Management and welfare of farm animals

Management and welfare of farm animals

THE UFAW HANDBOOK

Published by
Universities Federation for Animal Welfare (UFAW)
8 Hamilton Close, South Mimms, Potters Bar, Herts EN6 3QD, UK

First published 1971 by Churchill Livingstone as
The UFAW Handbook of the Care and Management of Farm Animals

Second edition 1978 and third edition 1988 published by Bailliere Tindall,
24-28 Oval Road, London NW1 7DX

Third edition reprint 1994 published by UFAW

Printed and bound in Great Britain by Henry Ling Ltd (The Dorset Press), Dorchester

British Library Cataloguing in Publication Data

Management and welfare of farm animals -
3rd ed
I. Livestock. Care
I. Universities Federation for Animal Welfare
II. Care and management of farm animals
636.083

ISBN 0 9007 6787 1

List of Contributors

R EWBANK MVSc MRCVS FIBiol — Director
Universities Federation for Animal Welfare, 8 Hamilton Close, South Mimms, Potters Bar, Herts EN6 3QD.

K R GOODERHAM BVSc DPMP MRCVS — Private Veterinary Practitioner
Marsh Lane, Hemingford Grey, Huntingdon, Cambridge PE18 9EN.

C M HAAN BSc(Agric) MS CBiol MIBiol — Senior Pig and Poultry Adviser
Ministry of Agriculture, Fisheries and Food, Great Westminster House, Horseferry Road, London SW1P 2AE.

P J HEARN NDP DipAgExt CBiol MIBiol (deceased)
Ministry of Agriculture, Fisheries and Food, 122a Thorpe Road, Norwich NR1 1RN.

J O L KING PhD MVSc BSc(Agric) FRCVS FIBiol — Emeritus Professor
Flat 6, Ashtree Farm Court, Willaston, South Wirral, L64 2XL.

I J LEAN BSc PhD CBiol MIBiol — Lecturer in Animal Production
Department of Agriculture, Horticulture and the Environment, Wye College, University of London, Nr Ashford, Kent TN25 5AH.

J D LEAVER BSc PhD FRAgS — Professor of Agriculture
Department of Agriculture, Horticulture and the Environment, Wye College, University of London, Nr Ashford, Kent TN25 5AH.

A MOWLEM FIAT CBiol MIBiol — Consultant
The Goat Advisory Bureau, 9 Pitts Lane, Earley, Reading, Berks RG6 1BX.

C NIXEY BSc(Agric) MAgrSc — Director of Technical Services
British United Turkeys Ltd, Hockenhull Hall, Tarvin, Chester CH3 8LE.

A J F RUSSEL BSc(Hons Agric) MAgrSc(Hons) PhD NDA — Principal Scientific Officer
The Macaulay Land Use Research Institute, Bush Estate, Penicuik, Midlothian EH26 0PY

D W B SAINSBURY MA PhD BSc MRCVS — Lecturer in Animal Health
Department of Clinical Veterinary Medicine, University of Cambridge, Madingley Road, Cambridge CB3 0ES.

A H SYKES MA PhD Reader in Animal Physiology
Department of Agriculture, Horticulture and the Environment, Wye College, University of London, Nr Ashford, Kent TN25 5AH

A J F WEBSTER MA VetMB PhD MRCVS Professor of Animal Husbandry
University of Bristol, Langford House, Langford, Bristol BS18 7DU.

H LL WILLIAMS BSc(Agric) MSc PhD FRAgS NDDH MIBiol
Reader in Animal Husbandry
The Royal Veterinary College, University of London, Boltons Park, Potters Bar, Herts EN6 1NB.

Contents

Preface

The Universities Federation for Animal Welfare was formed in 1938 in direct succession to the University of London Animal Welfare Society, which had been founded in 1926. From the outset it promoted the better treatment of all animals, working on the principle that one of the best ways of doing this was through education, i.e. by lectures, symposia and workshops, and more generally, by technical publications.

In 1947, after the war, UFAW's first major handbook was on the care and management of laboratory animals. However, UFAW was soon involved with the welfare needs of farm animals and submitted evidence to the Government's Technical Committee of Enquiry, chaired by Professor Brambell, which reported in 1965 on the welfare of animals kept under intensive livestock husbandry systems.

The First Edition of the *UFAW Handbook on the Care and Management of Farm Animals*, published in 1971, covered both old and new husbandry methods and also contained chapters on exotic species such as the water buffalo, Indian elephant and camel. The Second Edition, 1978, was reduced in size and concentrated on the animals commonly farmed in Europe.

Perhaps the most important innovation of this Third Edition is the inclusion of the word *Welfare* in the title. There was some hesitation in using this word in earlier editions, because of its possibly emotive overtones. That it may now be used is perhaps a reflection of the profound changes which are taking place in the attitudes of consumers, agricultural scientists and livestock farmers towards the ways in which the food animals are kept.

To achieve the necessary emphasis on welfare, every chapter has been completely restructured and rewritten. New authors, all of whom are experts in their own fields, have been employed in nearly every case and were asked to consider the welfare implications of the husbandry systems which they describe. In this way it is hoped to provide the veterinary and agricultural students, to whom the book is mainly directed, with a critical view of production practices which they might otherwise take for granted. It is important in the present social climate for everyone associated with the agricultural industry to be aware of the opinions which society in general has of modern husbandry methods.

In this rapidly-advancing technological age, there will be many developments in husbandry methods. Most of the changes will be directed to increase efficiency of animal production, but society is now demanding that developments should be seen to benefit the well-being of the animals and not merely reduce the cost of food to the consumers. With this in mind this new Edition, as well as being a comprehensive textbook of current animal husbandry, also aims to promote a humane attitude in the minds of all those responsible for the care, management and welfare of farm animals.

Acknowledgements

Mrs Phyllis Ray, MSc, formerly the Assistant Director of UFAW and latterly a Member of UFAW Council undertook on our behalf the lengthy tasks of planning this completely revised handbook, coordination and preparation of manuscripts and proof reading. For this tremendous effort, we are extremley grateful. Thanks are also due to Miss Clio Bromley who provided the necessary general administrative assistance. Our deep appreciation is expressed to the many contributors who have given us the benefit of their expert knowledge.

Also we wish to record the following organisations and individuals who have provided illustrations and copyright material.

British Denkavit
British United Turkeys Ltd
Buxted Ducklings Ltd
Dr H A Elson
Farm Animal Welfare Council
Her Majesty's Stationery Office
Liscombe Experimental Husbandry Farm
Meat and Livestock Commission
Milk Marketing Board
Ministry of Agriculture, Fisheries and Food
Macaulay Land Use Research Institute
Mr J C Sandford
Scottish Farm Buildings Investigation Unit

1 Animal welfare

R. Ewbank

Man uses animals in enormous numbers. He uses them for companionship, for work, for sport, for food and for the production of wool, hides and hair.

Livestock production systems evolved slowly over several thousands of years and a balance was struck, seemingly, between the needs of the human population and the degree of exploitation of the animals. Over the last 200 years, however, scientific methods have been increasingly applied to agriculture with a resulting rise in productivity, and during the last few decades world economic forces have changed the structure of farming so that many livestock units are now large in size and intensive in nature. At the same time, i.e. over the last 30–40 years, society has become increasingly critical of the ways in which animals are used.

These criticisms, which range from the practical to the philosophical, have led, among other things, to an increasing interest being taken by the general public, by scientists and by the farming community in the whole subject of animal welfare.

Welfare: definitions and explanations

What is meant by "welfare"

Many attempts have been made to define the term welfare as applied to animals. Two recent and widely used definitions are: "Welfare on a general level is a state of complete mental and physical health where the animal is in harmony with its environment" (Hughes, 1976), and "The welfare of an individual is its state as regards its attempts to cope with its environment" (Broom, 1986).

Both definitions refer in a general way to the balance which exists between the animal and its surroundings. They are not immediately helpful at the practical level in determining whether an animal is in fact enjoying a correct balance. For practical purposes, there is merit in simply replacing the word "welfare" by the terms "health" and "well-being" – both of which have strong positive components. Health is more than the mere absence of disease and well-being is more than the absence of discomfort and distress. This positive approach is to be welcomed because it inherently encourages high standards and it also plays into the natural pride of the good stockman in having contented, thriving and productive animals.

Possible causes of suffering in animals

Ill-treatment

This term covers those actions (or inactions) by man that cause animals to suffer. Ill-treatment can be divided into three main forms: abuse, neglect and deprivation (see Table 1.1). Some people would prefer to use the term cruelty rather than abuse.

Abuse and neglect result in changes in the signs of health and well-being of animals recognized by experienced stockmen and veterinary surgeons as being indicative of trouble. This results in individual biological inefficiency on the part of the animal (lowered growth rate,

Table 1.1
Consequences of ill-treatment of animals

Type	Symptoms/signs	Effect on production
Abuse (deliberate) →	Fear, injury, pain, distress, etc. (i.e. suffering) →	Individual biological inefficiency and financial loss
Neglect (occasional) through idleness, ignorance or overwork →	Malnutrition, disease, distress, etc. (i.e. suffering) →	Individual biological inefficiency and financial loss
Deprivation (built into some husbandry systems) of facilities to fulfil behavioural and/or physiological needs →	Changes in behaviour, occasional abnormal behaviours, etc. (suffering ?) →	? ? ?

Modified from Ewbank, (1985).

lowered milk yield, food conversion inefficiency, etc.) and financial loss to the farming enterprise. The stock keeper's traditional claim that his animals cannot be suffering because they are producing so well, is probably a good defence as long as he is only thinking of abuse and neglect.

Deprivation relates to situations where animals are prevented from fulfilling physiological and/or behavioural needs (see below). A denial of some of the facilities necessary to meet such needs is characteristic of several of the modern intensive systems, such as battery cages, sow tethers and stalls, and traditional veal crates. Animals kept in such accommodation sometimes show considerable behavioural abnormalities, compared with those kept in more extensive systems. Although such alterations in behaviour may or may not be signs of suffering, they are fairly certainly indicators that the animals are having to make an extreme behavioural adaptation to their environment. Deprivation is sometimes, but not usually, linked to depression in biological production – indeed, some so-called deprivation systems are highly productive, especially when measured in financial terms. High production alone is not necessarily a full defence against an accusation of deprivation!

In assessing the effects of changes in husbandry systems on animal production, care must be taken to distinguish between the effects on (a) biological production of individual animals and (b) the financial return from the whole enterprise. Although, for example, increases in stocking rate will usually lead to a lowering of individual daily live weight gains, or of productivity in terms of, say, individual egg production, the financial return from the enterprise – measured both as a total and on an individual animal basis – may be increased. This may be due to a sharing of the capital cost of the same building across a larger number of animals and also, perhaps, more importantly, to the greater number of animals being looked after by the same or even a reduced number of attendants in the new and often now semi-automated enterprise.

Stress, overstress and distress

Animals respond to challenges in their environment by employing a variety of interlocking physiological, biochemical and behavioural adaptation mechanisms. Potentially harmful stimuli which induce the main physiological and biochemical responses are called stressors and the response itself has been termed the *stress* or *stress response*. These changes are largely adaptive but are usually employed at a biological cost. In acute or prolonged stress the side-effects of some of the responses can be actually harmful to the animal. For an excellent but now somewhat dated account of the general aspects of animal stress, see Archer (1979).

Most users of the term stress have identified an increased level of the hormones secreted by the adrenal cortex (the corticosteroids) as the main measurable indicator of stress. This indicator is seemingly especially valid if it is accompanied by evidence of an increased metabolic rate and/or of a suppression of the animal's immune response (Barnett, 1987). A strong plea has been put forward by Fraser *et al.* (1975) that the term stress should only be used when an animal has to make an abnormal or extreme adjustment in its physiology or behaviour to cope with adverse aspects of its environment. This argument has considerable merit, but a case can be made (Ewbank, 1985) to subdivide this stress response and to use the term *overstress* for the adaptive, medium-level response which has been made at a biological cost, and which may be accompanied by some damage to the animal, and *distress* for the high-level response which has a high biological cost, is damaging to the animal, is probably sensed by the animal as unpleasant (i.e. causing it suffering), and which is outwardly expressed by recognizable changes in its behaviour.

This use of the term *distress* fits in with the legal terminology of the Agricultural (Miscellaneous Provisions) Act, 1968 (see below for details of this legislation) where the main offence is to cause "unnecessary pain or distress". Pain (an internal physical/psychological process causing suffering) in animals is usually noticed by the human observer when the animals show outward signs of distress. It is possible also that animals suffer from emotional distress which is not associated with pain. Most stock keepers would probably agree that their charges feel and probably suffer from fear. If farm animals experience one emotion, could they not have others? The ewe separated from its lamb shows physiological and behavioural changes indicative of distress. How does the ewe's reaction compare with the response of a human mother forcedly separated from her child? There are no real answers to these questions but practical experience, common sense and humaneness suggest that, in the absence of positive proof to the contrary, animals must be given the benefit of the doubt. If an animal is in a situation that would induce unpleasant emotions in humans, and shows behavioural signs of distress, then it is probably undergoing an unpleasant emotion process – in other words, it is probably suffering. It is interesting to note that emotional suffering in animals has been recognized in UK law for many years. The Protection of Animals Act, 1911 stipulates that it is an offence to "infuriate or terrify any animal" (see below for further details of legislation).

Physiological and behavioural (ethological) needs

The *European Convention for the Protection of Animals Kept for Farming Purposes* (Council of Europe, 1976) states several times that farm animals must be housed (managed, fed, etc.) according to, among other things "their physiological and ethological needs".

The concept of physiological needs is straightforward. If, for example, an animal is not given adequate quantities of a suitable food, it will probably not grow properly, or, in the case of a mature animal, be difficult to breed from; it may show signs of malnutrition, or vitamin or mineral deficiencies. If kept long enough on this inadequate diet it may even die. There is a physiological need for sufficient amounts of the correct kind of food. If animals are not allowed to exercise, pathological changes may be found in their joints, bones and muscles and, therefore, it can be argued that there is a physiological need for exercise (see discussion of sow stalls in Chapter 6). There is definite physical evidence of the adverse effects of deprivation, both of an adequate diet on the one hand, and of adequate exercise on the other.

Behavioural (ethological) needs are less readily understood. The concept, however, is central to many ideas in farm animal welfare and is perhaps best approached from a historical perspective. In 1964, the UK government, in response to the public concern which had been brought to a head by the publicaton of the book

Animal Machines (Harrison, 1964), appointed a Technical Committee (Brambell, 1965) to enquire into the welfare of animals kept under conditions of intensive husbandry. The deliberations of this committee were strongly influenced by contemporary behavioural ideas.

In the 1940s, 1950s and 1960s one of the main areas of interest for European ethologists was the study of animals' internal motivational states. The ethologists of that time took the view that most, if not all, behavioural patterns were largely controlled (motivated) by genetically determined internal drive mechanisms. It was believed that over a period of time, the level of the internal drive state would steadily rise until it exceeded a threshold and the animal would then perform an appropriate behaviour. External stimuli from the environment might alter (usually lower) the threshold; they might be required for the full expression of the resulting behaviour but they were not necessary either for the initial build-up or for triggering the behaviour. If it was made difficult for the animal to carry out the behaviour (e.g. by not giving it the required physical facilities, such as, in the case of the domestic fowl, nesting material for the pre-egg laying behaviour) then the animal might be expected to show so-called intention movements and/or inappropriate behaviours or even overtly abnormal behaviour. Many people consider these types of responses to be signs of frustration and a possible form of emotional distress. These ideas were largely accepted by the Brambell Committee, who further held that some intensively kept livestock housed in seemingly barren surroundings did indeed show behavioural signs of frustration.

The concept of behavioural needs was subsequently taken up by various groups of workers – agriculturalists, veterinarians and applied ethologists – who were investigating various welfare problems which were occurring in some intensive units. These problems were usually of a behavioural nature, and included such things as feather-pecking in yarded poultry and tail-biting in pigs in crowded pens. However, by this time, most mainstream "pure" ethologists were becoming increasingly interested in other behavioural matters and did not really follow up the further implications of behavioural needs. More recently, however, some ethologists have been pointing out that the internal drive theory is too simple an explanation of why animals perform certain behaviours at certain times and places. They are also suggesting that needs, if they indeed exist, would be better termed biological, and that they are satisfied through a mixture of physiological and behavioural mechanisms. For a critical yet sympathetic review of some of the modern ideas on behavioural needs, see Dawkins (1983).

The idea of behavioural needs, however, had been taken up by the legislators. In the late 1960s and 1970s, when the Council of Europe was drawing up its *Convention for the Protection of Animals Kept for Farming Purposes*, it took on board many of the points, including the concept of behavioural needs, contained in the Brambell Report. The UK has signed and ratified the Convention and has thus indicated not only that the government of the day agreed with the spirit of the document but that it also believed that its domestic legislation was sufficient to implement all the conditions (including the meeting of behavioural needs) laid down by the Convention. It seemed, from the point of view of the UK lawyers, that the important word was "need". A need is not a luxury or an option but a necessity. An animal deprived of a necessity will, in the view of most people suffer – showing signs of pain, discomfort and/or distress. Such signs could be the basis of a prosecution under the Protection of Animals Act, 1911 or under the Agriculture (Miscellaneous Provisions) Act, 1968. The lawyers, therefore, argued that the law indirectly ensures that animals' physiological and behavioural needs are likely to be met.

Another approach to ensuring, as far as is reasonable, that animals are able to fulfil their biological needs has been to draw up lists of basic husbandry requirements. The earliest and best known example is Brambell's (1965) so-called "five freedoms": "An animal should at least have sufficient freedom of movement to be able without difficulty, to turn round, groom itself, get up, lie down and stretch its limbs." On its establishment in 1979 the Farm Animal Welfare Council (FAWC) issued a press statement (No. FAWC/1) indicating that it intended to revise the Codes of Recommendation for the Welfare of Livestock, made under the Agriculture (Miscellaneous Provisions) Act, 1968, to provide farm animals with:

1. Freedom from thirst, hunger and malnutrition.
2. Appropriate comfort and shelter.
3. Prevention, or rapid diagnosis and treatment, of injury and disease.
4. Freedom to display most normal patterns of behaviour.
5. Freedom from fear.

The prefaces of the latest editions of the Codes expand these into a list of ten provisions which would allow the animals to fulfil their basic needs (see p. 6 for details).

Many of the points made in these lists can be supported by scientific, technical and field evidence, while some cannot. There are a few items which might be described as "self-evident" needs (or rights) for which, as yet, there is little scientific support. This is not to say that those items are necessarily wrong – it may be that they are being advanced in the expectation that factual evidence will sooner or later be found to back them up.

Animal rights

The ethical views as to how animals should be treated vary from the concept that man has an absolute dominion over the lower creatures, through a middle position that man has a duty to use his animals well (i.e. give them a good life while they are alive and a quick humane death when they have to be killed), to the suggestion that animals have total rights and that these rights must be fully respected.

The recent spread of animal rights ideas in the UK is largely due to the influence of three semi-popular but serious books written by moral philosophers (Singer, 1976; Regan, 1982; Rollins, 1981). The arguments put forward are often convincing within their own philosophical framework but at first reading seem to bear only a limited relevance to the practical situation. Some animal rights concepts have been taken up as beliefs by a number of animal welfarists and tend to be expressed in slogans which can vary from "animals have a right to be left completely alone" to "animals have a minimum right to be able to satisfy their physiological and behavioural needs".

Many animal rights campaigners tend to proselytize and be intolerant during discussions. This is a pity, because some of their ideas can contribute to an understanding of the complex ethical relationships that exist between man, animals and the environment.

The legal position in the UK

General animal welfare legislation

Certain aspects of this legislation have already been touched upon in the previous section, namely, the ways in which the present UK law is considered to be adequate to implement the conditions laid down by the *European Convention for the Protection of Animals Kept for Farming Purposes* for fulfilling animals' physiological and behavioural needs.

The Protection of Animals Act, 1911

This is the principal statute relating to the protection of domestic and captive animals. Offences amounting to cruelty include:

1. Cruelly to beat, kick, ill-treat, over-ride, over-drive, over-load, torture, infuriate or terrify any animal.
2. To cause unnecessary suffering by doing, or omitting to do, any act.
3. To convey or carry any animal in such a manner as to cause it unnecessary suffering.
4. To perform any operation without due care and humanity (in relation to which the provisions of the Protection of Animals (Anaesthetics) Acts are particularly relevant – see below).
5. The fighting or baiting of any animal or the use of any premises for such a purpose.
6. The administering of any poisonous or injurious drug or substance to any animal.

This Act is commonly grouped with several other related Acts and Amendments, known collectively as the Protection of Animals Acts, 1911–1964. Although these Acts do not apply to Scotland, there is separate legislation of a very similar nature applying specifically to that country.

The Agriculture (Miscellaneous Provisions) Act, 1968

This is the principal piece of legislation which applies specifically to the welfare of farm animals. It was enacted as a response to the Brambell Report and has four main points:

1. It makes it an offence to cause unnecessary pain or unnecessary distress to livestock being kept for farming purposes on agricultural land.
2. It gives authority for veterinary officers of the State Veterinary Service to inspect, on welfare grounds, farms where livestock are being kept.
3. It empowers the appropriate Minister to introduce Regulations to improve the welfare of livestock. To date, there have been four such Regulations:
   (i) The Welfare of Livestock (Intensive Units) Regulations, 1978, under which there is a duty for the stock keeper to inspect thoroughly the livestock and automatic equipment in intensive units at least once every day.
   (ii) The Welfare of Livestock (Prohibited Operations) Regulations, 1982, as amended in 1987 (see p. 7).
   (iii) The Welfare of Calves Regulations, 1987 (see Chapter 3).
   (iv) The Welfare of Battery Hens Regulatons, 1987 (see Chapter 9).
4. It authorizes the Agricultural Ministers to prepare Codes of Recommendations for the Welfare of Livestock – Codes have already been produced for cattle, sheep, pigs, rabbits, domestic fowls, turkeys and ducks. These Codes are of an advisory nature only (unlike the Regulations listed above), and failure to observe them is not in itself an offence. However, they do have legal standing: a person cannot claim ignorance of them, and he disregards their advice at his own risk. If he should be prosecuted under the Act for causing unnecessary pain or distress to livestock on agricultural land, his failure to observe the provisions of the Codes may be used in evidence.

The Prefaces to the Codes of Recommendations have already been mentioned in connection with physiological and behavioural needs. They list detailed recommendations which, if adopted, should go a long way towards ensuring that the stock keeper has met these needs. The provisions which they give as being necessary include:

1. Comfort and shelter.
2. Readily accessible fresh water and a diet to maintain the animals in full health and vigour.
3. Freedom of movement.
4. The company of other animals, particularly of like kind.
5. The opportunity to exercise most normal patterns of behaviour.
6. Light during the hours of daylight, and lighting readily available to enable the animals to be inspected at any time.
7. Flooring which neither harms the animals, nor causes undue strain.
8. The prevention, or rapid diagnosis and treatment, of vice, injury, parasitic infestation and disease.
9. The avoidance of unnecessary mutilation.
10. Emergency arrangements to cover outbreaks of fire, the breakdown of essential mechanical services and the disruption of supplies.

These Prefaces can be seen as guidelines which set the basic standards for the keeping of animals and which establish the attitudes which people should take towards their stock. The Codes of Recommendations themselves are detailed suggestions as to how these standards are to be practically implemented.

Control of minor surgery

The whole question of which minor surgical operations (for example, castration, docking, debeaking, dehorning) are allowed to be performed on farm animals, who can do them, and at what age anaesthetics must be used is very complex. The situation is covered by three interacting groups of legislation: the Welfare of Livestock (Prohibited Operations) Regulations, 1982, as amended in 1987; the Protection of

Animals (Anaesthetics) Acts, 1954 and 1964, with the various Orders made under them; and various Orders under the Veterinary Surgeons Act, 1966.

Under the Welfare of Livestock (Prohibited Operations) Regulations, a number of operations are prohibited, unless they are being carried out as first aid measures, or by a veterinary surgeon in the treatment of injury or disease. These are:

1. Penis amputation and other penial operations.
2. Freeze dagging of sheep.
3. Short-tail docking of sheep, unless sufficient tail is retained to cover the vulva in the case of female sheep and the anus in the case of male sheep.
4. Tongue amputation in calves.
5. Hot branding of cattle.
6. Tail docking of cattle.
7. Devoicing of cockerels.
8. Castration of a male bird by a method involving surgery.
9. Any operation on a bird with the object or effect of impeding its flight, other than feather clipping.
10. Fitting any appliance which has the object or effect of limiting vision to a bird by a method involving the penetration or other mutilation of the nasal septum.
11. Tail docking of a pig unless the operation is performed by the quick and complete severance of the part of the tail to be removed and either:
    (i) the pig is less than 8 days old, or
    (ii) the operation is performed by a veterinary surgeon who is of the opinion that the operation is necessary for reasons of health or to prevent injury from the vice of tail biting.
12. Removal of any of the antlers of a deer before the velvet of the antlers is frayed and the greater part of it has been sheared.

The 1987 Amendment totally prohibits tooth grinding in sheep, even by a veterinary surgeon.

The Protection of Animals (Anaesthetics) Acts are designed to prevent the infliction of unnecessary suffering during an operation. Certain minor operations may be performed without the use of an anaesthetic. They include, among others, the giving of injections, emergency first aid and the castration of a male animal before it has reached a specified age (see individual chapters for details).

The Veterinary Surgeons Act stipulates that, after a certain age, male animals may be castrated only by a veterinary surgeon. In the case of farm animals this age coincides with that after which an anaesthetic must be used.

For further details of the legislation covering animal welfare, see Cooper (1987) and Porter (1987a,b).

How to ensure good welfare

Recognition of health and well-being

Four interrelated groups of signs are used in deciding whether an animal or group of animals is in a state of health and well-being. These are physical appearance, behaviour, productivity and state of the environment. The actual details of the signs will vary between species, they may differ at different times of the year and they may be characteristic of a particular husbandry system.

There is often a total pattern of signs which may be recognized (sometimes it seems nearly subconsciously) by the stockman as being normal, denoting the healthy and productive state. As he works with his animals he (again it seems often subconsciously) tends to monitor this pattern and will be alerted to trouble if there is a deviation from the norm. Once alerted, he will undertake a thorough inspection; he will then probably come to a decision that the animals are either normal (healthy) or abnormal (unhealthy) or that he cannot really make a decision; he will then take action (if appropriate) and/or wait to see what happens and/or seek professional advice.

Once the welfare status of an animal (or group of animals) has been determined, the husbandry system can then be viewed against a positive/negative welfare scale (see Fig. 1.1). If the welfare is poor, i.e. the system lies in the negative part of the scale, it is likely that there will be obvious signs of pain, discomfort or distress and the stockman will take action to

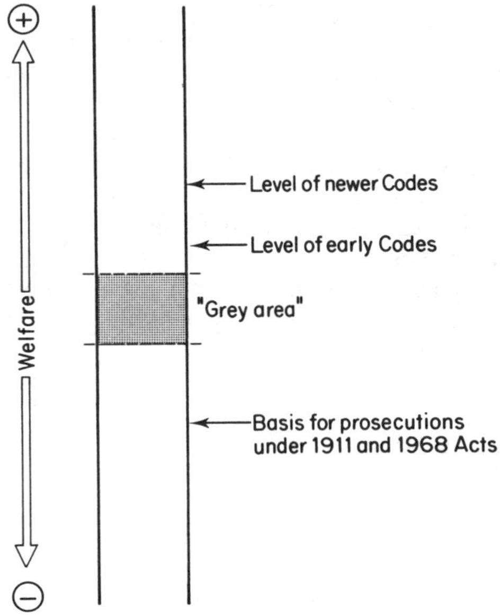

Fig. 1.1 Positive/negative welfare scale.

remedy the situation. An inspecting officer might decide that there is sufficient evidence of suffering to institute a prosecution. There will obviously be a variable, indeterminate "grey area" between the clearly positive and negative zones, but in most cases it will be clear to the stockman/veterinary surgeon/inspecting officer that the enterprise lies in one zone or the other. It can be argued that the original Codes of Recommendation for the Welfare of Livestock aimed at a welfare standard well above the "grey zone" and that the more recently revised Codes are set at an even higher level. As an animal (or enterprise) moves up the positive side of the scale there is usually an increase in individual biological production (growth rates, etc.) and thus an improvement in the profitability of the system. There may, however, be a limit (law of diminishing returns?) to the rises in production which will result from any further increments in the welfare status of the animals, i.e. there may no longer be a profit motive for further improvement by the farmer. If ethical or legal pressures push up welfare beyond the break-even cost/gain point then there may have to be a significant increase in the cost of the end-product (eggs, meat, milk) to the consumer.

Stockmanship

Stockmanship involves stock sense (a knowledge of, rapport with and ability to observe animals) and skill in stock tasks (the practical aspects of handling, care and manipulation of animals). A good stockman should be observant, patient, informed about animals and their needs, skilful in stock tasks, able to recognize health and disease states, and be knowledgeable about the workings of environmental control equipment and the measures to take when it fails. Most farmers and stockmen take a positive pleasure in having contented thriving animals under their care. This natural inclination should be respected and encouraged.

It is essential that stockmen should be well-trained, provided with the equipment and facilities to perform their work and have sufficient time to quietly and regularly observe their charges. One of the criticisms of modern intensive systems is that far too many animals are being looked after by far too few stockmen. It may be possible to perform much of the routine work with mechanized equipment but there may still be little time for individual attention to the animals themselves.

The stockman must be backed by management; there must be clear lines of communication. In large units the subdivision of responsibility can be a problem. In old-fashioned, small enterprises the stockman, the manager and the owner were often the same person. In modern large establishments, these three functions are often in separate hands and may be operating from different sites. The stockman can only perform his work well if he is provided with the correct physical facilities and if he has the continued support of the manager/owner.

For further full discussion of stockmanship, see UFAW (1983) and Seabrook (1987), and for an account of human/farm animal relationships, see Albright (1986).

Behavioural studies

The study of the behaviour of farm livestock has been, until recently, a somewhat neglected subject in agriculture. This is a pity because:

1. Changes in behaviour are often the first

signs of disease and the main signs of distress.

2. Behavioural changes can sometimes be used to assess the well-being (or otherwise) of animals (see Duncan, 1987).
3. Perhaps most importantly, the study of the behaviour of an animal (or group of animals) in its environment tends to produce a broad ecological/behavioural picture by which a husbandry system can be judged as to whether or not it is biologically sound.

The best practical introduction to farm animal behaviour is probably that by Fraser (1980). This should be supplemented by consulting Wood-Gush (1983) and Kilgour and Dalton (1984). Further advanced texts are Fraser (1985) and Hafez (1975).

Control of disease

Disease can lead to substantial suffering in animals as well as being an important cause of economic loss. Its control demands an intelligent cooperation between the farmer's own veterinary surgeon, specialist advisers and the stock keeper. It is crucial that disease control is thought of in the planning stage of any new animal enterprise. Well-planned preventive measures are not only more humane, in that the diseases do not develop and therefore the animals do not suffer, they are also more efficient in terms of productivity and costs. For further information, consult Sainsbury and Sainsbury (1987).

Humane killing of diseased and injured animals

It may be necessary, at times, for diseased, injured, deformed or surplus farm stock to be killed on the farm. The veterinary surgeon and licensed slaughterman may do this in the course of their routine duties but, in an emergency – to relieve acute suffering – a farmer or stockman may himself have to kill the animal(s). This emergency slaughter must be carried out as humanely as possible and it is essential that the farmer or stockman should give some thought in anticipation of the problem and prepare accordingly. He should consult his veterinary surgeon and make sure that the appropriate equipment is available at all times and that the farm staff have been trained in its use.

A welfare problem can arise, however, over the disposal of diseased or injured stock which are not suffering acutely. If they are sent while still alive (as a casualty animal) with an appropriate veterinary certificate to a slaughterhouse which is willing to accept them, the farmer will probably receive a greater financial return than if they had been killed on the farm and the carcass disposed of through a knacker's yard. There is, therefore, a temptation for some farmers to despatch animals while they are still alive, animals which should, on humane grounds, have been slaughtered on the farm. This welfare problem will probably only be overcome by a change in attitude of farmers, veterinary surgeons and slaughterhouse owners, by the firmer implementation of the laws which make it an offence to cause unnecessary pain, suffering or distress to animals and possibly by the adoption (compulsory?) of insurance schemes by which the financial loss resulting from the killing of an animal on humane grounds could be claimed for under the insurance policies.

For practical details of humane killing, see under the appropriate species, and for further general information, see UFAW (1973).

Extensive and intensive systems

Much of the concern about the welfare of farm animals centres around so-called intensive husbandry systems (traditional veal crates, battery cages, sow tethers and stalls). For a recent compilation of critical articles and photographs, see Animal Welfare Institute (1987). A strong supportive case for intensive systems has been advanced by Curtis (1986). Concurrently with the criticism of intensive systems has been an interest in the development of new extensive methods of animal keeping (e.g. UFAW, 1981).

Many intensive systems can be seen as restrictive and in some cases may not provide animals with their basic biological needs. It is therefore tempting to suggest that extensive systems which give animals more freedom must be better.

Husbandry systems can be classified as intensive, intermediate and extensive. Figure 1.2, for

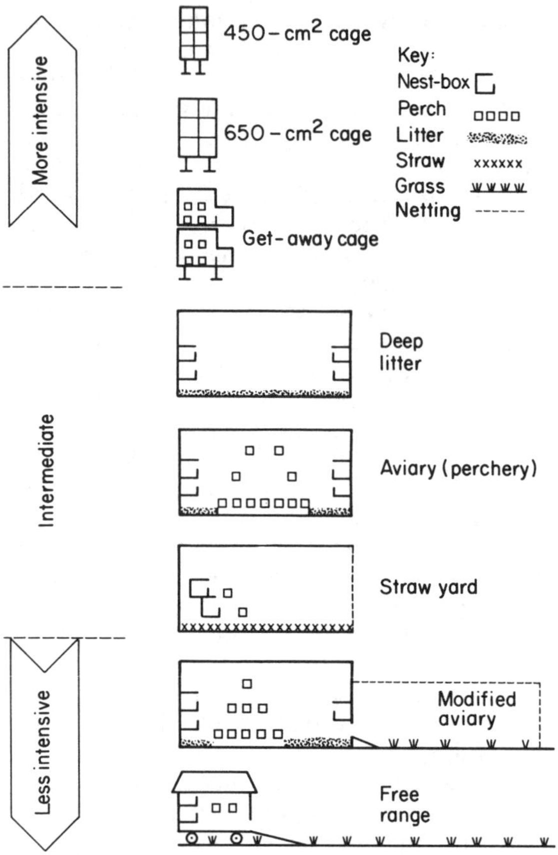

Fig. 1.2 Diagrammatic representation of housing systems for laying hens (not to scale). (Modified from Ewbank, 1981.)

example, illustrates the range of housing systems available for laying birds. Generally speaking, as systems become less intensive, they are more costly to run, and the end-product is more expensive to the consumer (although not as expensive, however, as some people would like to imagine – see Carnell, 1983); they demand a higher level (both in quality and quantity) of stockmanship; the animals are more exposed to the weather; and often there is a higher level of disease. The animals, however, do have more freedom, although at times this can appear as "freedom to get into trouble".

Extensive systems are not necessarily "better" and, conversely, intensive systems are not necessarily "worse". Each type has its good and bad points (see individual chapters for details). Rather we should realize that there are both good and bad units in each category, as measured by health, well-being and productivity. The aim should be to encourage good welfare, whatever the system.

Keeping animals in good health and well-being

For animals to be kept in good health and well-being there should be:

1. A full understanding of their biological (physiological and behavioural) needs.
2. The provision of facilities which allow the animals to fulfil these needs.
3. Sufficient time allowed for the stockmen to give attention to their charges.
4. The provision of round-the-clock back-up services so that, in case of breakdown of equipment, accidents or outbreaks of disease, the animals do not suffer unduly.
5. An attitude on the part of the stockman, the manager and the owner which ensures that the animals are given a good life when alive and a humane death when they have to be killed.

The way ahead

Most welfare criticism is directed against three intensive systems:

1. Traditional (single animal) crate-rearing of veal calves.
2. Stalling and/or tethering of pregnant sows.
3. Confinement of laying birds in battery cages.

Legislative action (The Welfare of Calves Regulations, 1987) has been taken by the UK government, which should in effect stop the crate-rearing of veal calves. It is interesting to speculate that this change in the law could only really be brought about because the farming industry itself had developed and largely changed to the alternative, yarded veal-rearing system.

It is likely that, sooner or later, legislation will be introduced to phase out stalling and/or tethering of pregnant sows. Many pig farmers realize this and, if they are replacing old buildings and worn out equipment, they are tending to put in the well-tried "small-group-of-sows-in-a-straw-yard" units or the newly-developed "large-group-of-sows-on-automatic-feed-dispenser" systems. Both these replacements are more costly to install/run and demand a higher level of stockmanship than stalls/tethers, but they have considerable welfare benefits for the animals. One difficulty may be that strains of pigs which are selected because they thrive under confinement systems, may not be behaviourally suited to social life in yards. An appropriate breeding programme should solve this.

The problem of battery cages seems less easily resolved. It could be both a welfare and an economic disaster to force the present population of battery birds out into the relatively untried alternative systems. Free range and intermediate systems (percheries, aviaries, straw yards, deep litter, etc.) do give the birds considerably more freedom but they are more difficult to run than battery units: they demand a higher level of stockmanship, the outdoor systems seem to have a higher level of disease, and the eggs cost more to produce. There is a real need for further research and development work on the intermediate and free range systems, in the hope that the problems will be solved and that it will finally be possible to phase out battery cages.

It has been suggested that the regular inspection and possibly the annual licensing of intensive farm animal units would be one way of improving welfare. This may be so – the manpower and administrative costs could be high but inspection and licensing might help to allay or remove some of the worst and sometimes unfounded fears of the general public.

Real progress mainly comes through changes in the attitude of those involved in the use of animals. Farmers, stockmen, veterinary surgeons, agriculturalists, etc. are not separate from the rest of the community; they are members of society and are influenced by what society wants. If society wants its animals treated in a particular way then sooner or later it will happen. And it will largely come about because the animal users themselves will eventually want it to happen, and will acquire the knowledge and skills needed to make the change.

Animal welfare is of increasing relevance to the farming industry. The literature is immense, and is perhaps best referred to via five important and comprehensive books/reports: Brambell (1965), Council for Agricultural Science and Technology (1981), Dawkins (1980), Fox (1984) and Sainsbury (1986).

It is vital both for the health and well-being of the animals involved and for the financial future of the farming industry that an increasing and critical interest should be taken in that mixture of economic, scientific, ethical, aesthetic and practical concepts which make up the complex subject of animal welfare, and that action should be taken on the new knowledge and ideas thus gained.

References and further reading

Albright, J. L. (1986). Human/farm animal relationships. In Fox, M. W. and Mickley, L. D. (eds), *Advances in Animal Welfare Science 1986/87* pp. 51–56. Washington, D.C.: Humane Society of the United States.

Animal Welfare Institute (1987). *Factory Farming: The Experiment that Failed.* Washington, D.C.: AWI.

Archer, J. (1979). *Animals Under Stress.* London: Arnold.

Barnett, J. L. (1987). The physiological concept of stress is useful for assessing welfare. *Australian Veterinary Journal* **64**: 195–196.

Brambell, F. W. R. (Chairman) (1965). *Report of the Technical Committee to Enquire into the Welfare of Animals Kept Under Intensive Livestock Husbandry Systems.* Cmnd 2836. London: HMSO.

Broom, D. M. (1986) Indicators of poor welfare. *British Veterinary Journal* **142**: 524–526.

Carnell, P. (1983). *Alternatives to Factory Farming.* London: Earth Resources Research Ltd.

Cooper, M. E. (1987). *An Introduction to Animal Law.* London: Academic Press.

Council for Agricultural Science and Technology (1981). *Scientific Aspects of the Welfare of Food Animals.* Report No. 91. Ames, Iowa: CAST.

Council of Europe (1976). *European Convention for the Protecton of Animals Kept for Farming Purposes.* European Treaty Series No. 87. Strasbourg: Council of Europe.

Curtis, S. E. (1986). The case for intensive farming of food animals. In Fox, M. W. and Mickley, L. D. (eds), *Advances in Animal Welfare Science 1986/87*, pp. 245–255. Washington, D.C.: Humane Society of the United States.

Dawkins, M. S. (1980). *Animal Suffering*. London: Chapman and Hall.

Dawkins, M. S. (1983). Battery hens name their price: consumer demand theory and the measure of ethological 'needs'. *Animal Behaviour* **31**: 1195–1205.

Duncan, I. J. H. (1987). The welfare of farm animals – an ethological approach *Science Progress (Oxford)* **71**: 317–326.

Ewbank, R. (1981). Alternatives: definitions and doubts. In UFAW, *Alternatives to Intensive Husbandry Systems*, pp. 5–9. Potters Bar: UFAW.

Ewbank, R. (1985). Behavioral responses to stress in farm animals. In Moberg, G. P. (ed.), *Animal Stress*, pp. 71–79. Bethesda, Maryland: American Physiological Society.

Fox, M. W. (1984). *Farm Animals: Husbandry, Behavior and Veterinary Practice*. Baltimore: University Park Press.

Fraser, A. F. (1980). *Farm Animal Behaviour*. London: Baillière.

Fraser, A. F. (ed.), 1985. *Ethology of Farm Animals*. Amsterdam: Elsevier.

Fraser, D., Ritchie, J. S. D. and Fraser, A. F. (1975). The term 'stress' in a veterinary context. *British Veterinary Journal* **131**: 653–662.

Hafez, E. S. E. (ed.) (1975). *The Behaviour of Domestic Animals*. London: Baillière.

Harrison, R. (1964). *Animal Machines*. London: Stuart.

Hughes, B. O. (1976). Behaviour as an index of welfare. In *Proceedings 5th European Poultry Conference, Malta*, pp. 1005–1012.

Kilgour, R. and Dalton, C. (1984). *Livestock Behaviour. A Practical Guide*. London: Granada.

Porter, A. R. W. (1987a). Animal welfare. In *Legislation Affecting the Veterinary Profession in the United Kingdom*, pp. 16–42. London: Royal College of Veterinary Surgeons.

Porter, A. R. W. (1987b). Practice of veterinary medicine and surgery. In *Legislation Affecting the Veterinary Profession in the United Kingdom*, pp. 111–116. London: Royal College of Veterinary Surgeons.

Regen, T. (1982). *All That Dwell Therein*. Berkeley: University of California Press.

Rollins, B. E. (1981). *Animal Rights and Human Morality*. Buffalo, NY: Prometheus Books.

Sainsbury, D. (1986) *Farm Animal Welfare*. London: Collins.

Sainsbury, D. and Sainsbury, P. (1988). *Livestock Health and Housing*, 3rd edition. London: Baillière.

Seabrook, M. (ed.), (1987) *The Role of the Stockman in Livestock Productivity and Management*. Report EUR 10982EN. Luxemburg: Commission of the European Communities.

Singer, P. (1976). *Animal Liberation*. London: Chapman and Hall.

UFAW (1973). *Humane Killing of Animals*. Potters Bar: UFAW (new edition in preparation).

UFAW (1981). *Alternatives to Intensive Husbandry Systems*. Potters Bar: UFAW.

UFAW (1983). *Stockmanship on the Farm*. Potters Bar: UFAW.

Wood-Gush, D. G. M. (1983). *Elements of Ethology. A Textbook for Agricultural and Veterinary Students*. London: Chapman and Hall.

2 Dairy cattle

J. D. Leaver

Cattle were first domesticated over 6000 years ago and have been a mainstay of agriculture in most countries. They have contributed greatly to human welfare, supplying milk, meat and hide, and in some countries draught power and fuel. Their ruminant digestive system has greatly extended man's food supply by its ability to convert inedible fibrous foods into edible energy and protein for human consumption in the form of meat and milk.

The two major species of domesticated cattle are *Bos taurus*, which is found mainly in temperate regions and includes the European breeds of cattle and *Bos indicus* which occurs in tropical regions.

The dairy industry in the UK

Since the early 1960s, rapid changes have occurred in the UK dairy industry. The trend towards fewer but larger herds accelerated in the 1960s and 1970s with no change in the size of the national herd (Table 2.1). By 1985 over one-third of dairy cows were kept in herds of over 100.

Table 2.1
UK herd structure 1965–1985

Year	UK milk producers	Dairy cows (million)	Average herd size
1965	124 688	3.186	26
1975	76 827	3.242	42
1985	48 827	3.133	64

Source: *Dairy Facts and Figures*, 1985, published by the Federation of Milk Marketing Boards.

Most milk producers are situated in the wetter, western half of the country, where farms are smaller than in the major arable areas, and where dairying relies to a large extent on grass and its conserved products.

Dairy breeds

Dairy cattle have a conformation designed for the production of milk. Viewed from the side, they are wedge-shaped, and cows of good conformation have strong suspensory ligaments for the attachment of the udder, and sound hind legs and feet. Viewed from above, the dairy cow is narrow at the shoulder, widening towards the hook bones of the pelvic girdle. For high yields of milk, the cow needs to mobilize body fat in early lactation and has to convert nutrients to milk rather than to body fat. Consequently, high-yielding cows are often quite thin.

The popularity of the Dairy Shorthorn and Ayrshire breeds has declined since the 1950s and, during this period, that of the Friesian and Holstein breeds has increased. The Dairy Shorthorn is a dual purpose animal providing a good beef carcass as well as having dairy attributes. The colours are red, white, red and white, and roan. The decline of the breed has been dramatic; pre-war, this was the dominant dairy breed, but there are now few herds remaining. In Scotland, the brown and white Ayrshire was the dominant breed until the 1970s, but the poor beefing potential, and the failure to progress significantly through the use of artificial insemination, have led to a continuing decline in breed numbers. In the rest of Britain, the Ayrshire has been a minority breed since the 1950s.

Cows of the Channel Island breeds – the Jersey and Guernsey – produce milk of a high fat and protein content, and are smaller than other breeds (apart from the Dexter which can be considered a rare breed). Most of the Channel Island herds still remaining produce milk for direct processing on the farm.

The Friesian originated in the Netherlands. In Britain, the breed was selected for dual-purpose characteristics, and this led to a small black and white animal with large amounts of subcutaneous fat. In North America, selection was for yield of milk and for size (height at withers). This led to a much taller and more dairy-like animal, the Holstein. The traditional British Friesian is now increasingly crossed with Holsteins from Canada, the USA and Europe to produce these dairy-like characteristics, and pedigree Holstein herds have also increased in number. The Friesian and Holstein breeds and their intermediate crosses now account for over 90 per cent of dairy cows in the UK. The higher yield of these breeds (Table 2.2) has been a contributory factor to the overall increase in milk yield per cow.

Table 2.2
Production figures for major breeds in recorded herds in England and Wales, 1984–1985

Breed	305-day yield (kg)	Milk fat (g/kg)	Milk protein (g/kg)
Ayrshire	5026	39.6	33.4
British Friesian	5556	38.7	32.3
British Holstein	6214	38.4	31.7
Dairy Shorthorn	4996	36.9	32.7
Guernsey	4005	46.7	35.5
Jersey	3835	52.7	38.0

Source: *Dairy Facts and Figures*, 1986, published by the Federation of Milk Marketing Boards.

The milk yield potential of cows within breeds has also increased through selection. Progeny testing of bulls and the use of artificial insemination (AI) have contributed significantly to higher production levels. The application of knowledge concerning the nutrition and management of cows has also been effective in increasing yields per cow.

Milk utilization

Milk that is produced surplus to liquid consumption is used for manufacture. In 1965, 70 per cent of milk produced in the UK was sold as liquid milk, whereas in 1985 only 45 per cent was consumed in this form (Table 2.3).

Table 2.3
Production and utilization of milk in the UK, 1965–1985 (million litres)

Year	Total milk supplied from farms	Liquid sales of milk	Manufactured milk
1965	10 710	7459	3251
1975	13 167	7761	5407
1985	15 242	6926	8301

Source: *Dairy Facts and Figures*, 1985, published by the Federation of Milk Marketing Boards.

The consumption of liquid milk declined from 2.8 litres/week per head of population in 1965 to 2.4 litres per week in 1985. Also in the 1980s, stimulated by the "health lobby", there was a swing to skimmed and semi-skimmed milk consumption at the expense of full-fat milk. A similar decline occurred in the consumption of butter.

The combination of expanding production of milk on farms, the declining consumption of milk and its products per capita, and the stabilizing of the population size, has thus led to surpluses of production in the EEC generally, as well as in the UK.

Milk marketing

Since 1933, marketing has been controlled by the Milk Marketing Boards in England, Wales, Northern Ireland and Scotland. These are producer organizations which have the exclusive right to buy all milk from producers, and to equalize the price paid (pool price), irrespective of the Board region in which they reside, or of the use for which the milk is intended.

Most milk is supplied by wholesale producers whose milk is collected from their farms by the Milk Board. The remainder is produced by producer-retailers and producer-processors who sell milk under licence, and by farmhouse cheesemakers who operate under contract to the Board.

The milk price paid to wholesale producers is based on the value of sales of milk for liquid consumption and for manufacture. There are monthly variations in price which are intended to depress production levels in spring and increase them in late summer (milk production reaches a peak nationally in May after cows have been turned out from their winter housing on to spring grass). The price is also subject to additions or subtractions which reflect the milk compositional quality (in particular fat and protein content). Premiums are paid for good hygienic quality measured on total bacterial counts (TBC): the detailed standards differ for each Milk Board, but premium payments are given when TBC's are below 15 000–20 000 bacteria per ml of milk. Deductions are made for antibiotic contamination, for water inclusion, and for the EEC co-responsibility levy which is used to help in financing the disposal of surpluses.

European Economic Community

The policy for dairying in the EEC is to manage the Community markets to obtain a target price for milk. This is determined by the Council of Agricultural Ministers and is not a guaranteed price. A complex system of intervention buying of butter and skimmed milk powder, and of subsidies and levies, is used to attempt to secure product prices.

These methods were successful in maintaining prices to farmers, but stimulated surplus production which accumulated in intervention stores as butter and skimmed milk powder. This situation led, in 1984, to an imposition of milk quotas in all EEC countries, with super-levies on over-production. Initial levels of quotas imposed were well above the levels of consumption of milk and dairy products and, consequently, a reduction in quotas has continued.

Biology

The modern dairy cow has been selectively bred to produce a large amount of milk relative to its size. This entails having an ability and willingness to consume large amounts of feed daily (high yielding cows have a daily dry matter intake of up to 3.5 per cent of their body weight) and, in addition, an efficient digestive system to provide the basic components from which milk will be synthesized.

Digestive system

The cow is a ruminant animal and can digest large amounts of fibrous material. The food is grasped by the powerful rough tongue in conjunction with the eight incisor teeth (in an adult cow) on the lower jaw and the dental pad on the upper. Mastication takes place by the molar teeth, and saliva is secreted (over 100 litres/day) on to the food. The saliva contains phosphate and bicarbonate, and is an important buffering agent for the rumen, maintaining the pH of the contents at about 6.5.

A unique aspect of digestion in the ruminant is the ability to ruminate (chew the cud). Boluses of coarse food are regurgitated from the rumen into the mouth and chewed for about 1 min. The bolus is then swallowed and the process is repeated. The cow spends about 6–8 h per day in rumination, the actual time depending on the amount of fibrous material consumed. To maintain normal rumen function, some fibrous feed is necessary. This should not be less than 0.5 kg DM/100 kg bodyweight as grass or forage.

The rumen has a capacity of about 150 litres and is subdivided into four compartments.* It acts as a large fermentation vat where the microorganisms digest the food. The bacteria in the rumen are assisted by protozoa and yeasts, and the balance of the various microbes varies with the type of diet eaten. It takes 1–2 weeks for the population to adjust to a large change in diet, such as from high forage to low forage. It is, therefore, essential that changes are made gradually, otherwise digestive upsets will ensue. For example, sudden large intakes of concentrates reduce the pH rapidly and, to avoid consequent digestive problems, concentrates should be fed little and often (maximum 3 kg per feed).

* For an illustration of the digestive system, see Chapter 3, Fig. 3.2.

In the rumen, carbohydrates such as cellulose and starch are hydrolysed to monosaccharides, which are subsequently fermented to the organic acids – acetic, propionic, butyric and some longer-chain acids. These volatile fatty acids (VFA) are absorbed through the rumen wall and pass into the bloodstream. High fibre diets lead to 60–70 per cent of the acids as acetic acid, 16–20 per cent as propionic acid and 7–12 per cent as butyric acid. A substitution of concentrates for forage leads to a reduction in acetic acid to about 50 per cent, with a balancing increase in the proportion of propionic acid. The proportion of VFAs produced has important effects on the metabolic hormone balance, particularly of insulin and growth hormone, and this has implications for milk production and body fat deposition. Diets producing a high proportion of propionic acid tend to reduce milk fat content and to encourage the partition of energy to body fat at the expense of milk production.

An important function of the rumen microbes is the synthesis from non-protein nitrogen, of microbial protein, which is subsequently digested in the stomach (abomasum) and small intestine as in monogastric animals. The microbes also synthesize B-complex vitamins.

The fermentation process produces large amounts of gas, particularly methane and carbon dioxide, and these are expelled from the rumen by the regular reflex action of belching.

The reticulum is a sac at the anterior end of the rumen. It has a honeycomb appearance and is concerned with the passage of food boluses up the oesophagus during rumination, and with the passage of digesta from the rumen to the omasum. In the pre-ruminant calf, a reflex action stimulated by sucking or drinking milk causes the closure of the muscular reticular groove, giving direct access to the abomasum, and allowing the milk to bypass the poorly developed rumen.

The omasum removes a large amount of water and organic acids from the digesta. It has a capacity of about 15 litres and, like the rumen and reticulum, has no secretions. The abomasum is the true stomach secreting the normal gastric juices, and further digestion similar to that in monogastric animals is carried out here and in the small intestine.

Mammary system

The mammary system of the cow is partially developed at birth, and development continues during the rearing period. Between 3 and 9 months of age there is a rapid multiplication of cells in the mammary gland, followed after puberty by increased proliferation of the duct system. The onset of pregnancy brings about rapid development of the gland under the influence of the pregnancy hormones, and during the final month there is accelerated development of secretory tissue. This mammary growth continues after parturition until peak milk yield is reached about 6 weeks after calving. Thereafter, the balance between production and loss of cells swings in favour of loss, and there is a reduction in cell numbers.

The udder of the cow has four quarters, with about 60 per cent of the milk produced in the hind quarters. Fibrous lateral suspensory ligaments and more elastic medial suspensory ligaments hold the gland in place. In older cows, these ligaments may stretch, giving rise to the pendulous appearance of the udder seen in some animals. One of the objectives of breeding management is to produce cows with a good udder support system.

The alveolar cells synthesize and secrete constituents of milk. These are discharged into the alveolar lumen, and from there into the ducts, sinuses and cisterns. A large amount of blood flows through the udder (about 500 litres per litre of milk produced) and some milk constitutents transfer directly from blood to milk. The major precursors of the constituents of milk are free fatty acids, triglycerides, amino acids, acetate and glucose. The main constituents of milk are outlined in Fig. 2.1.

The milk fat is made up of fatty acids derived both from synthesis in the udder and by transfer from the blood. The fatty acids which contain 4–10 carbon atoms are synthesized from acetate and β-hydroxybutyrate arising from the volatile fatty acids produced in the rumen. Those with 18 or more carbon atoms are directly transferred from blood triglycerides within the udder. The intermediate length fatty acids derive from both sources. Milk fat contains a mixture of both saturated and unsaturated fatty acids.

The lactose in milk (milk sugar) is derived

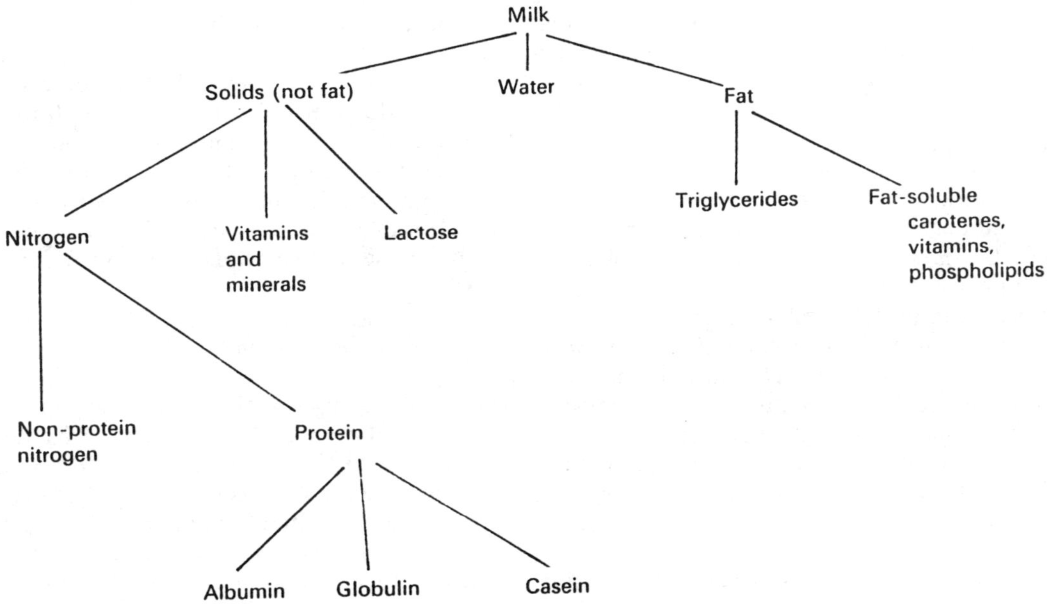

Fig. 2.1 Constituents of milk.

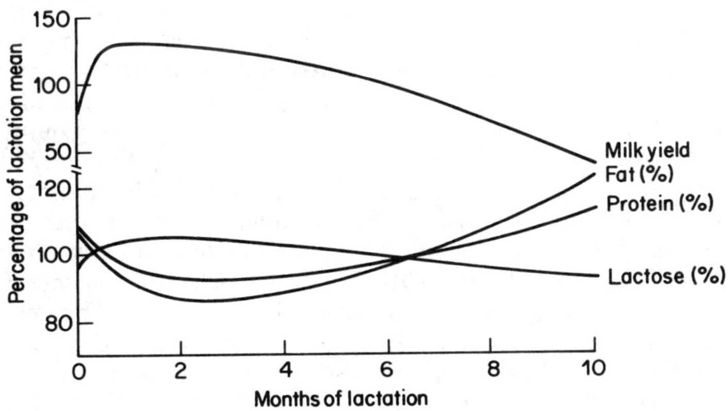

Fig. 2.2 Lactation curves of milk and its constituents.

mainly from blood glucose. The yield of milk is particularly dependent on the amount of water secreted in the udder, and this is associated with the water-soluble constituents, in particular lactose. Thus the milk yield is closely related to the amount of lactose synthesized in the udder.

The milk protein is derived mainly from the amino acids in the blood. The ash content is made up of the three major elements – calcium, phosphorus and magnesium – in addition to potassium, sodium, chlorine, and trace elements. The vitamins are absorbed from the blood, in particular vitamins A and the B complexes. Other vitamins include C, D, E and K.

Typical lactational trends in milk yield and composition are shown in Fig. 2.2. These stylized curves are modified by nutritional factors such as energy and protein intakes and the composition of the diet, and by non-nutritional factors such as genetic potential, age and disease.

Reproduction

Dairy cattle attain puberty when they reach about 40 per cent of mature weight. The age at which this is achieved depends on the plane of nutrition, but under good management conditions it will be at 10–12 months of age. There is a marked increase in follicular activity prior to puberty and, at puberty, normal oestrous cycles commence. These average 21 days in length, with the majority between 19 and 23 days. The cow is in oestrus for about 15 h, although this can range from 3 to over 24 h. Ovulation occurs 10–12 h after the end of oestrus.

If natural service is used, the cow will stand to be mounted by the bull during the oestrous period. When artificial insemination is used, the semen is deposited through the cervix into the uterus. Thus the optimum time to inseminate cows is in mid- to late oestrus. The normal gestation length is 283 days, but this varies between animals, and is particularly influenced by the sire of the foetus.

From the fifth month of pregnancy, which usually represents month 8 or 9 of lactation, milk yield begins to decline more rapidly, and it is normal practice to dry-off the cow 6–8 weeks before parturition is due, the simplest method being to stop milking. Cows giving large yields at drying off (over 15 kg/day) may have to be offered only straw and water for 2 or 3 days to reduce milk secretion. During the dry period the cow should continue to lay down fat in preparation for the next lactation, and should be fed accordingly.

After parturition, involution (return to normal size and function) of the uterus continues for 4–6 weeks, but reproductive cyclical activity normally recommences within 3 weeks of calving. Delays in resumption of cyclical activity are commonly caused by abnormalities of the reproductive tract following difficult calvings (dystokia), retained placenta, or endometritis, or are due to luteal or follicular cysts on the ovary. Underfeeding of cows can also lead to a delay in resuming normal oestrous cycles.

Normal behaviour patterns

Dairy cattle have been selected not only for their production potential, and therefore their ability to eat large amounts of feed and to convert this into milk, but also for their behaviour patterns in the farming environment in which they are kept. Cows which do not conform to the system tend to be culled from the herd.

Feeding and elimination

Feeding is a major activity of the cow. The feed intake of dairy cattle is determined by factors associated with the feed and with the animal. The feed factors concern the availability and type of feed. Animal factors include the genetic potential of the cow, the physiological state (pregnant, lactating, etc.) and behavioural factors. Intake during indoor feeding or during outdoor grazing can be considered as follows:

$$\text{intake} = \underset{(\text{min/day})}{\text{time spent eating}} \times \underset{(\text{g DM/min})}{\text{rate of eating}}$$

This simple relationship can be further broken down in behavioural terms. "Time spent eating" can be calculated as the number of bouts per day of eating multiplied by mean bout length (minutes). Similarly, "rate of eating" can be measured as the number of bites per minute multiplied by mean bite size. Typical feeding behaviour of dairy cows is shown in Table 2.4. Grazing behaviour is modified particularly by sward conditions. The maximum number of grazing bites per day appears to be about 40 000, which places a limit on herbage intake. With indoor feeding, restriction of feed availability leads to faster rates of eating and to a greater variation in intake between cows.

Rumination occurs in many bouts of differing length, but mainly when the cow is lying down, which is mostly after nightfall. Longer rumination times are associated with large intakes of fibrous feed. The frequency of drinking, and the amount consumed each time is determined by the dry matter content of the diet.

A cow will typically urinate on 8–10 occasions during the day and defaecate 10–15 times. The volume and number of these eliminations is determined by the amount and type of

Table 2.4
Typical feeding behaviour of lactating dairy cows

	Time spent feeding (min/day)	Rate of intake (g DM/min)	Total DM intake (kg/day)
Early season grazing	480	32	15.4
Late season grazing	560	19	10.6
Indoors *ad libitum* silage[a]	200	40	8.0
Indoors restricted silage[a]	100	60	6.0
Complete diet 60 : 40 ratio[b]	350	50	17.5

[a] In addition to 9 kg per day of concentrates.
[b] Ratio of concentrate DM to forage DM.

feed ingested. The amount of urine produced is often 10–20 kg/day, together with 30–50 kg of faeces.

Resting

Dairy cows generally lie down for 9–11 h/day, during which a high proportion of the rumination occurs. In grazing animals the diurnal activities of lying, standing and grazing are dictated by the hours of darkness and by weather conditions. In housed cattle a more regular diurnal pattern is seen, with lying and feeding activity dictated by the milking time and by the hours of darkness. Cows tend to lie in the same area of a building each time, although with cubicle housing, not necessarily in the same cubicle.

The amount of time spent lying is greater for young cattle – about 14 h/day for 3- to 6-month-old animals, and up to 20 h/day for younger calves.

Association

The social hierarchy in a group of dairy cattle is quickly established, and is for the most part linear, although it is usual for some triangles of dominance to occur within the general order. Quite often, the most dominant animals are not the most aggressive. Also, the cows leading the herd to and from the field for milking, and through the milking parlour, are not at the top of the dominance order, but usually from the middle. The most submissive animals do, however, tend to be the last animals through the parlour and, in competitive feeding situations, are the last to feed.

It is not clear what the maximum group size should be for dairy cow management systems. Herds of over 150 cows are now becoming increasingly common. However, for convenience of housing and management, such large herds are often split into groups of 50–100 during the winter months. Nevertheless, it is important that the groups are stable, because any interchange of cows between groups will upset the social hierarchy. Also, the optimum size of groups in which submissive cows can thrive is likely to be much smaller under housed than under grazing conditions.

The social hierarchy can be upset when cows are in oestrus or are sick. Submissive cows in oestrus can temporarily become dominant in encounters with cows higher in the order. Sick cows often lose their position, particularly if they are unable to isolate themselves from the rest of the herd.

Reproductive behaviour

At oestrus the cow shows signs of excitability, and there is increased grooming and occasional bellowing. The main behavioural signs are her attempts to mount other cows, which prompts them to mount her. Confirmation that she is in oestrus is given if she stands to let other cows mount. The duration of oestrus is from 3 to

24 h, the length depending on the time of year and the nutritional status of the cow. Duration is shorter in winter and for cows in poor body condition.

A bull is capable of detecting pre-oestrus in cows. He noses the perineum and probably smells the urine. An olfactory reflex is sometimes exhibited in which the neck is extended, the nostrils contracted and the upper lip curled. If the cow remains standing after nudging and courtship, the bull mounts the cow and mating takes place. A pumping action of the tail is normally shown, together with a forward thrust when ejaculation takes place. After mating, the bull often remains with the cow, and further mating might take place.

The pregnant cow generally shows no reproductive behavioural signs, but when parturition approaches she becomes restless and, if in a field, will often find a sheltered position away from the herd. Calving occurs most frequently at night, but there is some evidence that late-night feeding can delay parturition until the following day.

Most births occur without complications. In the first stage of labour, uterine contractions commence and continue for several hours and eventually the water bag and anterior feet of the calf appear. During a normal birth, the second stage, in which the calf is born, lasts about 1 h. The third stage, involving the expulsion of the placenta, takes up to 8 h.

Other behaviour patterns

Dairy cows are docile by nature and only change their behaviour patterns when environmental or management changes are introduced. When turned out to grass in the spring after being housed all winter, they exhibit play activities of vocalizing, together with galloping, kicking, pawing, and some playful fighting.

Grooming is a natural body care activity of the cow. Those parts of the body which cannot be licked are often rubbed against walls, fences, trees, etc. One cow may groom another by licking its head and neck, the grooming animal often being slightly below the other in the social hierarchy.

Under the circumstances of being put into a different building or field, some exploratory behaviour is exhibited. Territorial activity is not normally seen in dairy cattle.

Welfare considerations

There is now a general consensus that the suitability of systems of animal husbandry should not be judged simply on whether they affect animal performance. Due consideration has to be given to whether animals can fully exhibit the normal behaviour patterns described above, and this can lead to the safeguard of dairy cow welfare under a variety of management systems. Pressures to improve the welfare of dairy cows arise from both within and without the industry. As dairy cow potential increases through selection, and as economic pressures on farming increase, the need to have highly productive, profitable, long-living cows will be a priority. This can only be achieved where welfare is given a high profile. Public disquiet with some "intensive" animal systems will also help to ensure that welfare is safeguarded.

Problems of welfare are more likely to arise during the housing period, which in the UK accounts for between 4 and 8 months of the year, depending on the geographical location. Important requirements are:

1. Adequate space for freedom of movement and to exhibit normal behaviour patterns.
2. Daily exercise for cows which are tied.
3. The provision of a clean dry lying area.
4. Ventilation which prevents build-up of gas and minimizes smell.
5. Natural light during daytime and low-level lighting overnight.
6. A non-competitive feeding system.
7. Separate bedded areas for sick or calving cows.
8. Handling facilities for cows requiring treatment.
9. Emergency arrangements in the case of fire or disruption of supplies and services.

These factors are outlined in the Preface to the *Codes of Recommendations for the Welfare of Livestock*: *Cattle* (MAFF, 1983a) and they should receive full consideration in the design of dairy cattle facilities.

Genetic improvement

For the past 200 years there has been a gradual improvement in the performance of dairy cows and, with the introduction of artificial insemination (AI), and progeny testing of bulls, rapid progress has been made in the last 40 years within breeds. So far little progress has been made in developing cross-breeding schemes to exploit hybrid vigour, as has been done in other species of farm animals.

Genetic improvement is brought about by selection, i.e. choosing the parents of the next generation. The genotype or genetic merit of a cow represents the maximum performance which is possible under optimum management conditions. The phenotype is the actual performance achieved on the farm and is a function of the combined effects of genotype plus environment.

Most traits associated with milk production are quantitative, being controlled by a large number of genes. Those genes controlling milk yield, for example, are concerned with food intake, digestion and metabolism, blood supply to the udder, the amount of secretory tissue and udder size. Thus it is not surprising that there is considerable variation between cows in their ability to produce milk.

The heritability of a trait determines the rate at which it can be improved by selection. Milk, and milk fat and protein yields have moderate heritabilities of about 0.25, whereas milk composition (fat, protein and lactose content) has a higher heritability of about 0.50. Traits concerned with viability and health, such as fertility, disease resistance and longevity, have low heritabilities of less than 0.10, and therefore are difficult to improve by selection. In these cases improvement has to be brought about by management.

Only about 6 per cent of genetic progress is derived from female selection and this emphasizes the overwhelming importance of sire selection. Currently, young potential AI sires derive from matings of cows of high genetic merit with progeny tested bulls, and are themselves progeny tested. After 750 straws of his semen have been used, the young sire is laid off until an adequate number of his daughters have production records available from their first lactation. This takes 4–5 years. Only 10–20 per cent of such young sires are selected to join the AI stud for subsequent widespread use.

Young sires are progeny tested by the Milk Boards and by private companies. The first lactation daughter records used for the test are obtained from milk records on commercial farms by the official milk recording associations. Their records are compared with those of contemporaries (daughters of other sires) within the herd. This is termed a "contemporary comparison".

In future, the practice of multiple ovulation and embryo transfer (MOET) is likely to be used increasingly to speed up the rate of genetic progress. The production of a large number of brothers and sisters from the same mating allows the transmitting ability of young sires to be assessed more quickly from the performance of their sisters, instead of from their daughters as in the progeny test. MOET schemes will involve setting up nucleus herds in which testing takes place. An added advantage over progeny testing is that other traits, such as food conversion efficiency can also be measured and selected for.

Breeding

Reproductive efficiency

The reproductive efficiency of a dairy herd is a major factor affecting profitability. A high annual milk production per cow is dependent on regular calving at about 12-month intervals. Poor reproductive efficiency leads to longer calving indices and a reduction in milk sales per cow per annum. It also leads to an increase in the number of cows culled for failing to conceive. On average, in the UK, only about 85 cows calve per annum for every 100 cows available for breeding. Of the 15 per cent loss, 8 per cent is due to cows failing to breed and being culled, and 7 per cent is due to calving intervals of over 12 months (the average being 13 months).

The calving index achieved in a herd can be wholly accounted for by three management factors: time from parturition until the time when the decision is made to start serving; oestrous detection rate; and pregnancy (or conception) rate per service. Average and target values for these factors are shown in Table 2.5.

Table 2.5
Target and average values for reproductive performance

	Target	Average
Calving to start serving (days)	60	70
Oestrous detection rate (%)	80	60
Pregnancy rate to all services (%)	65	55
Calving index (days)	365	395

In well managed dairy herds, delay in the onset of oestrous cycles *post partum* is not a problem, with the majority showing cyclical activity by 21 days, and almost 100 per cent by 42 days *post partum*. High yielding cows may show a slight delay in resumption of these cycles, but rarely for more than 10–14 days. Clinical problems such as endometritis, and luteal and follicular cysts are much more likely to be the cause of delays, and any cows exhibiting these symptoms should be examined by the veterinary surgeon.

Oestrous detection

The oestrous detection rate (ODR) is the number of cows observed in oestrus durng a 3-week period, in relation to the number of cows expected to be observed. The average ODR on farms is about 60 per cent. It has been found that over 90 per cent of those cows not observed to be in oestrus are cycling, the remainder being acyclic. Thus the problem is mainly one of stockmanship. The duration and intensity of oestrous activity is variable and in winter, when activity is less, many oestrous periods are not detected.

The presence of a bull enhances oestrous activity, and the siting of a bull pen adjacent to the housing area or collecting yard can increase detection rates. Heat-mount detectors fixed to the back, or paint applied to the tail-head of the cow, are aids to an improved detection rate. When another cow mounts a cow in oestrus, the pressure on the cow's back triggers off a colour change in the heat-mount detector, and with tail paint, the paint is rubbed off. In both cases the attention of the stockman is drawn to cows showing these changes.

One alternative is to synchronize the oestrous period of several animals, by controlling the length of the luteal phase of the oestrous cycle with prostaglandin or progesterone. Injection with prostaglandin F2α or an analogue causes the corpus luteum to regress, and ovulation follows in 60–84 h. As only about 50 per cent of cycling cows are in the luteal phase at any one time, two injections are given at 11-day intervals. The cows are then given either a single insemination at 78 h or two inseminations at 72 and 96 h after the second injection. As prostaglandin F2α is an abortive agent, it should not be administered to animals which might be pregnant. It is also dependent for its action on the cow exhibiting normal cyclical activity, so treatment of non-cycling cows is wasteful.

Progesterone is administered as an impregnated intravaginal coil implanted for 9–12 days. Following withdrawal, the cow is inseminated 48 and 60 h later. This treatment, unlike prostaglandin, is also beneficial for cows which have not resumed normal oestrous cycles, as it stimulates ovarian activity.

Unfortunately, synchronization techniques, while overcoming the need to detect oestrus, do result quite often in poor conception rates. Their use, therefore, is not normally recommended, except in individual problem cases where the veterinary surgeon has confirmed that the cow has normal reproductive function, but oestrus has not been detected.

Conception

Although inseminations carried out at the optimum time (mid- to late oestrus) result in about 90 per cent of ova being fertilized by the spermatazoa, the average pregnancy rate (diagnosed 6–8 weeks post-insemination) is only about 55 per cent. Embryo losses account for the difference, most of them occurring within 14 days, with the cow returning to oestrus after the normal 21-day period. About 10 per cent of embryos are lost after 14 days and most losses occur before implantation is complete at about 30 days. These losses are characterized by a delayed return to oestrus, and are more common in older cows.

Natural service by bulls is still used on many farms, but the risk of slowing down genetic progress, and the inherent danger of handling bulls (in particular those of the dairy breeds), has persuaded most dairy farmers to rely mainly on AI.

Semen for AI is collected from bulls which are generally kept at AI stations. The semen is examined microscopically after collection to determine the proportion of live spermatozoa. A diluent is then added and the mixture is put into 0.25-cc straws and frozen in liquid nitrogen. Each straw contains about 20 million spermatozoa. They are stored in tanks of liquid nitrogen until the time for insemination. Most inseminations are carried out by trained operators from the Milk Boards. With the insemination technique, the operator inserts a gloved hand into the rectum and grips the cervix through the rectum wall. A catheter with the straw of semen is then inserted into the vagina, and passed through the cervix. The semen is deposited at the anterior end of the cervix. Careful training and hygiene are necessary for the technique to be effective.

On some farms, do-it-yourself (DIY) AI is carried out by trained members of staff to reduce costs, and as an attempt to increase pregnancy rates. The semen is stored on the farm in tanks, which are topped up with liquid nitrogen at intervals. Cows can be served at selected times of the day according to the stage of oestrus, whereas the AI Services only visit the farm once daily. Nevertheless, this potential advantage of DIY has to be set against the relative inexperience of the on-farm operators compared with full-time inseminators.

Appropriate handling facilities, such as AI stalls, are necessary for inseminations to be carried out. The cow should only be held in such stalls for a minimum period, as the deprivation from food and water and from the rest of the herd could be stressful.

Pregnancy diagnosis is carried out 6–8 weeks post-insemination by the veterinary surgeon, who palpates the reproductive tract *per rectum*. The development of on-farm tests which measure the progesterone levels in milk means that much earlier tests can be carried out at 21–26 days post-insemination, at which time levels in the pregnant animal will be markedly higher than in the non-pregnant, cycling cow.

Recording

If high levels of reproductive efficiency are to be attained, it is essential to have a satisfactory system for recording breeding data, particularly dates of calving, oestrous periods, services, pregnancy diagnoses, and veterinary treatments. Circular and horizontal wall charts are used on many farms to record the reproductive progress of individual cows. For large herds, computerized recording systems are being used increasingly.

Nutrient requirements

A large proportion of the feed eaten annually by the dairy cow is in the form of grazed grass and forage, and its digestive system is uniquely adapted for this purpose. The remaining part of the diet is made up of concentrate feeds which enable high energy intakes to be achieved. In total, a high yielding cow will eat over 5.5 tonnes of feed dry matter (DM) per annum. Energy, protein, minerals, vitamins and water are required for maintenance of the body and for productive purposes including lactation, reproduction and growth.

Energy

In the UK, the energy requirements of the dairy cow are measured as metabolizable energy (ME). This is the amount of energy available for metabolism after subtracting the energy losses in faeces and urine, and as gas from the rumen (Fig. 2.3).

The ME requirements of dairy cattle, and the ME contents of feeds can be found in MAFF (1984). The gross energy (GE) of concentrate feeds is greater than for forages, but in the total diet GE is normally in the range of 18–19 MJ/kg DM. The main variation between diets is the faecal energy loss (and hence digestible energy content). Losses in the faeces range from under

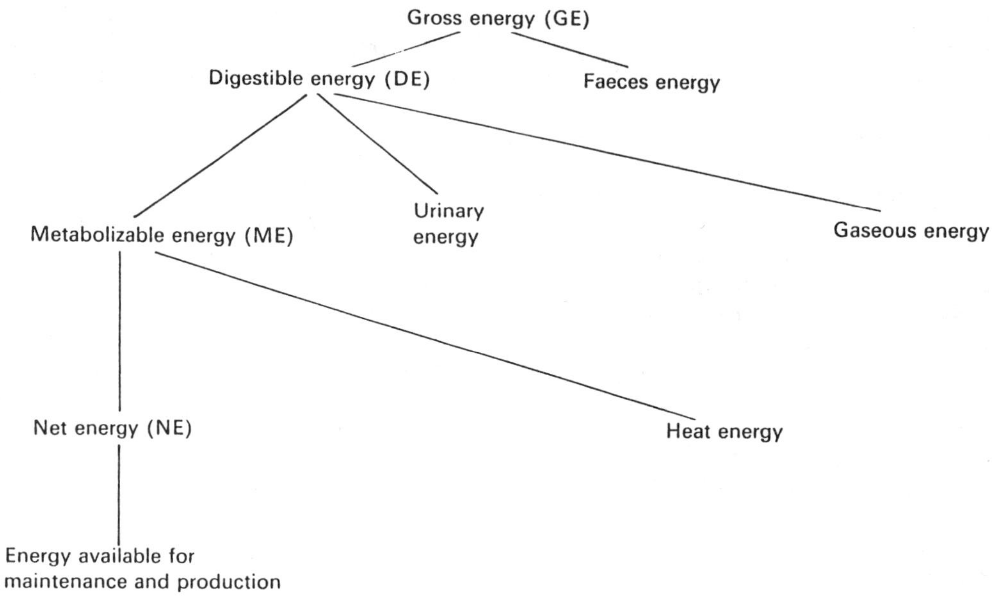

Fig. 2.3 Energy losses during the digestion and metabolism of a feed.

15 per cent in some concentrates to over 60 per cent in some straws. Urinary energy losses are normally in the range of 2–8 per cent of GE and gaseous losses from 6–10 per cent. Thus, ME can be estimated from digestible energy (DE) with some accuracy using a factor of 0.81.

The average efficiency with which the ME of the diet is utilized for different body processes is as follows: maintenance (0.72), milk production (0.62); live-weight gain in lactation (0.62), live-weight gain in the dry period (0.3–0.6, depending on diet metabolizability); and body tissue mobilization for milk production (0.82).

Food intake

A major factor affecting the performance of dairy cows is the voluntary food intake. This is determined by both animal and feed factors.

Food intake changes as lactation progresses, with peak intake occurring at about 15 weeks *post partum*. This peak occurs earlier in, and is greater for, thin compared with fat cows. Intake increases by about 1 kg DM/50 kg live-weight, and by about 1 kg DM/5 kg milk.

The most important feed factor affecting intake is the ME concentration of the diet (MJ/kg DM). Diets which have a low energy density and high fibre content (such as low-quality forages) are broken down slowly by the micro-organisms of the rumen. As a consequence the rate of passage of material out of the rumen is slow and intake is limited. Increasing the energy concentration of the diet by substituting concentrates for forage leads to a faster throughput of the rumen and higher intake. At very high energy concentrations the main limitation to intake is not the physical capacity of the rumen and the rate of passage of digesta, but the metabolic control mechanism, which acts in the same way as in the monogastric animal. In this situation the ME intake is not increased by further concentrate input (and DM intake declines).

Protein

The cow requires protein for both maintenance and production purposes. Formerly, protein requirements were expressed as digestible crude protein (DCP), with maintenance requirements of 350 g/day and production requirements of 0.55 g/kg milk for a 600-kg cow. It is now accepted, however, that this system has limitations and does not take into account how protein in the diet is digested and metabolized. The protein proportion of the diet is made up of true protein and non-protein nitrogen (NPN). Part of the true protein of the diet and the NPN are degraded to ammonia by the microorganisms in the rumen. The protein which is not degraded passes into the abomasum and small intestine, and is digested by enzymes. Much of the ammonia in the rumen is utilized by the microorganisms for growth, and this microbial protein also passes along the alimentary canal for normal enzymatic digestion. The ammonia which is surplus to the requirements of the microorganisms is absorbed into the bloodstream and transported to the liver where it is converted to urea. Some urea is recycled to the rumen via the saliva and the remainder is excreted in the urine.

The feed can therefore be considered to have a rumen-degradable protein (RDP) content and an undegradable protein content (UDP). For high yielding cows in early lactation, microbial protein production is inadequate to supply their protein requirements and a source of UDP is required in the diet. The choice of protein is thus of greater importance in early lactation. The degradability of feeds ranges from 100 per cent for NPN down to about 30 per cent for fish meal. The digestibility and amino acid make-up of the UDP are additional factors affecting the supply of protein to the tissues of the animal.

Matching protein supply to protein requirements is a complex process, because degradabilities of feeds are affected by associative effects with other feeds and by the feeding level, which affects the rate of passage through the rumen. A further factor in protein nutrition is the effect of the crude protein (CP) content of the diet on voluntary feed intake. Total DM intake increases as the CP content of the diet increases up to about 190 g/kg DM.

The protein requirements of dairy cows, and protein supply from feeds are outlined in ARC (1980, 1984).

Minerals and vitamins

Both minerals and vitamins are necessary for the basic functioning of the body. The requirements are fully discussed by Underwood (1981). Calcium and phosphorus account for about 70 per cent of the ash content of a cow's body, and are closely interrelated. The requirements for calcium are about 15 g/day for maintenance and 1.7 g/kg milk, and for phosphorus 13 g/day for maintenance and 1.5 g/kg milk. Magnesium is not easily mobilized from the bones where it is stored, and feeding levels of 11 g/day for maintenance plus 0.7 g/kg milk are essential. In spring, and occasionally in autumn, grazed grass is low in magnesium, and supplementation is required to prevent the onset of hypomagnesaemia (grass staggers; see p. 40).

Trace elements are required by cows in small quantities and the need for supplementation depends on the feed offered.

Water

Non-pregnant, non-lactating cows require about 4 kg water/kg feed DM eaten, supplied from the feed and drinking water. At low temperatures (0°C and below) this falls to 3.5 kg/kg and at high temperatures (over 27°C) it increases to 5.5 kg/kg feed DM. Cows in late pregnancy require about 50 per cent more water, and lactating cows have an additional requirement of about 0.87 kg/kg milk produced. Thus, a 600-kg cow yielding 30 kg milk/day, fed silage and concentrates, will drink about 70 kg water/day.

Feeding systems

Winter feeding

The choice of strategy for feeding the dairy herd depends on the available resources of the farm, in particular the land, labour, capital and milk quota. Thus the farmer has to decide how many cows to keep at what milk yield level to meet his quota target. The choice of few cows at high yield or more cows at a lower yield will be dictated by the land and labour availability, cost, and the opportunity for other profitable enterprises.

Table 2.6
Approximate concentrate allowances and forage requirements in early and mid-lactation for different yield levels and forage qualities

Lactation yield (kg)	Daily concentrates (kg/cow)[a]	Total forage requirement (tonnes DM)[b]	
		9.7 ME	10.5 ME[c]
4400	2		2.6
	4	2.2	
5000	4		2.4
	6	2.0	
5600	6		2.2
	8	1.9	
6100	8		2.1
	10	1.8	
6500	10		1.9
	12	1.6	

[a] For first 20 weeks of lactation.
[b] For a 200-day winter.
[c] Forage ME (MJ/kg DM).

Forage

In the UK, the traditional basal winter feeds of hay and roots have gradually been replaced by silage, produced mainly from grass, although in the south-east maize is also grown for silage. The amount of grass silage produced per hectare of grassland depends not only on the rainfall, soil type and fertilizer policy, but also on the frequency of cutting. Earlier and more frequent cutting leads to an increase in quality (digestibility and ME) but a decrease in quantity. The cow's requirement for forage for winter feeding is determined both by the quality of the forage and by the amount of concentrates fed – both these factors affect forage intake. Table 2.6 indicates the average amount of daily concentrates required for winter feeding of herds of different yield level, using silages of differing quality.

The "quality" of silage as it affects intake depends not only on the digestibility and ME, but also on the fermentation quality. Silages which produce a butyric acid fermentation with a high pH and high ammonia nitrogen (over 100 g/kg total nitrogen) content, are not particularly acceptable, and feed intakes are low.

If the forage produced on the farm is deficient either in quantity or quality, feed has to be purchased to make up for the deficit. Hay, straw and brewers' grains are commonly used as substitutes if silage is in short supply; if the available silage is of poor quality, more expensive concentrates have to be purchased to supplement it.

Concentrates

Concentrates offered to dairy cows comprise a whole range of materials. These include straights such as barley, sugar beet, maize gluten, soyabean meal and fishmeal; protein concentrates, which are mixtures of high protein straights fortified with minerals and vitamins together with some cereals; supplements such as vitamins, minerals and pharmaceutical additives in conjunction with a carrier, which are included at a low rate; and the most commonly used compounds containing mixtures of straights plus supplements.

The mean daily amount of concentrates to feed for different forage qualities and milk yield levels is shown in Table 2.6. The choice of protein level for the concentrate will depend on the protein content of the forage. Target crude protein levels in the total DM of the diet should be at least 16 per cent in early lactation and 14 per cent in late lactation.

The system used to allocate concentrates to individual cows is of secondary importance compared with the choice of the overall amount of concentrate for the herd, which is a major factor affecting profitability. Traditionally, concentrates were offered in proportion to the milk yield of the cow (e.g. 0.4 kg concentrates/kg milk). However, experiments carried out in the 1970s and 1980s with *ad libitum* forage showed that feeding according to yield gave no advantage in milk production of the herd compared with feeding all the cows the same daily amount. These conclusions apply to the early and mid-lactation period (weeks 1 to 24) and where the same total amount of concentrates is fed to the herd. As a consequence of these studies, many farmers use a simple flat-rate system in which all cows in early and mid-lactation are fed identical concentrates at levels similar to those in Table 2.6. This simple system alleviates the need for regular milk yield recording for feeding purposes. For cows in later lactation (post 24 weeks of lactation) the concentrate level can be reduced at a rate of 1 kg every 4 weeks without adversely affecting the persistency of lactation.

In larger herds, where more straight than compound concentrates are used, mixed diets of concentrate and forage (often termed complete diets) are fed. This system also places less emphasis on individual cow feeding. The normal practice is to offer mixed diets with a ME concentration of 11.5–12.0 MJ/kg DM in early lactation, 11.0–11.5 in mid-lactation and 10.5–11.0 in late lactation.

Feeding methods

In herds of less than 120 cows, self-feeding remains a common method of feeding silage. Cows graze the silage face, and a barrier or electrified wire is used to control the feeding and prevent waste. If the height of the silage is greater than 2 m, consideration should be given to removing the top material mechanically and feeding it in troughs, otherwise the cows have difficulty in pulling silage from the face, due to its density.

When the silage is available for about 20 h per day, a feed face width of 150–200 mm per cow is adequate. In systems where low amounts of concentrates are fed (6 kg/day or less), or where the time available for feeding is restricted, an increased feed face width must be allowed (e.g. up to 300 mm per cow). If this is not possible, some forage should be offered in troughs. The feed barrier or fence should be maintained at a distance of 400–600 mm from the feed face, to prevent a restriction in intake and, in the case of an electrified wire, the height above the ground should be about 800 mm.

In trough, bunker or passageway feeding systems, the silage is transported from the silo with a fore-end loader on the tractor, or by a forage box. The feed-face width per cow can be similar to that provided for self-feeding, providing other feeds such as concentrates or roots are not also fed. In these cases, all cows need to have access to feed at the same time and a minimum width per cow of 650 mm is necessary.

Fully mechanized forage storage and feeding systems have not been widely adopted. The storage of high DM silage (over 30 per cent DM) in concrete or metal towers, combined with automatic filling and emptying is a high capital and maintenance cost system. It allows weighed amounts of forage, with added concentrates, to be transported directly to troughs, with a minimal labour requirement.

A large proportion of the concentrates offered to dairy cows is fed in the milking parlour. Dispensers, which are often programmed for individual cows, transfer the concentrates into the trough in front of each milking stall. The amount of concentrates which can be fed at each milking depends on the rate of eating, and the milking parlour size and work routine. The maximum amount which can consistently be eaten is, however, about 3 kg.

Developments in mechanization and computerization have led to concentrate dispensers being sited in the housing area (out-of-parlour feeders). Cows are individually identified by transponders attached to their collars and, when they enter the feeder, they are recognized, and a programmed amount of concentrates is dispensed. This type of dispenser normally divides the daily allocation between four 6-h periods so that all the concentrates are not eaten at once. It is essential with such dispensers that adequate protection is given to the cow which is feeding, to prevent dominant cows from gaining access to the allocated concentrate. A stall with a self-closing rear gate is often used for this purpose.

If individual allocation of concentrates is not

required, then out-of-parlour concentrates can be mixed with the silage, either by putting a layer of concentrates in or on the silage in the forage box, or by using a mixer wagon. The latter method results in a more homogeneous mixture, the mixing mechanism being either by augers or paddles. Feeding forage and concentrates as mixtures has the financial advantage of allowing the use of a wider range of feeds, including straight concentrates and silage substitutes. Frequent feeding of concentrates, either mixed with the silage or through out-of-parlour feeders, has the nutritional advantage of placing less stress on the rumen. Large amounts of concentrates (over 3 kg/feed) taken in a single feed result in depressions in rumen pH. Potential problems resulting from this practice include cows going off feed, reduced total feed intakes, and reduced milk fat contents.

Summer feeding

Grazed herbage is the lowest cost feed available on the farm, but its efficient utilization often presents problems to the farmer. Management requires both the efficient production and utilization of herbage, while at the same time maintaining cow production levels.

Herbage intake

The performance of grazing dairy cows depends on a range of feed and environmental factors which influence the cows' willingness and ability to harvest the grass (Fig. 2.4).

Under poor grazing conditions, the rate of intake (g DM/min) of herbage declines and, as the time spent grazing (min/day) increases only slightly, the total herbage intake decreases. This

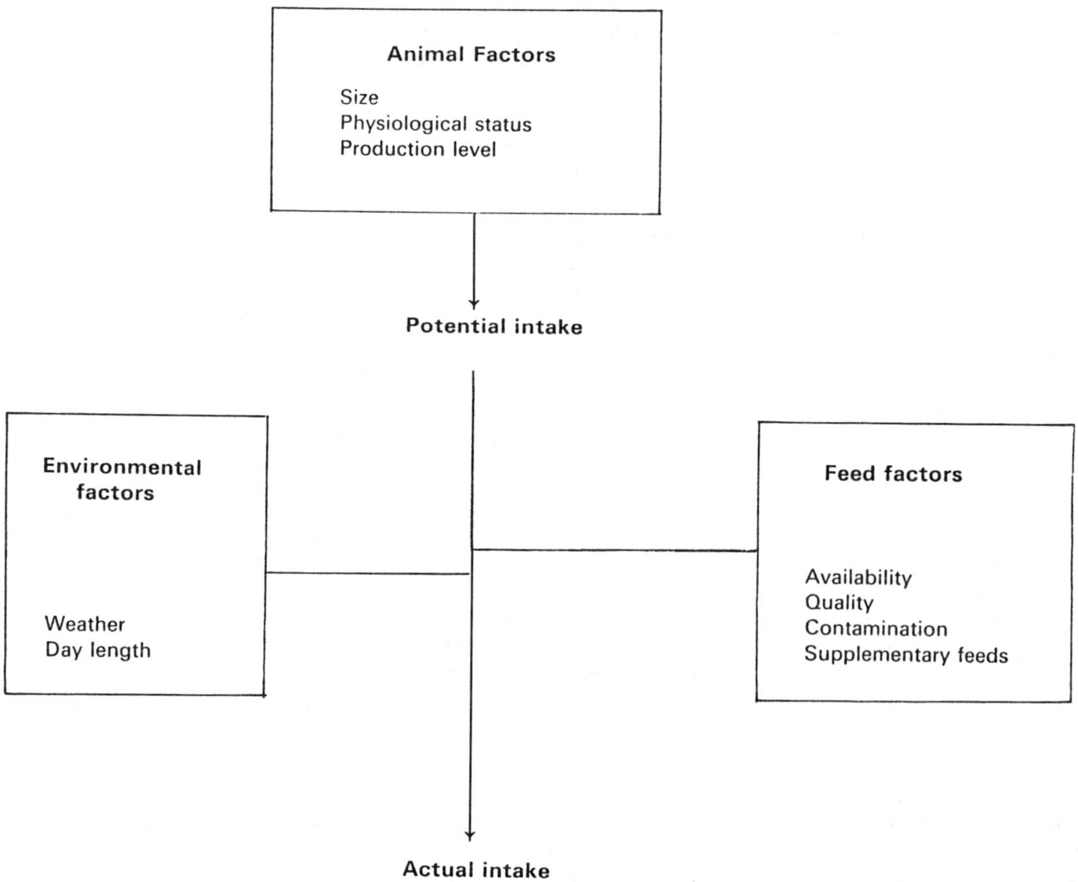

Fig. 2.4 Factors affecting the intake of grazed herbage.

is commonly seen when herbage availability and quality decline in late season. The rate of intake is a function of rate of biting (bites/min) and bite size (g DM; see also p. 18). Grazing studies have shown that rate of biting varies to only a small extent and consequently the major factor determining intake is bite size. Pasture management aimed at providing high intakes of herbage per day must therefore ensure that a large bite size is possible. Typical intakes and behaviour measurements over the grazing season are shown in Table 2.7.

Table 2.7
Grazing behaviour and typical grass intakes of dairy cows

Season	Intake[a] (kg DM)	Grazing time (min/day)	Rate of biting (bites/min)	Bite size (g DM)
Early	16	550	58	0.50
Mid	13	590	62	0.35
Late	11	610	64	0.28

[a] Intake = grazing time × rate of biting × bite size.

Supplementation

Supplementary feeding of grazing cows with concentrates is a common practice, although the economics are often questionable. The mean response in milk yield averages only 0.3 kg/kg extra concentrates, as the cow tends to compensate by consuming less herbage. Under good grazing conditions (at bite sizes of over 0.45 g DM) the time spent grazing is reduced by over 20 min/kg concentrate DM. The best responses to concentrates are therefore achieved where sward conditions are causing a limitation in herbage intake through a reduction in bite size. Cows of high milk yield potential also require supplementation if their potential is to be exploited.

When drought or overstocking lead to deficiencies in herbage availability, forages can be used to supplement intakes. A system of buffer feeding, in which silage, hay or a straw-based mix is offered *ad libitum* for a short period daily, allows the cows to decide for themselves how much to eat, thus making their own comment on the adequacy of the sward. The forage can be fed for 30–60 min after milking or, in more extreme circumstances, cows can be housed overnight and offered forage, and grazed intensively during the day.

Grazing systems

Rotational systems of grazing, such as strip-grazing or paddock grazing, impose a control on the amount of herbage made available. They require the regular application of fertilizer. Such systems lead to efficient utilization of grassland, but involve the added costs of fencing and watering compared with set-stocking.

Strip-grazing requires the daily or twice-daily movement of an electric fence, the decision on how far to move the fence being dictated by the length of stubble remaining on the grazed area. Similarly, with paddock grazing, the length of the grazed stubble indicates when the cows should be moved to the next paddock. Guidelines for the optimum stubble height are 6–8 cm for dry cows and 7–10 cm for lactating cows, the lower values reflecting spring and the higher late summer grazing. Rotational systems normally involve rotation lengths of 14–21 days in spring, increasing to 28–35 days in late summer.

Set-stocking (or continuous stocking) allows the cows free access to the whole grazing area, and this produces a short, dense sward, providing the pasture is not understocked. If this occurs, a mosaic of grazed and ungrazed areas develops. Advantages of the system are the increased longevity of the sward due to the high tiller population density, and the reduced labour, fencing and water trough requirements. Optimum sward heights are 5–8 cm for dry cows and 6–10 cm for lactating cows, with lower values for spring, and higher values for late summer grazing.

Zero-grazing involves housing cows in the summer, and cutting fresh grass each day to feed indoors. While the system has advantages in improved grassland utilization, it has the disadvantages of higher labour and mechanization costs. An alternative system is storage feeding; this involves year round indoor feeding of forage, which obviates the need for daily cutting of fresh grass. Both these systems face possible welfare problems, because the cows are housed all year round. Unless great attention is paid to ventilation, to cleanliness of the lying area with frequent scraping of concrete floors, and to cow space allocation, increased problems of lameness and mastitis may ensue. For such systems to be acceptable, an outdoor exercise paddock should be used for a minimum of 3 h daily during the summer. This will also be beneficial for oestrous detection purposes.

Stocking rate

Grass production is dependent on a variety of factors, including rainfall, soil type and fertilizer input and, because of this, it varies between geographical areas, between years and even between adjacent fields. The choice of stocking rate (number of cows per hectare of grassland) is therefore a more influential decision on performance than the actual grazing system.

High stocking rates increase both the amount of grass utilized, and cow productivity per hectare, but reduce individual cow performance. One method of overcoming this problem is to have high stocking rates to enhance grass utilization, but provide daily access to a buffer feed (forage or forage substitute) to allow the cows to top up any feed shortages from the pasture.

About 60 per cent of grass production occurs in the first 2 months of the grazing season. In eastern areas, where mid-summer drought is a problem, the proportion of the crop produced at this time is greater, while in the wetter western areas better suited to grass production, it is less. The stocking rate therefore should be high in early season (5.5–6.5 cows/ha), but should be reduced in mid- and again in late season (down to 2.5–3.5 cows/ha) in order to match grass production to cow requirements.

Monitoring and control of performance

Milk output

The performance of a dairy herd requires close monitoring and control. Within a quota system, month-by-month comparison of actual versus predicted milk sales is an integral part of management. This prediction is normally produced from a computer program which uses the predicted calving dates of cows to assess the mean lactation yields which are necessary, with that calving pattern and number of cows, to produce the annual quota.

The winter feeding programme is based on the quantity and quality of forage available, and on the required lactation yields. A ration formulation program can be used to assess concentrate inputs, whether they are to be fed on a flat-rate or on a feeding-to-yield system.

The most important recordings are, therefore, of total daily milk sales compared with predicted sales. A graph in the dairy or farm office shows at a glance whether changes in feeding management are necessary. If daily sales are below predicted levels, remedial action is necessary. The average response to an additional 1 kg of concentrates is 1 kg of milk. A greater or lesser response can be obtained, depending on a variety of factors. Greater responses in milk yield are obtained when forage is restricted in quantity (less than 90 per cent of appetite), when forage is of poor quality (less than 10 MJ of ME/kg DM) or when concentrate feeding levels are already low (less than 0.25 kg/kg milk). Extra concentrates may also lead to an increase in live-weight and to a reduction in forage intake. In examining the economics of responses to changes in concentrate input, therefore, the effects on milk yield, milk composition, live-weight change and forage intake must be taken into account.

Body condition

A further method of assessing the adequacy of feeding is the body condition of the cow. This can be estimated on the basis of fat cover around the tail-head (in beef cattle it is more common to assess the fat cover over the spinous processes in the loin area). The scale of scoring runs from 0 (extremely thin) to 5 (very fat) and has been described by Mulvany (1977). Most dairy cattle are in the 1.5–3.5 range of scores. Target condition scores at calving are 3.0–3.5 for the traditional British Friesian type and 2.5–3.0 for the more dairy-like Holstein type.

After calving, dairy cows will mobilize body fat, as they are in negative energy balance at this time (the intake of ME is less than the ME requirement for maintenance and milk production). Cows of high genetic potential will mobilize more fat than cows of low potential, and more fat will be mobilized on low levels of nutrition. The lowest condition score will coincide approximately with peak milk yield and also with the lowest milk protein content. After peak milk yield is reached, the intake of ME will exceed ME requirements for maintenance and milk production, and the body condition score and milk protein content will increase.

Because of the high energy requirements of the foetus, the dry cow often does not increase in body condition. Thus it is beneficial for the condition score achieved by day 250 of lactation to be within 0.5 of the target score at calving. If the mean group or herd score is not within 0.5 of a score at that time, feeding levels should be increased.

Condition scoring can have advantages over the weighing of cows. Live-weight fluctuates widely each day, due to milk extraction, drinking, feeding, urinating and defaecating. This disadvantage can be partially overcome by more frequent weighing, such as with an automatic weighing platform placed at the exit of the milking parlour. Results of such weighings are linked to a computer to give mean live-weights, and also live-weight changes over a period of time. However, even this sophisticated method does not define nutritional status accurately, as heifers and second parity cows in particular, are still growing as well as mobilizing or laying down fat. In early lactation, such animals are often mobilizing fat (reducing in condition score), but at the same time increasing in live-weight. Thus, a simple monthly assessment of condition scores in a herd can provide a very useful assessment of nutritional status.

Housing

In the UK, dairy cows are often housed for at least 6 months of the year. The overhead costs associated with housing and mechanization have increased as herd sizes have increased, and as systems have become less labour intensive. There is a need to ensure that this transformation in dairy cow management is associated with due consideration to animal welfare.

Environmental requirements

Dairy cows have a ruminant digestion and metabolism system which generates heat and which makes them very resistant to cold stress. This resistance is enhanced by tissue insulation and by the length of the coat. The heat production of a cow depends on its food intake, its size and its activity. Therefore, high yielding cows have a large heat output and are more resistant to cold than, for example, dry cows fed at maintenance level.

Dairy cattle in general have a lower tissue insulation than beef cattle, mainly because of the reduced thickness of skin and subcutaneous fat. The variation between animals within breeds is nevertheless quite high, due to differences in body condition. In cold environments the coat is shed less readily and becomes thicker and longer. However, high yielding dairy cows tend to have thin coats in winter because of the high level of heat production from digestion and metabolism.

The temperature range in which metabolic heat production is independent of air temperature is termed the zone of thermal neutrality. The range is not fixed, but will vary according to the metabolic state of the animal and to other environmental factors such as wind and rain (see also p. 62 and Table 3.13). The lower limit is the lower critical temperature, below which the animal has to increase heat production by shivering and other means. This critical temperature is raised by exposure to wind and rain.

In the UK, heat stress is not a major problem. If it does occur, it is likely to affect high yielding more than low yielding or dry cows, and a reduction in food intake is the normal response. However, adaptation does occur and the air temperature at which a reduction in food intake is seen depends on the temperature to which the animal has been accustomed. Heat is dissipated by sweating and panting, and solar radiation is to some extent absorbed and reflected by the coat.

The milk yields of cows decline below an air temperature of about −5°C, which is well above the lower critical temperature, which may be due to vasoconstriction and a consequent reduction in the blood flow to the udder. In European dairy breeds, milk yields and milk fat content are reduced at temperatures over about 25°C. The performance of dairy cows is therefore affected by less severe intensities of cold and heat than those affecting body heat production (at temperatures within the zone of thermal neutrality).

It may be concluded that the major justifications for housing cattle in winter are for the convenience of management and to avoid poaching damage to the land. For cows kept outdoors, potential heat stress in summer is overcome by the provision of adequate shelter from the sun. In winter, shelter from the wind and adequate levels of nutrition are necessary to ensure cattle are maintained above their lower critical temperature.

Housing requirements

The requirements of a housing system include those necessary for the well-being of the cows and those needed to provide an economic workable system for management. The housing area is the central component of the dairy unit complex, and has to be integrated with those other components associated with feeding, milking, handling and effluent disposal (Fig. 2.5).

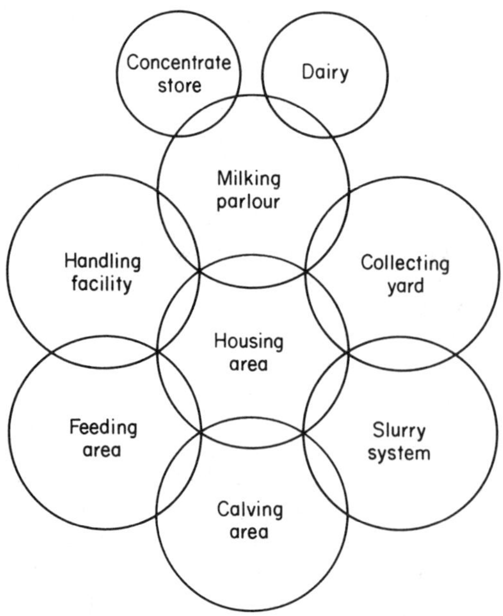

Fig. 2.5 Components of a dairy housing system (from Leaver, 1983).

The traditional system was to house and milk cows in cowsheds. This system was labour-intensive and not ideal for cow welfare, because exercise was limited. The increase in herd sizes, and the swing to silage as the main forage, has led to a predominance of loose housing systems. If managed well, these systems allow greater cow comfort and more exercise, and permit the mechanization of feeding and cleaning.

The cubicle system of housing is now the most common system. For dairy farms in non-arable areas it has the advantage of economy of bedding. Capital cost is high, and a satisfactory slurry removal, storage and spreading system is needed. It is illegal to permit contaminated water to enter a water course, and great emphasis has to be placed on containing slurry from cubicle housing systems, milking parlours and collecting yards, and silage effluent from silos. The legislation affecting the application of manures and wastes to land is included in several Acts of Parliament and is summarized in MAFF (1983b).

The most important aspect of cubicle division design is the prevention of injury to the cow. The lower rail should not catch the cow during lying and rising, and there should be ample lungeing space in front (with head-to-head cubicles) or to the side (with wall-facing cubicles) for when the cow is rising. The dimensions of the cubicle are the most important feature. These are shown in Table 2.8. A headrail attached to the top cubicle division rail, or a brisket rail placed on the cubicle floor, should be used to prevent dunging and urinating on the bed, by not allowing the cow to move too far forward in the cubicle.

Table 2.8
Recommended dimensions for cubicles for dairy cows

Live-weight (kg)	Length (m)	Width (m)
600–700	2.30	1.15
500–600	2.15	1.10
350–500	2.00	1.00

It is essential that ample clean, dry bedding is used in cubicles to prevent injury, particularly to the hocks, to provide comfort and insulation, and to prevent mastitis. The most widely used bedding materials are sawdust and wood shavings, with usage per cubicle of 1.0–1.5 kg/day. Damp sawdust and shavings should not be purchased, because they are likely to contain large concentrations of coliform bacteria, which can cause severe mastitis. Straw (usually chopped) is also a useful bedding for cubicles with a usage per cubicle of 1.5–2.0 kg/day. Cow mats made of rubber or a similar material reduce the abrasive surface of the cubicle bed and provide insulation. However, they are expensive, and some bedding material is still needed to absorb any slurry carried on to the bed by the cows' feet.

Strawyards provide a relatively low-cost housing system in arable areas. A major advantage of these is that the manure is handled in a solid form and can be easily stored until spreading. The amount of straw needed ranges from 2.5 to 5.0 kg/cow/day, depending on whether the whole or only part of the housing area is bedded.

The main types of dairy buildings are:

1. Framed (steel, concrete or timber) buildings used for cubicle or strawyard housing, with concrete block walls, space boarding above the walls and roofs of corrugated asbestos cement or aluminium sheets.
2. Prefabricated timber kennel buildings containing cubicles whose uprights support the sheeted roof.

Framed buildings have a greater life-span and lower maintenance costs and are more adaptable, whereas the kennel type has a lower initial cost and tends to be warmer.

The general requirements of housing systems relating to animal welfare are outlined on p. 20. *A British Standard Code of Practice for the Design of Buildings and Structures for Agriculture* (BSI, 1981) is available. The stockman should give particular attention to space, lighting, ventilation and the provision of feed and water.

Space allowance

In cubicle systems, an individual cubicle should be provided for each cow. The loafing/feeding area should provide a minimum of 3 m^2/cow, including a minimum passageway width between cubicles of 2 m and feeding passageway width of 3 m. In strawyard systems, the bedded area should provide a minimum 5 m^2/cow plus a loafing/feeding area of at least 2 m^2/cow.

Lighting

Adequate lighting is necessary, not only for the cattle to go about their normal routine of lying, standing, walking, feeding, drinking, grooming, etc., but also to allow the herdsman to inspect them at regular intervals throughout the day and evening. In buildings with insufficient roof lights, natural lighting should be supplemented with artificial lighting during the daytime. Experiments have shown that milk production benefits where artificial lighting is used to extend "day length" to 16 h when the natural day length is shorter than this. Thus, "full lighting" for 16 h, plus "minimal lighting" for the remaining 8 h to allow cattle to move around, is recommended.

Ventilation

There is a need for effective ventilation without draughts. Where possible, buildings should be designed for ventilation to be provided naturally. This not only reduces costs, but is generally more effective, and obviates the need for emergency cover in case of breakdown. Poor ventilation results in condensation, which increases the problem of keeping beds dry, and can increase the incidence of environmental mastitis. Where slurry is stored within the building in slatted tanks, extreme caution should be used in agitating the slurry during mixing or emptying. The gases generated are fatal to man and cattle, and additional ventilation, such as the opening of doors, should be considered during these processes.

Feed and water

Where cows are fed in or adjacent to the housing area, ample space (discussed earlier, p. 27) should be provided to allow all cows, irrespective of their place in the dominance order, to receive their daily requirement of feed. Cows require to drink four or five times a day (more on dry forages than on silage), and an adequate supply of clean water in troughs or bowls is necessary.

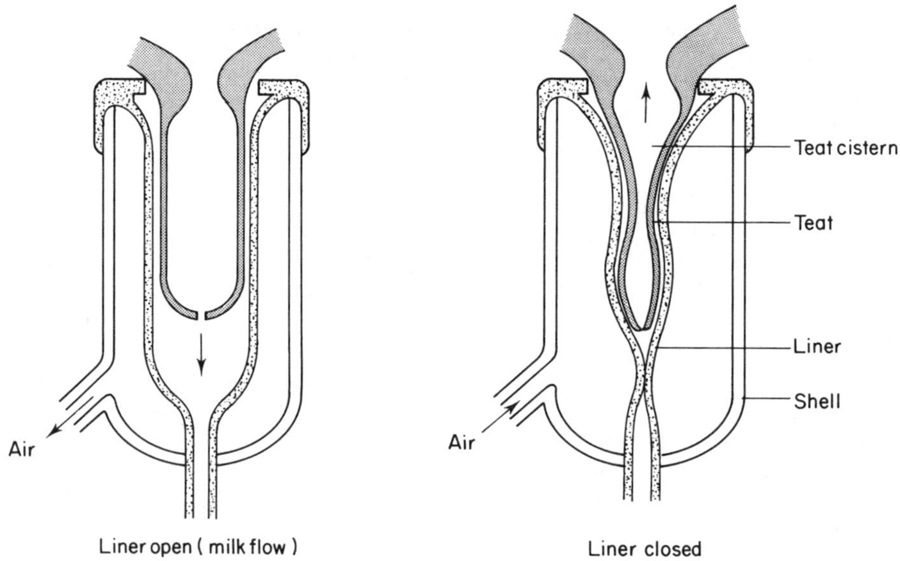

Fig. 2.6 Extraction of milk from the teat during machine milking.

Milking management

The efficiency of the milking process has become more important as herd sizes have enlarged, with more cows being managed per man. Hygienic extraction of milk and transfer to the bulk tank, with little stress on cow and man, are the main objectives.

Milk secretion

Milk is secreted continuously by the alveolar cells in the udder, and a reduced secretion rate does not occur unless the interval between milkings is increased to over 15 h. Thus, except in high yielding herds (over 6500 kg milk/cow), there is little benefit in having a more even time interval between twice daily milking than 14 h and 10 h. Thrice daily milking increases milk yields by about 15 per cent at all stages of lactation compared with twice daily. This suggests that it is not a reduced intramammary pressure which increases yields, but a stimulatory effect of the additional milking.

Milking machines

Milk is extracted by applying a vacuum to the teat end within the teat-cup liner (Fig. 2.6). The resulting pressure difference between the open liner and the teat sinus causes the teat sphincter to open, and the milk to flow out. A pulsation of alternate opening and closing of the liner prevents the damage to the teat end which would arise with a continuous vacuum. Milking machines work at a vacuum level of 45–50 kPa (13–15 inches of mercury). In the pulsation cycle, the liner is open for 50–75 per cent of the time, and there are 45–70 cycles per min.

The teat-cup liners are made of rubber or synthetic rubber, and an important characteristic is that the liner does not slip down the teat during milking. The milk passes from the liner into a claw-piece where the four liners meet, and this normally has an air-bleed which enhances the speed of removal of milk down long milk tubes into individual-cow, glass recorder jars (in milking parlours), or directly to the bulk milk tank.

Milking parlours

With the increase in herd size, the majority of cows are now milked in milking parlours, although many small herds are still milked in cowsheds, with a milk line transferring the milk to the bulk tank. The most common milking parlour in operation is the static herringbone (Fig. 2.7). Rates of throughput average about 30 cows per man hour for cowsheds, 55 cows per man hour for static herringbones and over 100 cows per man hour for rotaries.

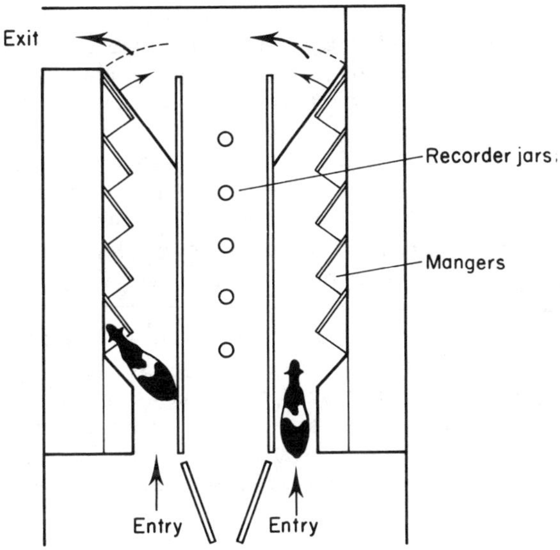

Fig. 2.7 A 10-stall, 5-milking-unit herringbone parlour.

The milking performance (cows milked per man hour) is determined by the time spent on (a) the work routine per cow, and (b) the time spent waiting by the operator for cows to milk out. The work routine is the time taken per cow to carry out all the operations associated with milking. The waiting time depends not only on the work routine, but also on the length of time the cows take to milk out, and the length of the parlour. Longer parlours (more cows per side) reduce the waiting time per cow because the operator spends longer carrying out the work routine on the side where clusters are being applied, before returning to the other side where the cows are already milking.

The total bacterial count of milk (TBC) which is carried out by the dairy authorities is mainly affected by the cleanliness of the cows' teats at milking and by the cleanliness of the milking plant. Washing teats with clean water (containing disinfectant), followed by drying with individual paper towels, leads to a reduction in the TBC level, which in well-managed herds is less than 15 000/ml. Milking parlour plants are cleaned automatically at the end of each milking by circulation cleaning or by acidified boiling water (ABW) cleaning. In the former system, which is the most common, there is a pre-rinse of warm water, then a circulated hot wash with detergent solution, followed by a cold water rinse. The process uses water at 85°C for the middle wash and takes about 15 min in total. The ABW system takes only about 6 min but requires more water at a temperature of about 95°C. The acid is added in the first half of the wash and the water is not circulated. (For measures associated with control of mastitis, see p. 42.)

Rearing dairy replacements

On most dairy farms the decision is made to rear replacements for the dairy herd rather than to purchase them from the market. The main reason is the job satisfaction achieved from influencing genetic progress in the herd through the choice of sires. Also, being self-contained prevents the introduction of disease, and training and preparation of heifers can take place prior to these entering the dairy herd.

Very often, however, rearing is done inefficiently, with this enterprise taking up land, labour and capital which could be more profitably used for other enterprises. Most dairy farmers also rear too many replacements, leading to high culling rates in the herd. In an efficient business the culling rate is only about 20 per cent and, if heifers are reared to calve at 2 years of age, there is a maximum requirement of about 50 youngstock per 100 dairy cows. None the less, a large proportion of herds carry over 100 female youngstock per 100 cows.

It can be considered good management in most dairy enterprises if no more than 60 per cent of cows are served with a dairy-bred sire, the remainder being to a beef sire to enhance calf values (see Chapter 3). Where possible, cows of high genetic index should be selected for serving with a dairy sire. Thus, a reduction in calving age to about 2 years, combined with a replacement rate of less than 25 per cent, will lead to many financial advantages compared with conventional calving at 2.75 years at a replacement rate of over 25 per cent. The advantages are reduced land requirement for youngstock, reduced working capital requirement and associated interest charges, reduced labour, mechanization and housing requirements, and an increased rate of genetic progress.

To calve heifers at 2 years of age at target live-weights (Table 2.9) does not require intensive rearing, and much of the total diet is based on grass and forage (Table 2.10). There is clear evidence that intensive feeding, producing very high pre-puberty growth rates, can be detrimental to lifetime milk production. The rearing period which is critical appears to be the 6-month period prior to puberty, when growth rates should not exceed about 0.8 kg/day, otherwise mammary development may be adversely affected, leading to permanent harmful effects on performance.

Table 2.9
Target live-weights (kg) for heifers of different breeds

	Birth weight	Target weight		Mature weight
		Mating	Pre-calving	
Holstein	43	360	550	660
Friesian	40	325	510	600
Ayrshire	32	280	430	510
Guernsey	27	260	390	450
Jersey	24	220	340	380

Newborn calves

If the calving takes place indoors, a clean dry pen should be made available. After birth, any mucus should be removed from the nose and mouth of the calf, and the navel should be dressed with iodine solution or with an antibiotic spray. The calf should suckle the dam within 6 h of birth to ensure an adequate absorption into the calf's bloodstream of immunoglobulins contained in the colostrum. The ability of the alimentary tract of the calf to absorb the immunoglobulins declines after about 6 h, and is negligible by 24 h after birth. The immunoglobulins play an important role in preventing *Escherichia coli* infections in the first few weeks of life, and also prevent the onset of other diseases.

If the calf is not seen to suckle in the first 6 h, then colostrum should be removed from the dam or another recently calved cow, and fed to the calf by a teat and bottle, or by using a stomach tube to transfer it directly to the abomasum. A minimum of 3 litres should be given.

Calf rearing systems

The calf should remain with its dam for at least 24 h. It can then be transferred to an individual pen and either trained to drink from a bucket or be fed by teat. For days 2 to 4 inclusive, colostrum should continue to be fed at a minimum of 5 litres per day in two or more feeds.

Table 2.10
An example of a rearing system for autumn-born Friesian heifers

Month	Live-weight (kg)	Expected LW gain (kg/day)	Total concentrates	Bulk feed
September	40			
		0.50	14 kg milk sub. + 50 kg conc.	Hay
November	70			
		0.65	350 kg conc.	Silage
April	180			
		0.70	100 kg barley	Grazing
October	310			
		0.50	250 kg barley	Silage
April	400			
		0.70	—	Grazing
September	510			

Table 2.11
Example of milk substitute feeding systems for calves following 5 days of colostrum feeding

	Once-daily bucket	Twice-daily bucket	Cold[a] *ad lib.*	Warm[b] *ad lib.*
Age at weaning (days)	35.0	35.0	35.0	35.0
Amount milk powder (g)[c]	500.0	250.0	1000.0	1000.0
Amount water (litres)[c]	3.0	2.5	7.0	7.0
Milk powder/calf to weaning (kg)	15.0	15.0	35.0	40.0
Live-weight gain (kg/day)	0.5	0.5	0.7	0.9

[a] Fed through a teat connected to a container of milk substitute.
[b] Fed through a teat connected to a dispensing machine of milk substitute.
[c] For once- and twice-daily bucket, amounts refer to amount per calf at each feed, and for *ad lib.* systems, amounts refer to concentration of milk substitute on offer.

Starting at 5 days of age, a milk substitute can be fed as an alternative to cows' milk. The four most common methods of feeding are shown in Table 2.11 (see also Chapter 3, p. 66). In all systems, clean water, fresh concentrates and a forage (hay or straw) should be made available. Weaning can take place abruptly, providing the calf is eating at least 650 g/day of concentrates. After weaning, calves should be housed in groups and offered forage, together with an appropriate level of concentrates (1–3 kg/day depending on forage quality) to produce a growth rate of 0.65–0.80 kg/day.

Routine procedures

All calves must be marked by 14 days of age with an approved ear tattoo or metal ear-tag bearing the herd letter and number, together with the individual calf number. If the calf is sent to market, it must be identified irrespective of age, but such identification is not necessary if it is sent direct from farm to slaughterhouse (*Tuberculosis Orders, 1964*).

Disbudding of calves should take place at 1–3 weeks of age using a gas or electrically heated cauterizing iron. A local anaesthetic is necessary for this operation (see Chapter 3, p. 76). At the same time, supernumery teats of heifer calves should be removed with a sharp pair of sterile scissors and iodine should be applied to the area.

General care and handling

Good stockmen are said to have personalities which can be defined as "confident introverts". As dairy herds become larger, the need has increased for self-reliant stockmen who not only do the routine jobs well, but also have the intelligence and drive to spend time observing and caring for their animals.

Cattle are creatures of habit, both individually and as a herd. Regular observation of cows and youngstock going through their normal routine alerts the stockman to any abnormal behaviour patterns of individuals. This helps to identify sick animals, cows about to calve, cows in oestrus, etc. Abnormal behaviour patterns, such as restlessness and bellowing in the herd, will also indicate where there are problems of feed availability and environmental problems. A good stockman observes such anomalous behaviour and knows when and how to react. He/she considers the welfare of the animals as important as their productive performance.

Handling and restraint

A cattle crush is an essential piece of equipment on the dairy farm for the control of cows for treatment. The crush can be usefully sited adjacent to the milking parlour and should incorporate a gathering pen, a forcing funnel, a race incorporating a footbath, the crush, and a shedding gate into two collecting yards (Fig. 2.8). Such a system can be used for most veterinary treatments involving dosing, injecting, blood sampling or rectal palpation. For the treatment of lame cows, the crush must have the ability to restrain the animal satisfactorily, and should incorporate a yoke, a bellyband and a means of securing the leg to be treated.

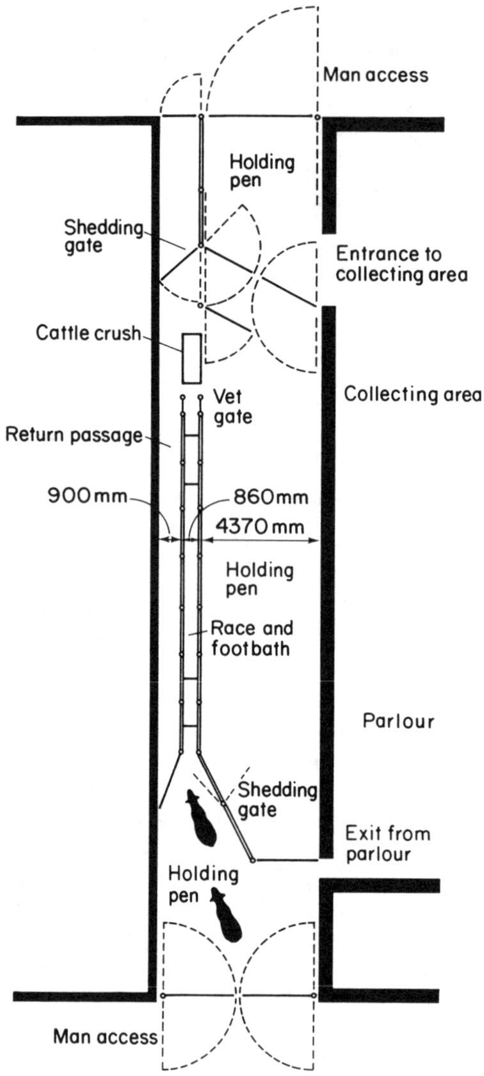

Fig. 2.8 Example of a handling facility for dairy cows (Crichton Royal Farms, Dumfries).

If a standard crush is used for inseminating cows, it is beneficial for the floor level to be at the same height for the cow and the inseminator. An alternative is to have separate AI stalls about 2.3 m long by 0.65 m wide.

The provision of self-locking yokes in a feeding passage is a useful facility for some veterinary treatments, for AI and for routine treatments such as tail trimming. Also, locking the cows in the yokes daily is a useful aid to oestrous detection, allowing a detailed examination of the vulva and tail area of each cow.

Calving and isolation areas

An area for calving cows and for the housing of sick or injured cows is a necessary requirement. A calving box of 12–25 m^2 is the most common facility on farms. This has the disadvantage of a small size, giving problems of manoeuvering when assistance is needed at calving. There is also the problem in a restricted area, that calves may be trampled or laid on. The difficulties of cleaning out such boxes can lead to a build-up of disease organisms, causing endometritis or coliform mastitis in the newly calved cow, and enteric disease or navel ill in calves.

Larger straw pens, in which a number of cows can be housed together, are preferable and have benefits if the stocking density is maintained at a low level (over 10 m^2/cow). Such pens are more easily cleaned out mechanically, and therefore are more likely to be maintained in a clean and dry condition. Calving areas should have wall surfaces which can be easily cleaned and disinfected, and should contain ample feeding space and drinking water.

For sick or suspect animals, isolation pens are required which should have a minimum size of 12 m^2, with no wall less than 3 m in length. The pen should be situated away from healthy stock, it should have adequate ventilation and a separate foul drainage system, and all surfaces should be able to be cleaned. A means of handling isolated animals is needed, and a tying ring and a crush gate are the simplest arrangement. Feeding and watering facilities should be incorporated.

Bull pens

If a bull is kept on the farm, a specialized bull pen should be available. This should have a housing area of at least 1 m^2/60 kg live-weight, with ample ventilation. An open exercise area should be provided, twice the size of the housing area, and with walls at least 1.5 m high. A service pen 3.3 m long by 1.2 m wide should be sited adjacent to it.

For safety reasons, provision for escape from bull pens must be made via escape gaps, railings or refuge walls. The pen should have a warning notice on the door. Health and Safety Guidelines on bull handling are available from HM Agricultural Inspectorate, Health and Safety Executive.

Transport

The movement of animals from one environment to another creates stress, and any transportation of animals should take full account of their welfare. In addition to the psychological stress, minor injuries are common in cattle transported by lorry or trailer when there is insufficient bedding material or inadequate restraint. These difficulties can be overcome by good stockmanship.

The Transit of Animals (Road and Rail) Order, 1975 lays down detailed requirements regarding vehicles used for the transportation of animals, and for the welfare of the animal during loading, carriage and unloading. Horned cattle must be separated during transit from those without horns, and calves must not be mixed with older animals (although a cow and her calf or calves may remain together). Food and water must be provided at intervals not exceeding 12 h, although this period can be extended to 15 h when the journey can be completed within that period of time. Calves are particularly susceptible to stress and disease during transportation, especially if mixed with other calves from different sources. They should have a minimal period in transit, should not journey through other markets, and dehydration should be avoided.

There is a great responsibility on stockmen, attendants at markets and hauliers to ensure that animals have a comfortable journey without suffering. The transport of sick or injured animals requires special attention, particularly during loading and unloading. Low-loading vehicles are beneficial for this purpose. In cases where the animal is unable to walk, loading should take place under veterinary supervision.

For severe injuries or sickness, the veterinary surgeon may arrange for slaughter of the animal on the farm, before transportation takes place (see Chapter 3, p. 78).

Health and disease

Dairy herd health is important, both in terms of the financial loss to the dairy farmer resulting from ill health in the herd, and in the suffering it may cause to the animals. Financial losses arise from deaths, culling, reduced milk production and from the costs of treatment. It is rare for deaths in a dairy herd to average more than 3 per cent per year, the most common causes being hypomagnesaemia (staggers), hypocalcaemia (milk fever), and coliform mastitis. In calves, mortality rates often exceed 5 per cent if management is not of a high standard. The major problems are calf scour and pneumonia.

Prevention of problems is more cost beneficial and causes less animal suffering than relying on the cure of sick animals. Advice in preventive medicine taken from the veterinary surgeon and animal husbandry advisers, and implemented on the farm, is an essential component of good farm management.

Reproductive problems

Dystokia

Difficult calvings (dystokia) arise in about 5 per cent of births in dairy cattle, although in first-calving heifers the incidence may be over 10 per cent. The result may be a stillbirth, particularly if prompt assistance is not given. The sire of the calf has a significant effect on both calf size and gestation length, and there is considerable variation in sires both between and within breeds (see also Chapter 3, p. 53 and Table 3.7). For maiden heifers, sires should be chosen which produce a known low incidence of dystokia. Heifers which are overfat at calving are more liable to dystokia. The plane of nutrition in late pregnancy should, therefore, be controlled to prevent too much fat accumulating around the reproductive tract.

Cows should be examined internally if labour has been in progress for several hours without the water bag or feet of the calf appearing. Internal examination should also take place if the labour does not progress within 2 h of the appearance of the water bag. These examinations should be carried out aseptically. Assistance with the calving can be given by attaching ropes to the legs of the calf and pulling in conjunction with the normal contraction of the uterus. If this assistance does not result in progress, or if the calf is in an abnormal position, veterinary assistance should be called.

Retention of the placenta

Any malfunction in the complex endocrine control of the third stage of labour can lead to retention of the placenta. This tends to occur following premature births (which are common with twins), and there is a high incidence of retention in herds with brucellosis infections. In many situations, a retained placenta will be expelled naturally after 6–10 days but, if there is any sign of endometritis, the veterinary surgeon should be called to remove it, and treat the infection.

Endometritis

Endometritis is an inflammation of the uterus which occurs after calving, and which leads to a discharge from the vulva. In severe cases, the cow becomes very ill, goes off her feed and loses condition. A reduced milk yield and poor fertility may ensue. Prompt treatment by the veterinary surgeon is necessary. To reduce the incidence of this infection, calving in hygienic conditions and the washing and disinfecting of hands and arms, and calving equipment during assisted calvings, is necessary.

Metabolic disorders

These disorders are a result of an imbalance between nutrient input and nutrient output.

Hypocalcaemia

The most common disease is hypocalcaemia or milk fever which most commonly occurs in cows in their third lactation or later, within 24 h of calving, although many cases are seen from 1 day before to 3 days after calving. Clinical signs are unsteadiness in walking, followed by the cow lying quietly and then being unable to rise. A continuing reduction in blood calcium leads to a cessation of intestinal movement, and muscle tremors. Eventually, if untreated, the cow rolls on to her side, and death can occur from the pressure of gas in the rumen. Treatment with calcium borogluconate administered subcutaneously and, in severe cases, intravenously, brings about a speedy recovery. Preventive measures include feeding a low calcium diet in the dry period and introducing a high calcium diet just before the calving date, or drenching with or injecting vitamin D, 6–10 days before calving, and feeding a high forage diet indoors for 2–3 weeks before calving.

Hypomagnesaemia

Hypomagnesaemia, or low blood magnesium, occurs mainly in animals on spring grass, which has a low magnesium content, and occasionally on autumn grass. As the body has virtually no stores of magnesium, an adequate daily intake is required. The symptoms of this condition are increased excitability followed by falling over with spasms in the legs. Death due to heart failure can follow quickly. Subcutaneous administration of magnesium sulphate solution should be given immediately the condition is recognized, and the veterinary surgeon called. Intravenous therapy can be fatal and such treatment requires professional expertise. Hypomagnesaemia can be prevented by feeding 60 g/day of calcined magnesite mixed with the feed. The magnesium can also be supplied in drinking water using a metering device which delivers a measured amount into the water trough.

Acetonaemia

Acetonaemia, or ketosis is a problem of high yielding cows in early lactation. The cow has a reduced appetite and becomes dull and lethargic. Milk yield falls, and there is a sweet smell from the breath and in the milk, due to the formation of ketone bodies. These are produced in the liver from acetate when, because of inadequate energy intake, there is insufficient propionate available from the rumen. Treatment of the disease can include intravenous injection of glucose, although this effect is only transient. The condition can be triggered by other clinical problems such as mastitis, lameness, or change of diet, or any problem which causes the cow to reduce its feed intake. The disease can usually be prevented by having available good quality forage fed on an *ad libitum* basis.

Mineral deficiencies

For dairy cattle fed mixed diets, mineral and trace element deficiencies are rare. However, on all-grass or all-forage diets, deficiencies in some trace elements may occur in lactating cows in some geographical areas. The most common problems are associated with low levels of copper, cobalt or selenium in the grass or forage. Blood analyses arranged by the veterinary surgeon to examine the energy and protein status of cows, can also be used to examine their trace element (and major mineral) status.

Disorders of the digestive tract

The most common disorders are those relating to the rumen, and these are mainly caused by the type of and changes in diet.

Bloat

Bloat is the result of the failure to expel, by belching, the gas formed during the normal fermentation process. With some diets, a stable foam which prevents the escape of gas is produced in the rumen. Other causes are the cessation of ruminal contractions, such as in cows with hypocalcaemia, and blockages of the oesophagus. The pressure which is built up in the rumen causes the animal to lie down on its side, and death can result from heart failure, or from choking following inhalation of rumen contents. In severe cases, the veterinary surgeon uses a trochar and cannula to gain entry to the rumen through the side of the animal, which allows the gas to escape. As a preventive measure, antifoaming agents such as peanut, linseed and paraffin oils, and synthetic detergents can be administered as a drench, in the feed or in drinking water, or by application to the flanks for licking, or by spraying the pasture. Bloat is most commonly seen on legume pastures. Incidence of this condition can be reduced by feeding long roughage.

Scouring

Scouring (loose faeces) in cows is commonly seen when diets contain a high protein level and/or a low fibre level, as with spring grass. It is an indicator of digestive problems, if not linked with this type of diet. Possible causes are liver fluke, Johne's Disease and, rarely, parasitic gastroenteritis. If such scouring is seen, the veterinary surgeon should be called.

Acidosis

Acidosis is a digestive disorder resulting from a lowering of the rumen pH from its normal level of 6.0–6.5. It is usually a consequence of the consumption of large amounts of starchy concentrate. The lower pH reduces rumen motility, and feed intake is reduced or ceases altogether. At very low levels of pH (4.5 and below), caused by gorging on concentrates, the lactic acid concentration in the rumen leads to a reduction in microbial activity, and toxin production. This may cause liver damage and, in some cases, death. More typically, cows on high-concentrate and low-forage inputs show a more chronic acidosis, which is characterized by low milk fat levels and laminitis.

Preventive measures include provision of high quality forage offered *ad libitum*, feeding concentrates little and often (or mixed with the forage), ensuring that concentrates are not all of a starchy type, and feeding sodium bicarbonate (as 1–2 per cent of the concentrate).

Displacement of the abomasum

Around the time of calving, the abomasum occasionally becomes displaced, migrating under the rumen from the right side to the left. Subsequent gas formation in the abomasum causes the cow to go off her feed, and the milk yield declines. Acetonaemia may follow as a result. The abomasum can be returned to its normal position, either by surgery or by manipulation through the abdominal wall with the cow lying on its back. The feeding of a high forage diet in the dry period reduces the incidence of the condition.

Ingestion of foreign bodies

Cows with apparent indigestion may have eaten a foreign body such as a piece of wire. The symptoms are grunting, an arched back and forward kicking of the hind legs. The veterinary surgeon should be called to deal with such cases, which may require surgery.

Lameness

Lameness problems are usually associated with the winter months. Cubicle housing with slurry systems, silage feeding, high planes of nutrition, and genetic factors are all possible predisposing causes. The lameness represents the symptom of a variety of possible underlying problems, although most cases are caused by foot disorders. About 90 per cent of cases are associated with the outer claws of the hind feet.

A large proportion of foot problems result from laminitis, which occurs in early lactation, and results in poor quality horn growth. The disorder is characterized initially by tenderness in the feet, followed some weeks or months later by foot problems such as solar ulcers, as the poorly developed horn reaches the sole surface. Both high-concentrate and high-protein diets have been found to be predisposing causes. The incidence of the problem is reduced by not "lead feeding" concentrates in early lactation, by ensuring the total diet has not more than 18 per cent crude protein in the total DM, by regular scraping (minimum twice daily) of concrete passageways, and by regular hoof trimming to prevent overgrowths (toe length from hair-line to point of toe should be about 700 mm). Severe claw problems can be alleviated by gluing a wood block under the unaffected claw to take the load-bearing from the affected claw.

In summer and winter, foul of the foot, caused by *Fusiformis necrophorus*, can reach a high incidence. The symptoms are a swelling above the hoof, severe lameness and a putrid smell between the claws. Footbathing in formalin, copper sulphate or zinc sulphate solutions will prevent the condition. Clinical cases can be treated with sulphonamide drugs given subcutaneously, or by antibiotic injection administered by the veterinary surgeon.

Lameness in dairy cattle leads to a loss of cow condition and a reduced milk yield, and may necessitate increased culling. This economic loss, combined with the suffering of affected animals, highlights the regular attention which stockmen have to give to the problem.

Mastitis

Mastitis, or inflammation of the mammary gland, is caused by the action of disease organisms which have entered the udder via the teat orifice. Many infections remain at a subclinical stage and clots are not seen in the milk, although adverse effects on milk yield occur. Such infections may be eliminated naturally, while others become clinical some time later. The presence of clots (and of clinical disease) can be detected by taking a foremilk sample from each teat prior to milking. Such clots are formed from precipitation of milk proteins by leucocytes and epithelial cells. In severe cases the quarter may become swollen and, if the infection gets into the bloodstream, a raised temperature and death can occur.

From 10 to 90 per cent of cows in a herd may be infected at any one time, although clinical cases are normally under 5 per cent. Depressions in milk yield average about 15 per cent, and this is associated with a reduced milk fat and lactose content.

Most infections are contracted during the milking phase, and are transferred by the teat cups from diseased to susceptible quarters. Common organisms are *Streptococcus agalactiae*, *S. dysgalactiae*, *S. uberis* and *Staphylococcus aureus*. Good hygiene at milking will maintain low levels of these infections. Preventive measures include washing and drying of udders prior to attaching the cluster, and teat disinfection immediately after removal of the cluster by dipping or spraying the teats with an iodine, chlorhexidine or hypochlorite solution. A routine check on the milking plant, in particular the reserve vacuum level, is also indicated as a preventive measure. Antibiotic therapy administered by infusing all quarters at drying off is an established technique of reducing the overall herd level of infection. During lactation, clinical cases of mastitis should be detected early by the use of foremilking at each milking. Treatment is by intramammary infusion, or by injection of antibiotic or other preparations.

Other organisms may be picked up between milkings. *E. coli* infections generally originate in the housing area. They are often seen in early lactation in herds with low cell counts in milk. Contamination of the teat orifice with dirt is the main cause. This type of infection is often severe, with no subclinical phase, and affected quarters, even if treated successfully, often fail to produce milk subsequently. Very severe cases can result in fever and death. Prevention through keeping cows clean and dry should be a major objective.

Antibiotic therapy for clinical mastitis in lactation given by intramuscular or intramammary methods must be associated with the discarding of milk for the prescribed period recommended by the manufacturer. Milk is tested for antibiotic content at the dairy receiving milk from the farm, and farmers selling milk containing antibiotics are penalized. Milk is also tested by the Milk Board for its white cell numbers, which indicate mastitis levels in the herd. Average herd levels are about 350 000 cells/ml. In well managed herds cell counts are below 250 000 cells/ml.

Summer mastitis

Summer mastitis in maiden heifers and dry cows occurs in July and August. It is caused mainly by *Corynebacterium pyogenes*, which is transmitted by flies, and invariably results in the loss of a quarter. The problem is controlled by keeping flies away from the udder by treating at intervals with fly spray or Stockholm tar, and by long-acting intramammary dry-cow antibiotic therapy.

Health problems of youngstock

Most calf deaths occur in the first 4 weeks of life, mainly from septicaemia and scours (see also Chapter 3, p. 76).

Septicaemia

Septicaemia arises from *E. coli* bacteria being absorbed through the gut wall into the bloodstream. This usually takes place soon after birth, in calves which do not receive adequate amounts of colostrum, and which, therefore, have little or no passive immunity in the form of immunoglobulins in the blood. The result is often sudden death.

Scouring

Scouring in calves (diarrhoea) is caused by localized infections of the intestine. This results in faeces which are usually white and pasty in appearance, but sometimes dark and watery. Dehydration results and, if the calf is not treated, death can occur. Treatment should include a reduction in the concentration of the milk substitute, accompanied by the feeding of electrolyte solution. Calves which do not respond to these treatments should be seen by the veterinary surgeon.

Pneumonia

Following weaning, the major problem in housed calves is pneumonia caused initially by viruses, but often with secondary bacterial infections. The predisposing factors are inadequate air space per calf resulting from low roofs, high stocking densities, and too little ventilation. Draughts should be avoided. In older calves which are ruminating, low air temperatures are not of importance. Naturally ventilated buildings, such as the monopitch design, generally result in a low incidence of pneumonia. Vaccines are available which give a degree of protection from some of the viruses causing pneumonia.

A good indicator of calf well-being is the rectal temperature, which is taken with a clinical thermometer. Any calves which are reluctant to feed, which have sunken eyes, starey coat, hunched back or loose dung should have their temperature taken. Any animals which show a deviation of ± 1°C from the normal temperature of 101.4°C should be examined immediately by the veterinary surgeon.

Parasitic diseases

In summer, two nematode parasites common to

calves in their first grazing season are lungworm (*Dictyocaulus viviparus*), causing parasitic bronchitis (husk), and stomach worms (*Ostertagia ostertagi*), causing parasitic gastroenteritis (see also Chapter 3, p. 71).

Lungworm infection results in a husky cough, bronchitis and pneumonia, together with a severe weight loss 3–4 weeks after infective larvae, picked up from the pasture, reach the lungs. The severity of the disease is dependent on the amount of larvae ingested, and high mortality rates can occur. The most dangerous period is from June to November. The disease can be treated by anthelmintics, but the best procedure is to vaccinate calves with an oral vaccine containing irradiated larvae before they are turned out to grass (two doses with a 4-week interval between).

Stomach worm infection is characterized by diarrhoea and a rapid loss of weight of calves in their first grazing season, usually from July onwards. It causes clinical disease in a high proportion of animals, but mortality is low. Larvae which overwinter in the grass sward are eaten in the spring, and these develop into adult worms in the abomasum and, as with the lungworm, the eggs pass out in the faeces and develop on the sward into infective larvae. Vaccination against the disease is not possible, but grassland management, in conjunction where necessary with the use of anthelmintics, can control the disease.

One method of control is to treat the calves with anthelmintic in July to remove the adult worms, and then move the calves on to clean pasture (one that has been cut for silage or hay, or has been grazed, either by cattle in their second or later grazing season, or by sheep.) A second method is to integrate calves, older animals and grass cutting (for silage or hay) into a rotational system, to reduce the numbers of infective larvae on the sward (Fig. 2.9), because clinical disease only occurs when there is a high concentration of these.

A second type of stomach worm disease can occur in the late winter months, and is caused by larvae which have been ingested in autumn becoming inhibited in the wall of the abomasum. These develop into adult worms in late winter and produce similar symptoms to the summer disease, but with low morbidity and high mortality. This disease is controlled by grazing calves on clean aftermath in late summer and autumn.

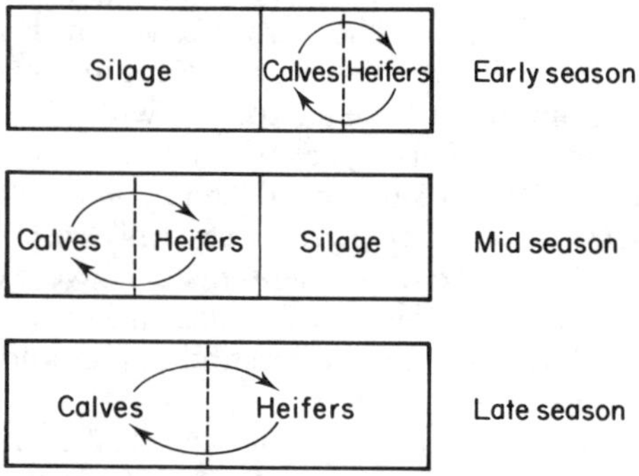

Fig. 2.9 Grazing system for dairy youngstock.

Notifiable diseases

The Ministry of Agriculture has the authority to control the movement of livestock and regulate imports when these measures are necessary for the control of certain diseases which include, in cattle, anthrax, brucellosis, enzootic bovine leucosis, foot-and-mouth disease, bovine tuberculosis, and warble fly infestation. It is the legal requirement of the person in charge of an animal which is suspected of having one of these diseases to report it to the Ministry of Agriculture or to the police. Legal powers are conferred in Orders under the Animal Health Act, 1981. Zoonotic diseases, such as salmonellosis and brucellosis (which are transmissible from animal to man), must also be reported as required by the Zoonosis Order, 1975.

It is a legal requirement for owners to record movements of cattle to and from the farm, to identify animals from birth with an ear-tag, and to report all cases of abortion and sudden death.

The way ahead

The economic pressures on dairy farming will ensure that the number of herds in the UK reduce and herd sizes increase. It is unlikely, however, that there will be any swing back to labour-intensive systems. Technological developments will continue to produce labour-saving husbandry methods. This need not be detrimental to animal husbandry because, given good management, the removal of drudgery from the daily work routine should allow more time for the stockman to oversee his animals.

New breeding developments, including multiple ovulation and embryo transfer, will speed up genetic progress in milk yield potential. But good management will be essential if this potential is to be realized in individual cows, and maintained over a number of lactations.

Feeding systems will need to provide *ad libitum* access to feed for much of the lactation cycle. Thus, buffer feeding of forage or forage substitutes in the summer, and free access to good quality forage in the winter, should ensure that undernourished cows, and associated fertility and metabolic problems, are less of a problem in the future. Mixed diets of forages, by-products and straight concentrates are likely to be used increasingly.

Environmental pressures on the spreading of slurry may encourage a swing back to straw-bedded systems of housing dairy cows in arable areas. This may also have benefits in cow comfort. The possible development of automatic milking cluster attachment has wide-ranging implications for management. This could allow the elimination of the milking parlour, because cows could at their leisure enter stalls to be milked automatically. More frequent milking would increase production and place less stress on the udder. An important benefit would be to allow the stockman to spend more time observing and tending his animals and less time on routine laborious work.

Pressure from consumers of animal products is likely to ensure that these are free from artificial additives. This may prevent the use of stimulants to enhance milk production, even when their presence in milk is negligible or undetectable. Interest in farm animal welfare by the general public is also increasing. This awareness should lead to a continuing improvement in the standard of management of dairy cattle.

References and further reading

Agricultural Research Council (1980). *The Nutrient Requirements of Ruminant Livestock*. Slough: Commonwealth Agricultural Bureaux.

Agricultural Research Council (1984). *The Nutrient Requirements of Ruminant Livestock*, Suppl. No. 1. Slough: Commonwealth Agricultural Bureaux.

Blowey, R. W. (1985). *A Veterinary Book for Dairy Farmers*. Ipswich: Farming Press.

British Standards Institute (1981). *British Standard Code for the Design of Buildings and Structures for Agriculture, BS 5502, Section 2.2*. London: BSI.

Broster, W. H., Phipps, R. H. and Johnson, C. L. (eds) (1986). *Principles and Practice of Feeding Dairy Cows*. Technical Bulletin 8. Reading: NIRD.

Federation of Milk Marketing Boards (1985 and 1986). *Dairy Facts and Figures*. Thames Ditton: Federation of Milk Marketing Boards (published annually).

Fraser, A. F. (1980). *Farm Animal Behaviour*. London: Baillière Tindall.

Leaver, J. D. (1983). *Milk Production – Science and Practice*. London: Longman.

Ministry of Agriculture, Fisheries and Food (1983a). *Codes of Recommendations for the Welfare of Livestock – Cattle*, L 701. London: HMSO.

Ministry of Agriculture, Fisheries and Food (1983b). *Farm Waste Management. General Information*, B 2077. London: HMSO.

Ministry of Agriculture, Fisheries and Food (1984). *Energy Allowances and Feeding Systems for Ruminants*, RB 433. London: HMSO.

Mulvany, P. M. (1977). *Dairy Cow Condition Scoring*. Paper 4468. Reading: NIRD.

Thiel, C. C. and Dodd, F. H. (eds) (1977). *Machine Milking*. Reading: NIRD.

Underwood, E. J. N. (1981). *The Mineral Nutrition of Livestock*. Slough: CAB.

3 Beef cattle and veal calves

A. J. F. Webster

The beef industry in the UK

The British beef industry is closely integrated with the dairy industry. There are about twice as many dairy cows as beef cows (Tables 3.1 and 3.2). Each dairy cow will be kept, on average, for four to five lactations which means that 20–25 per cent of calves born to dairy cows enter the dairy herd as replacement heifers. Most of the rest – bulls, castrated bulls (steers) and heifers – are reared for prime beef and killed at 1–2 years of age. A small proportion of calves from the dairy herd (about 8 per cent) are either killed at a few days of age as "bobby" veal or reared for about 4 months for quality veal. The majority of these are exported and finished on the continent of Europe (Table 3.2).

During the years from 1975 to 1985, beef cows (or suckler cows), whose only saleable products are their calves and eventually their own carcasses, have declined in numbers relatively more than dairy cows (Table 3.1), whose prime saleable commodity is milk. However, the imposition of quotas on dairy production in 1984 may reverse this trend. Beef production in the European Economic Community (EEC) currently exceeds demand, although in the UK it is in almost exact balance. Table 3.1 shows that the majority of suckler beef cows are kept in herds of less than 50, which implies that they

Table 3.1
Statistics relating to the UK beef industry in 1975 and 1985

	Year 1975	1985	1985 : 1975 ratio
Livestock numbers (thousands)			
dairy cows	3242	3131	0.96
beef (suckler) cows	1899	1339	0.70
total cattle and calves	14 717	12 847	0.87
Beef cow units, by size (thousands)			
1–19 cows	72.5	51.4	0.70
20–48 cows	20.8	13.9	0.66
over 50 cows	9.1	6.6	0.72
total	102.4	71.9	0.70
Meat supplies (tonnes, thousands)			
UK production	1125	1115	0.99
imports	263	203	0.77
exports	110	189	1.71
total new supply	1278	1129	0.83
UK production as % new supply	88	99	1.12

Source: MAFF (1976, 1986).

Table 3.2
Sources of UK production of beef and veal, 1985

	Numbers (thousands)	% total
"Prime beef", UK-bred young cattle		
from the dairy herd	2278	52
from the beef herd	1031	23
imported Irish store cattle	171	3
total prime beef	3480	80
"Cow beef"		
dairy cows	345	8
beef cows	252	6
Veal		
slaughtered in UK	99	2
exported	153	4

Adapted from *Beef Yearbook* (MLC, 1986).

are either on small, low-income farms or a relatively minor enterprise within a larger mixed farm, there to add a touch of class to what might otherwise be a rather humdrum existence.

The product

The objective of beef rearing systems is to produce beef of the highest possible quality at the lowest possible cost. The quality of beef, essentially its taste, tenderness and juiciness, is determined by age at slaughter, position on the carcass (e.g. rump, shoulder, etc.) and fat concentration. Generally, cow beef contains much more mature connective tissue than that from prime beef carcasses, and is therefore tougher. Prime cuts, such as fillet, sirloin and rump, fetch relatively high prices, partly because of their tenderness, and partly because they contain little or no gristle (connective tissue) or *inter*-muscular strips of fat. The *conformation* of a beef carcass describes its shape in terms of muscle : bone ratio and the proportion of muscle in the higher-priced cuts.

Most prime beef animals are slaughtered when the fat concentration in their carcass is close to 20 per cent, although in recent years entire bulls have been slaughtered at a fat concentration of about 15 per cent. About half the fat is trimmed off before the beef is sold to the final customer. This appears wasteful but it is because quality beef requires a sufficient quantity of *intra*muscular fat (or "marbling"), and this is only achieved after the animal has deposited substantial amounts of "waste" fat within the abdomen – kidney, knob and channel fat (KKCF) – and in the subcutaneous tissues. The animal deposits fat for its own ends, not ours.

The procedure adopted within the EEC for assessing the conformation and fatness of a beef carcass is illustrated in Table 3.3. Conformation is marked within seven classes: E (excellent), U+, U, R, O, O−, to P (very poor). Fatness is also marked within seven classes; 1 (leanest), 2, 3, 4L, 4H, 5L, 5H (fattest), 4L being about optimal. Table 3.3 illustrates the distribution of castrate males (steers) within the conformation and fatness classes. In general, young bulls are leaner and females (heifers) fatter at the same conformation class. Cull cows tend, on average, to have poorer conformation than steers but to be fatter at slaughter. The distribution matrix for cull cows is therefore further down and to the right of that illustrated for steers in Table 3.3.

The age and weight at which a young beef animal is deemed to be ready for slaughter in terms of conformation and fatness is governed by its breeding and sex (which together determine its phenotype) and its plane of nutrition. Cattle are said to be early maturing if they fatten at a relatively young age and light weight.

Table 3.3
Carcass classification within the European Economic Community and overall incidence for steers (per cent)

Conformation	Fatness: 1 (leanest)	2	3	4L	4H	5L	5H (fattest)	Overall
Excellent, E								0.2
U+				0.9				2.3
U			1.8	5.4	4.0	1.0		12.5
R		0.8	7.1	19.8	12.1	2.2		42.3
O		0.8	6.9	14.1	6.8	1.0		29.9
O−		1.0	4.0	4.7	1.4			11.6
Very poor, P								1.2
Overall	0.5	3.2	21.6	45.3	25.2	4.6	0.6	100

Blank spaces indicate an incidence of less than 0.5 per cent.
Trends: heifers, →; young bulls, ←; cull cows, ↘.

The Hereford is an earlier maturing breed than the Charolais; the heifer matures earlier than the bull. The animal (within any phenotype) that consumes nutrients at the greatest rate during growth also deposits fat at the greatest rate relative to lean tissue, and so reaches the degree of fatness deemed optimal for slaughter at the lightest weight.

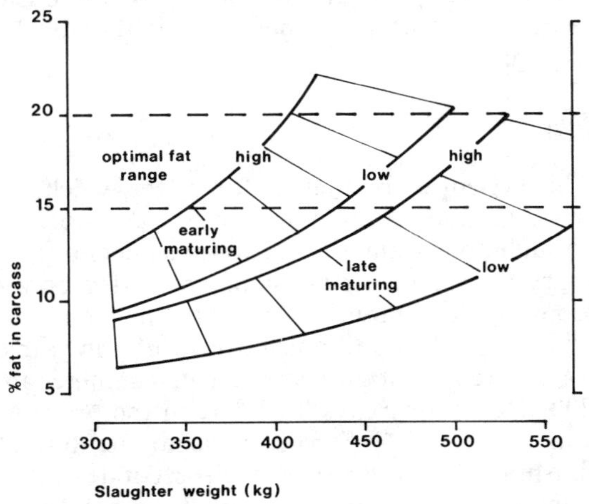

Fig. 3.1 Effect of type (early versus late maturing) and plane of nutrition (high versus low) on the relationship between slaughter weight (kg) and percentage fat in the carcass.

Figure 3.1 illustrates the effect of plane of nutrition on optimal slaughter weight for an early-maturing animal like a Hereford × Friesian steer and a late-maturing animal like a Charolais × Friesian bull. Given a highly nutritious diet containing a large proportion of cereals, the Hereford × Friesian steer will be ready for slaughter at 400 kg in perhaps 10 months. On a diet consisting almost entirely of fresh and conserved grass it may be 460 kg before it is ready for slaughter, and may have taken twice as long to get there. The larger Charolais × Friesian bull, fed a high quality diet, reaches the right degree of finish at a weight of 500 kg in perhaps 12 months. On a low-quality diet, it may be very difficult to achieve a minimum fat concentration of 15 per cent. Optimal strategies for matching feeds to breeds will be considered in more detail later.

Prior to the imposition of a ban in 1986, it had become common practice to implant growing steers and heifers with anabolic steroids (testosterone, oestrogen and synthetic compounds with similar actions) in order to stimulate lean tissue growth and increase slaughter weight, making them, in effect, more like bulls. Steroid implants were undoubtedly cost-effective in practice. They had no perceptibly deleterious long-term effect on the welfare of the animals and, if administered in accordance with the law, left no residues in the carcass. The decision to ban steroidal implants for beef cattle was political rather than scientific, but none the worse for that. If the consumer wishes beef without added anabolic steroids then it is up to the producer to provide it.

Table 3.4
Semi-intensive beef production systems for calves from the dairy herd

	Winter finishing	Summer finishing
Season of birth	autumn to early winter	late winter
Body weight (kg)		
weaning	65	65
turnout, year 1	180–200	100–120
yarding	320–360	200–240
turnout, year 2	↓	290–330
slaughter	400–500	440–540
Age at slaughter (months)	14–18	20–24

Rearing systems

Beef from the dairy herd

The 75 per cent of calves born to dairy cows, who are destined to become prime beef, are removed from their mothers shortly after birth and reared artificially. This involves providing them with a liquid milk-replacer diet until such time as they are eating enough dry food to support maintenance and growth. This usually occurs at about 5 weeks of age. Thereafter, the rearing system of choice is determined by phenotype (breed and sex), the cost and availability of food (e.g. grass and cereals) and season of birth.

Semi-intensive beef systems

The two most popular rearing systems in the UK are semi-intensive: the cattle spend their first summer on pasture, then are finished either in yards at the end of their second winter (winter finishing), or after a second summer at grass (summer finishing). Male calves are normally castrated and reared as steers. Table 3.4 illustrates the target weights for each system, the range being determined by breed and sex.

Winter finishing

Winter finishing at 14–18 months usually involves calves who are born in the months of September to December, which is the peak calving period for dairy cows. Such animals are able to gain approximately 0.6–0.8 kg/day over the first winter on a mixture of forage and cereals, so that they weigh 180–200 kg when the spring comes and they can be turned out on to grass. Their ruminant digestive system is sufficiently mature to enable them to gain about 0.8–0.9 kg/day during the summer, so that they are 320–360 kg when yarded again in October or November. If they are to finish before the following spring, they will require a high plane of nutrition, based usually on grass silage and an amount of cereal appropriate to the breed and sex. A Hereford × Friesian heifer may finish by January at a weight of 400 kg. A Limousin × Friesian steer may achieve a weight of 500 kg by April.

Summer finishing

This system is used either for calves which are born in mid- to late winter, so are not big enough to obtain the maximum benefit from grass during their first summer, or for calves from big, late-maturing bulls like the Charolais which are difficult to finish over winter in yards. Target weights at the end of the first summer are 200–240 kg for calves born late in the previous winter (Table 3.4). Since there is no intention of finishing the cattle over their second winter, they may receive only silage with minimal supplementation of essential minerals and vitamins. They undergo what is known as a store period, gaining about 0.6 kg/day so that they end the winter bigger but less fat than they entered it. They are turned out to grass for their second summer at 290–330 kg, well grown but lean and hungry (in store condition), and can then achieve growth rates well in excess of 1 kg/day on low-cost summer pasture. Because they have been grown more slowly, summer-finished cattle are heavier (440–540 kg) at slaughter than winter-finished cattle and will have consumed a smaller amount of expensive cereals. On the other hand, they will have occupied the pastures for two summers rather than one, and cattle prices per kg are always lower in the autumn than in the spring because of the large number of animals being finished off grass.

Intensive beef systems

Intensive beef systems are those in which calves are reared in confinement from the time of birth to the time of slaughter. In the 1950s, when cereal prices were very cheap relative to the price of beef, very many calves (especially Friesian bulls) were reared on a diet consisting largely of barley with minimal access to any forage. This system of *cereal-* or *barley*-beef (Table 3.5) achieves a slaughter weight of 420–440 kg at 10–11 months of age, an average weight gain of 1.25 kg/day. The decrease in the price differential between beef and barley has made this system less popular except for those beef farmers who grow their own barley. Improvements in recent years in the nutritional quality of grass preserved as silage have made it possible to grow cattle almost as fast and more economically on silage/cereal mixtures in which more than half the nutrients are provided by the silage.

Intensive beef systems require that the animals be confined throughout their lives at a high stocking density. This, incidentally, increases the risk of infectious diseases such as pneumonia relative to systems which permit the animals to spend half the year at pasture. These problems are particularly acute for calves in the first 3 months of life. Specialist contract-rearing units have therefore been developed to provide calves for the intensive beef units at an age of about 12 weeks and a weight of about 115 kg (Table 3.5). The health and welfare implications of these intensive and semi-intensive systems will be considered later (see p. 77).

Beef from the suckler herd

Calves born to beef cows stay with their mothers until weaning in late summer or autumn, when they are sold or moved into yards for winter feeding and finished out of yards in the spring or at pasture the following summer, once again according to size, breed and sex. Most beef farmers house their cattle over winter, not so much to protect them from the cold (since their cold tolerance is phenomenal – Table 3.13) but for convenience of feeding and to protect the pastures from poaching by the cattle. Calving is normally concentrated either in early spring, usually before turn-out, or at pasture in the autumn. When calves are born in the spring, the peak period for lactation in the cows coincides with the peak period for production of grass, so concentrate feeding to cow and calf is minimized (Table 3.6). The calves are weaned in the autumn at a weight of about 230 kg. Autumn-born calves are housed over winter with their mothers, and given access to a special creep area where they may rest and receive concentrate food without competition

Table 3.5
Intensive beef production systems for calves from the dairy herd

	Cereal beef bulls	Silage/cereal beef	
		Bulls	Steers
Body weight (kg)			
on entry at 12 weeks	115	115	115
at slaughter	420–440	490	450
Age at slaughter (months)	10–11	14–16	14–16
Weight gain (kg/day)	1.25	1.0	0.9

Table 3.6
Physical performance of upland and lowland herds calving in spring and autumn

	Lowland herds		Upland herds	
	Spring	Autumn	Spring	Autumn
Live calves reared (per 100 cows)	81	83	76	90
Age of calf at weaning (days)	217	244	209	303
Weight of calf at weaning (kg)	233	256	231	293
Weight gain to weaning (kg/day)	0.89	0.89	0.92	0.83
Feeding, concentrates				
to cow (kg)	102	167	92	132
to calf (kg)	35	122	35	135

Source: *Beef Yearbook* (MLC, 1986).

from the adult cows. They are turned out to pasture in the spring and weaned in mid-summer at weights of 250–290 kg. Table 3.6 presents figures for suckler beef production on both lowland and upland farms (over about 300 m). The differences between the two are small, which illustrates the point well that marginal land is perfectly adequate for beef cows and their calves.

Veal production

The calves used for veal production are those which are surplus to the needs of the dairy and beef industry, mainly male calves of pure dairy type like the Holstein or Jersey. The cheapest of these calves may be killed within 2 weeks of birth as bobby veal and used for little more than pies and calf-skin products. Others will be reared for quality veal on a diet that has, up to now, consisted entirely of a liquid milk-replacer diet containing an abnormally low concentration of dietary iron, in order to produce a calf weighing 160–200 kg when slaughtered at 14–16 weeks of age. Dietary iron intake is restricted to produce white veal, i.e. muscle deficient in myoglobin, the pigment that gives beef its cherry red colour. The welfare implications of this system and the search for alternatives are considered later (see p. 69).

The cattle

The husbandry of beef cattle, or any farm animal, should be based upon a fundamental understanding of how the animal works, especially in relation to factors which are economically important – digestion, growth and reproduction. It should also recognize the animal as a sentient being with a right to a reasonably comfortable life and a gentle death. This requires an equal understanding of the behaviour of the species and its response to the environment.

Digestion

Cattle are ruminant herbivores. In the natural state, wild ruminants subsist largely by grazing the leaves and stems of grasses and browsing the leaves of shrubs and trees where available. The main constituents of cell walls of these plants are cellulose, hemicellulose and lignin, compounds which are almost completely indigestible by man. When a cow grazes at pasture, all plant material enters the large paunch, or reticulo-rumen (Fig. 3.2) where it is mixed, diluted with saliva and subjected to microbial fermentation. The end-products of microbial fermentation of plant carbohydrates (sugars, starch, cellulose and hemicellulose) are absorbed, largely across the rumen wall, as the volatile fatty acids – acetic, propionic and butyric acid – which form the major source of dietary energy to the ruminant. The digesta that leave the reticulo-rumen and pass into the abomasum (which corresponds to the true stomach in man) contain large amounts of microbial protein. These are subjected to acid digestion and form the ruminant's principal supply of amino acids.

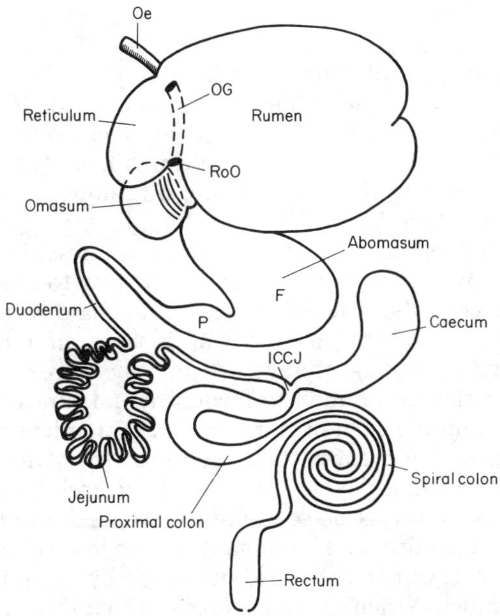

Fig. 3.2 The digestive tract of the adult ruminant (from Webster, 1987). Oe, oesophagus; OG, oesophageal groove; RoO, reticulo-omasal orifice; F, fundus of abomasum; P, pylorus; ICCJ, ileo-caeco-colic junction.

A cow normally consumes grass or conserved forage (e.g. hay) at speed, with little preliminary mastication. This enables her to harvest a lot of food in a short time, which has obvious survival value when she is faced by competition or predators. When the food is long and fibrous, she may then spend up to 8 h a day ruminating, i.e. regurgitating boluses of food and chewing them over at her leisure before re-swallowing them. Rumination usually occurs while the animal is at rest, and can even take place during light sleep. The duration of rumination is, however, directed strictly by the physical form of the food and the amount of comminution it requires before it can leave the rumen. If fed a milled diet, cattle hardly ruminate at all. The need to ruminate is physiological, not psychological.

The adult ruminant derives most of its nutrients from reactions in the rumen. The small intestine is the major source for absorption of amino acids and minerals. The large intestine acts as a second fermentation chamber, but normally contributes less than 8 per cent to nutrient uptake.

When the young calf sucks milk from its mother, this high-quality food is directed straight into the abomasum by reflex closure of the oesophageal groove (Fig. 3.2), thereby bypassing the fermentation chamber in the rumen. The protein, fat and carbohydrate (lactose) in milk are subjected to acid digestion and absorption in the same way as in a simple-stomached animal. This elegant arrangement of the digestive tract permits the calf to direct the different components of its food into the site most appropriate for their digestion, the rumen for grass and the abomasum for milk. When the artificially-reared calf sucks milk slowly through a teat, the oesophageal groove reflex is completely effective and liquid food passes directly into the abomasum. When it drinks rapidly from a bucket, closure is partial and some liquid food almost always enters the rumen. Since fats cannot be fermented in an anaerobic environment and can inhibit microbial activity, this is a bad thing.

Reproduction

The cow normally has one calf per year after a pregnancy of 9 months. There are small genetic differences in the duration of pregnancy, which are largely determined by the genotype of the calf. Table 3.7 lists the duration of pregnancy and the incidence of calving difficulties associated with the use of different sire breeds on the UK Friesian/Holstein cow. Charolais bulls carry a high incidence of calving difficulties, partly because they are large relative to the Friesian, and partly because they extend the gestation period by 2 days. The Limousin bull provokes an even longer gestation but less calving difficulties because the calves are not so large at birth.

Cattle have no discrete breeding season. A sexually mature cow or heifer will, if not mated, show oestrus or sexual activity (come "on heat", or "bulling") at intervals of about 21 days throughout the year. The duration of oestrus or heat varies with season, ranging from as short as 8 h in mid-winter to 36 h in summer. In the early stages, the cow will be restless and seek out the bull if present. She will become increasingly attractive to the bull and to other cows, and she will allow herself to be mated by the bull or mounted by other cows if there is no bull in the vicinity to chivvy them off. Most dairy cows are fertilized by artificial insemination, which places the responsibility for oestrous detection upon the stockman (see Chapter 2, p. 22). When beef cows are at pasture or on range, natural service by a bull running with the herd is practically obligatory. If the cows are confined, and there are good genetic reasons for using artificial insemination, it is possible to induce and synchronize oestrus and ovulation by the use of hormones. There are two approaches, both of which rely on controlling the onset of pro-oestrus, i.e. the period of rapid follicular development. They are injection with prostaglandin, or insertion of progesterone-releasing intravaginal devices (see Chapter 2, p. 22). The probability of fertilization following synchronization of oestrus and a single artificial insemination is unlikely to exceed 65 per cent, so the beef farmer who practices synchronization requires a back-up bull.

Table 3.7
Effect of sire breed and duration of gestation on the incidence of calving difficulties in Friesian/Holstein cows and heifers

Sire breed	Gestation length (days)	Calving difficulties (%)	
		Cows	Heifers
Friesian	281	2.7	5.7
Aberdeen Angus	279	—	1.4
Hereford	282	1.2	2.7
Charolais	284	3.4	6.7
Limousin	287	2.4	3.2

Behaviour

Whenever man dictates the environment for any other animal species, he must consider how this affects the natural behaviour of that species, and so determines whether the environment is appropriate and, indeed, humane. The essential maintenance behaviours of cattle, such as eating, drinking, ruminating, defaecating, etc., are seldom likely to be compromised by domestication, but social, sexual and exploratory behaviour are inevitably distorted. Cattle in the wild form large, stable herds whose size appears to be limited only by the availability of pasture. Being part of a herd reduces the possibility of capture for an animal that cannot hide. The cow's logic is, in effect, "The lion has got to eat someone but if I am in a herd it is less likely to be me than if I am standing out alone on the open plain."

Individual cattle become distressed if removed from the herd. One exception to this is the cow about to calve, who will isolate herself until such time as her calf is born and has become strong enough to join the herd, and experienced enough to recognize and run to its own mother. On open range and given plenty of space, cattle form stable sub-groups. A dominance hierarchy is established, whereby each cow knows its place, more or less, and adopts, as appropriate, a dominant or submissive attitude towards its neighbour, thereby avoiding conflict. When cows or steers are closely confined, aggression can occur, especially if the accommodation is so designed that the submissive animal cannot escape. Bulls are aggressive for reasons that are genetically sound but managementally difficult.

The exploratory behaviour of cattle involves a constant conflict between curiosity and timidity. Consider Fig. 3.3a. A novel object (a balloon) has been introduced to a calf in the familiar environment of its pen. Having decided that the balloon poses no immediate threat, the calf investigates it with nose and tongue. In Fig. 3.3b the calf meets the balloon in the less familiar environment of a yard. Here it is still curious but more apprehensive. Note the position of the feet in both pictures. In Fig. 3.3a the calf is in a relaxed stance, Fig. 3.3b it is poised for retreat.

Cattle breeds

The first thing to emphasize about cattle breeds is that they are entirely artificial, a creation of man motivated in part by a desire to improve productivity, in part by the desire to make money through the sale of allegedly superior stock, and in part by the fun to be had at livestock shows.

Fig. 3.3 Exploration of a novel object (a) in familiar surroundings and (b) in unfamiliar surroundings (from Webster, 1985).

Friesian/Holstein

Since most UK beef comes from the dairy herd, most mothers of beef calves are black-and-white, high-yielding dairy cows (Table 3.8). The breed originated in Northern Holland (Friesland) and Germany (Schleswig-Holstein) and has been selected primarily for its high milk yield. The Dutch and British Friesians have traditionally retained a reasonably good beef conformation because of the interdependence of the beef and dairy industries. The North American Holsteins have been selected exclusively for milk because of the large suckler beef cow population in the USA and Canada, and have become larger but less beefy than the traditional Friesian. Importation of Canadian Holsteins into the UK has increased milk yield per cow at the cost of some loss of conformation or beef type. At present, the British Friesian and Canadian Holstein breed societies are separate, but they may amalgamate in the near future. Whatever happens officially, the distinction between the two types will continue to blur. The male, relatively beefy Friesian-type calf is currently (1987) worth about £80–100 at birth, and is suitable for barley beef or 16–18 month finishing out of yards, given a reasonable intake of cereals. The very leggy Holstein calf is worth perhaps £40 less, and may be destined for barley beef or veal.

Table 3.8
Distribution of cows and heifers artificially inseminated by the Milk Marketing Board according to breed of bull and cow during the year 1 April 1986 to 31 March 1987

	Total no. of inseminations	
	Cows served	Heifers served
Bull breed		
Friesian/Holstein	1 033 133	37 116
Aberdeen Angus	33 375	27 198
Charolais	203 191	1366
Hereford	200 272	25 269
Limousin	380 071	24 255
South Devon	3 742	115
Simmental	73 059	1140
Others	178 043	9783
Cow breed		
Friesian/Holstein	1 937 694	106 168
Jersey	40 129	3405
Hereford	26 472	5009
Others	100 491	11 660
Total	2 104 786	114 582

Source: Milk Marketing Board (1986/1987).

Other dairy breeds

Table 3.8 shows the extent to which the Friesian/Holstein cow dominates the dairy industry. The only other dairy breeds that remain in substantial numbers are the Channel Island breeds (Jersey and Guernsey), which yield milk with a much higher concentration of solids. Pure-bred male calves of dairy breeds like the Ayrshire, Guernsey and especially Jersey are too small and light-muscled to make satisfactory beef animals. Having little cash value, many of them receive little care and attention during their short lives and end up as bobby veal.

Hereford

The Hereford was, for many years, the most popular sire for the production of beef from the dairy herd. On average, the mature Hereford bull is of moderate size (*c.* 900 kg), although it must be said that the range within this and every other breed is larger than the difference between average mature weights for the principal beef breeds. The breed has an excellent conformation, grows fast and fattens easily due to a large appetite for fibrous foods relative to its nutrient energy requirement for maintenance. The bulls carry a low risk of calving difficulties (Table 3.7). The gene that gives the Hereford its white face is dominant and transmitted to its calves. The beef producer seeing a white-faced calf in market knows that it was sired by a Hereford bull. The breed has currently been supplanted by the Limousin as the most popular beef bull for artificially inseminating dairy cows (Table 3.8). Hereford bulls, however, carry the added advantage of a remarkably equable temperament (as bulls go). Because of this, and the low incidence of calving problems, many Hereford bulls are kept on farms to serve heifers and to act as a back-up for breeding cows. Thus there are still more Hereford × Friesian calves than Limousin × Friesians. The male, black and white Hereford × Friesian calf may currently (1987) fetch £140 at market at 10 days of age because it is ideal for any of the semi-intensive systems. It is too expensive for barley beef. The Hereford × Friesian heifer calf is only worth about half as much. If reared for beef, she tends to fatten at a light weight (about 400 kg). Some Hereford × Friesian female calves are reared for veal and others become (very good) suckler cows.

Aberdeen Angus

These cattle are black and polled (i.e. naturally without horns) and both these characters are dominant. The breed was developed to produce high-quality beef, with conspicuous marbling, on the good land but in the short growing season of north-east Scotland. Many Aberdeen Angus cattle tend to fatten at unacceptably light weights in modern feeding systems, so their popularity has declined except as a safe sire for heifers (Table 3.8). The importation of some very large Aberdeen Angus bulls from Canada may reverse this trend.

Limousin

The Limousin is a red-coloured breed originating from central France and is currently the most popular beef sire for beef from the dairy herd. Its most attractive features are leanness and excellent carcass conformation with a high killing-out percentage and a high proportion of meat in the expensive cuts (Table 3.9). It is thus the butcher's favourite. Calves from a Limousin bull usually fetch more than £150 (1987 prices). Calving problems are acceptably low. However, selection for leanness and a high killing-out percentage (i.e. small gut) inevitably tends to select against appetite, so that Limousin-cross beef cattle may be hard to finish. They are best suited to intensive or semi-intensive systems in which they are yarded and slaughtered by 18 months of age. Limousin × Friesian bull calves are usually not castrated, and can be difficult to handle.

Charolais

The Charolais is a very large (mean 1100 kg), light yellow-coated breed with a rapid lean tissue growth rate, good beef conformation and the ability to reach a large size without putting on excess fat. It was for some years the most popular non-British sire for beef from the UK dairy herd, but has fallen back in popularity as

Table 3.9
Carcass composition of calves from sire breeds crossed with Friesian/Holstein cows

	Sire breeds			
	Angus	Charolais	Hereford	Limousin
Weight at slaughter (kg)	393.0	494.0	410.0	454.0
Feed conversion ratio (g gain/kg feed)	86.0	82.0	88.0	85.0
Killing out (%)	52.5	54.8	52.3	54.7
Saleable meat in carcass (%)	72.5	72.7	71.9	73.3
Saleable meat in expensive cuts (%)	44.1	44.8	44.1	45.4
Fat trim in carcass (%)	9.6	9.0	9.7	9.2

Source: Southgate, in More O'Ferrall (1982).

that of the Limousin has increased, partly because of the high incidence of calving difficulties and partly because the biggest calf is not necessarily the most efficient at producing beef. Charolais × Friesian calves are also worth more than £150 (1987 prices) at 1 week of age. They are slow to finish, and best suited to 18–24 month systems where they can achieve slaughter weights up to 550 kg.

The Charolais bull is, however, an excellent sire for beef from the suckler herd. Whereas the dairy cow earns her living chiefly as a producer of milk, with her calf as a bonus, the calf is the main source of income to the beef industry. To maximize efficiency, the weight of the calf at slaughter must be as large as possible relative to the weight of its mother to maximize income relative to the costs of maintaining the breeding herd. It is necessary, therefore, to cross-breed suckler cows with bulls of a large breed, despite the increased risk of calving difficulties. The Charolais bull is a most effective sire for beef production from a moderately large beef cow such as the Hereford × Friesian.

Other sire breeds

Other big, beefy sire breeds include:

1. The Simmental which looks rather like a large Hereford and can be used to increase size in commercial Hereford cattle without noticeably changing their colour and aesthetic appeal.
2. The Blonde d'Aquitaine which may be thought of as a Charolais by anyone but a member of either breed society.
3. The South Devon, a red-coated British breed, comparable in size to the Charolais.

Top-quality bulls from any of these large breeds can make excellent sires for beef production from the suckler herd.

Suckler cow breeds

Most suckler cows in the UK are not pure-bred but first crosses between two pure breeds. This is, in part, because genetic traits relating to maternal behaviour and calf survival are much more likely to be improved by heterosis (hybrid vigour) than by selection within breeds, and partly because suitable half-bred cows such as the Hereford × Friesian are available.

Over the last 200 years, the Hereford has become the most common breed of beef cow in the world outside the tropics. In the UK, however, most beef farms developed in the hills and uplands of Scotland, Wales and northern England using local breeds like the Aberdeen Angus, Shorthorn, Galloway or Welsh Black. Cross-breeding the black Angus or Galloway with the Shorthorn produces the Blue-Grey, a small, hardy cow able to outwinter with little supplementary feeding and to produce calves for 10 years or more.

The other most common beef cow is the Hereford × Friesian emerging from the dairy herd. Table 3.10 compares some aspects of the maternal performance of the Blue-Grey with the Hereford × Friesian. The latter is a little larger and milkier and, on lowland pastures, produces a calf about 14 kg heavier at 200 days of age. This advantage declines to 8 kg on the hills,

Table 3.10
A comparison of Blue-Grey and Hereford × Friesian suckler cows

	Blue-Grey	Hereford × Friesian
Average cow weights (kg)		
mature	450	475
at first mating	280	325
Lactation yield (kg)	1600	1940
Calf 200-day weights (kg)		
lowland	203	217
hill	190	198
(lowland – hill)	13	19

Adapted from Allen and Kilkenny (1985).

suggesting that the Hereford × Friesian is a little less hardy than the Blue-Grey. It is also worthy of note that the average Hereford × Friesian calf is removed from its mother at birth and artificially reared, yet manages to handle the responsibilities of motherhood on range very well.

Nutrition and feeding

Animals require nutrients and eat food. Nutrition is the science that equates the requirements of animals for energy, protein, minerals and vitamins to meet the needs of maintenance, activity, growth, pregnancy and lactation with the capacity of food to provide these specific nutrients.

Metabolizable energy requirement

Before any animal can utilize nutrients for metabolism, it must first digest and absorb them from the gut. By far the greatest quantity of nutrients absorbed from the gut is used as a source of energy for the work of maintenance and production. This is defined as metabolizable energy (M/E), where metabolizable energy = gross energy in food minus energy lost in faeces, urine and methane (from fermentation).

The ME requirement of an animal is expressed in MJ/day and the ME concentration in the dry matter (DM) of food is given in MJ/kg DM (M/D). The DM intake of a ruminant is stimulated by its requirement for nutrients, especially energy, and restricted by the amount of unfermented matter it can carry in the rumen. Thus a growing calf or lactating cow will have a greater appetite per unit of body weight than an adult, non-pregnant, non-lactating animal, and the less digestible the food (i.e. the lower its M/D), the more difficult it becomes for the animal to meet its energy requirements within the constraint of maximum possible DM intake. If we exclude young calves for the moment, we can say that feeding beef animals involves feeding small, controlled amounts of relatively expensive, cereal-based concentrate feeds with a high M/D, and then allowing the animal to eat forage to appetite so that the mixture of forage and concentrate achieves the M/D required to match ME intake to requirement for maintenance, growth or lactation.

Table 3.11
Daily allowances for metabolizable energy (M/D) for growing cattle

Liveweight (kg)	M/D (MJ/kg DM)	Daily liveweight gain (kg) 0.25	0.75	1.25
100	10	22	—	—
	12	21	33	—
250	8	40	—	—
	10	38	57	—
	12	37	52	75
450	8	61	—	—
	10	59	83	—
	12	57	78	108

Table 3.12
Metabolizable energy (M/D) and crude protein (CP) concentrations of some common feeds for beef cattle

	Dry matter (g/kg)	Composition of dry matter	
		M/D (MJ/kg DM)	CP (g/kg DM)
Pasture grass	200	11.0	175
Ryegrass silage			
excellent	200–300	10.8	170
moderate	200–300	10.0	150
poor	150–250	9.0	120
Maize silage	250–300	11.4	100
Meadow hay	850	8.5	90
Barley straw	850	5.8	40
Barley	900	13.4	110
Maize (corn)	900	14.0	100
Soyabean meal	900	12.5	500

The ME requirement for maintenance (ME_m) of any mammal is related to its metabolic body size or body weight (kg)$W^{0.75}$. For cattle, ME_m may be taken as 0.5 MJ/kg$W^{0.75}$. Thus for a 460 kg Blue-Grey cow

$$ME_m = 50 \text{ MJ/day}$$

At peak lactation, the ME requirement is about twice maintenance, i.e. 100 MJ/day, allowing for a loss of about 0.5 kg/day body weight as the newly-calved cow milks off her back.

The ME allowances needed for growing cattle to meet a range of target weight gains are given in Table 3.11 (for more detailed figures, consult Allen and Kilkenny, 1985). The blanks in Table 3.11 indicate live-weight gains that cannot be achieved because, at any given M/D, ME requirement exceeds appetite. Thus a diet with an M/D of 10 can sustain a gain of no more than 0.25 kg/day in a 100 kg calf but 0.75 kg/day at 250 or 450 kg. A diet with an M/D of 12 should be able to achieve a weight gain of 1.25 kg/day in an animal genetically capable of growing that fast.

Table 3.12 gives the M/D and crude protein concentration (g/kg) of some typical feeds for beef cattle. Note that the M/D of grass and forages ranges from 11 for pasture grass and 10.8 for excellent silage down to 8.5 for meadow hay and 5.8 for barley straw. Cereals and other concentrate feeds have M/D values between 13 and 14 MJ/kg DM. Summer pasture (M/D = 11) can provide sufficient ME to sustain a weight gain of 1 kg/day or more in growing steers. The amount of cereal required to achieve comparable weight gains during winter feeding depends on the M/D of the conserved forage. If it is very high (over 10.5) then many cattle will gain 1 kg/day without the need for cereal supplementation. Forages, such as poor silage or hay, with an M/D of 9.5 or less, may require up to 3 kg/day of concentrate supplement (depending on the breed of animal) to achieve optimal weight gains.

Feeds for breeds

It is necessary now to consider the interaction between feed quality, as defined by M/D, and genotype. Figure 3.4 provides a somewhat schematic illustration of the effect of increasing M/D on the efficiency of food utilization for growth, defined by gram live-weight gain per MJ ME for three types of beef cattle fed to appetite.

Given a poor quality diet of hay (M/D = 8) the Friesian achieves only 4 g gain per MJ ME, rising to 12 g per MJ ME on a barley-beef ration (M/D = 12.5). It is able to consume more energy from the barley-beef ration relative to its needs for maintenance, and therefore converts food more efficiently, although not necessarily more cheaply, since barley is more expensive than hay per unit of ME. Relative to the pure Friesian, the Hereford × Friesian is more efficient at converting low-quality foods, because it has a relatively lower heat production and so a lower ME requirement for maintenance. Given diets with an M/D of 10.5 or above, the difference between the two types is small, although the Hereford × Friesian does tend to get too fat too soon on rations containing a high proportion of cereals.

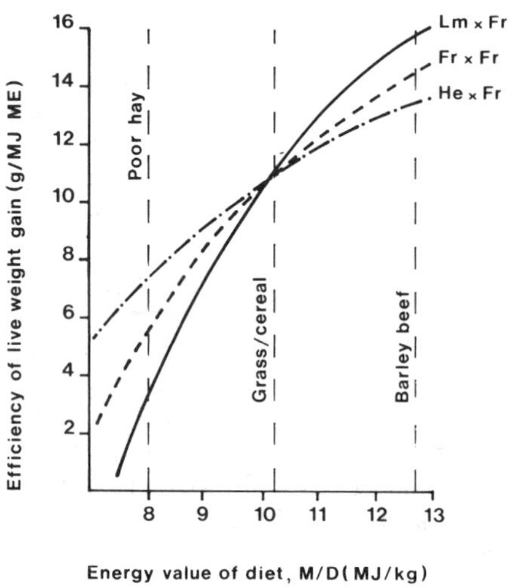

Fig. 3.4 A schematic relationship between diet quality (M/D) and food conversion efficiency (g gain/MJ ME) for three types of beef cattle.

The Limousin × Friesian has a relatively high potential for lean tissue growth but a relatively low gut capacity for indigestible fibre. Thus it is less efficient on poorer quality diets but becomes progressively more able to express its genetic potential as feed quality is improved. It cannot be said that the Limousin × Friesian is better or worse than the Hereford × Friesian, they are best suited to different diets. The Hereford- ×Friesian steer can be grown and finished with little or no cereal supplement, whereas the Limousin × Friesian requires some cereal, but having got it, yields more lean meat (Table 3.9).

Protein and other nutrients

Ruminants require protein for microbial protein synthesis in the rumen, and as a source of amino acids for protein synthesis within the body. The rumen microbes require about 120 g crude protein/kg DM. For beef cattle weighing more than 250 kg, this should also provide enough amino acids for growth. Younger animals require a diet with a higher protein concentration. Concentrate feeds for calves under 250 kg may contain 140–180 g/kg crude protein, depending on the balance between concentrate and forage in the ration (Table 3.12).

Diets based wholly or largely on fresh and conserved grass are likely to provide sufficient calcium, but phosphorus and magnesium may be marginal. In certain cases the diet may be deficient in copper, cobalt or selenium, and all these deficiencies may stunt growth. Minerals can be fed in a controlled fashion by incorporating them into the concentrate ration. They may also be offered free-choice in mineral licks, but this can present two problems: intake is very erratic and not related to requirements, and selenium (and possibly copper) are toxic when consumed to excess. Thus some cattle may eat too little and some too much.

Cattle have no dietary requirement for vitamin C, and are able to acquire B vitamins as an end-product of microbial synthesis in the rumen. Green grass and silage are good sources of carotene, the precursor of vitamin A. Sun-cured hay is a good source of vitamin D and whole cereals are good sources of vitamin E. In general, adult cattle are unlikely to suffer from deficiencies of the fat-soluble vitamins A, D and E, but body reserves of these vitamins drop during the winter. Calves are born with almost no fat-soluble vitamins in their body and normally acquire them by drinking colostrum. The colostrum of beef cows in late winter may contain very low concentrations of fat-soluble vitamins, which gives their calves a poor start and makes them particularly vulnerable to infectious diseases of epithelial surfaces, e.g. diarrhoea and pneumonia.

Water

The *Codes of Recommendations for the Welfare of Livestock: Cattle* (MAFF, 1983) state that "cattle should have access to sufficient fresh, clean water at least twice daily". This is necessary to maintain the water content of the body tissues in the face of inevitable and regulatory losses of water in the urine, faeces and by evaporation of sweat from the skin and respiratory tract. They also require water for milk production and to sustain the continuous culture of microbes in the rumen. Published figures for the water intake of cattle range between 50 and 150 g/kg body weight/day. It is greater during lactation and hot weather and less when the food is very wet (e.g. grass or silage). The first response of cattle to a water shortage is usually a drop in food intake. Ideally, cattle should have access to clean water almost continually. If they are only allowed access to water twice daily they should be free to drink as much as they can on each occasion.

Feedstuffs

Grass

This is almost the perfect food for beef cattle. The most nutritious part of the plant for ruminants is the young leaf, so its digestibility is greatest before the emergence of seed heads. Grazing management should be directed towards ensuring that the grass is long enough (at least 70 mm) to ensure that the cattle get enough to eat, but not so long that it becomes stemmy and fibrous. Well managed grassland provides sufficient ME and protein for maintenance and lactation in beef cows and sufficient for maintenance and growth in beef calves receiving milk from their mothers. It cannot sustain economically acceptable weight gains in calves from the dairy herd which are receiving neither milk nor concentrates before these reach a weight of 200 kg in the case of British breeds, or approaching 300 kg in the case of the larger continental breeds. Cattle which receive all or nearly all their food from fresh and conserved grass may, in certain areas, suffer debilitating deficiencies of copper, cobalt, selenium, and possibly phosphorus, and may be more susceptible to an acutely fatal attack of hypomagnesaemia or grass staggers.

Conserved forages

The grass crop may be sun- or air-dried and conserved as hay or compacted into an air-tight clamp and conserved anaerobically as silage. Well-made silage has a greater nutritive value than hay because the crop is cut at an earlier, leafier stage (Table 3.12). When making silage for dairy cows, it is important to cut the grass very young to achieve the highest possible quality (M/D = 10.5). For beef cattle, the crop can be left a little longer to achieve a greater yield and an M/D of about 10.0. The anaerobic fermentation of grass sugars in the silage clamp increases the acidity of the crop and prevents further microbial breakdown. In effect, the crop is pickled. The ME of silage is contained almost entirely in the cell-wall fraction. Silage fed alone tends to ferment and leave the rumen relatively slowly. Supplementation of grass silage with a little starch or sugar (1–2 kg DM from barley or a root crop) and/or very small quantities of high quality protein (e.g. 250 g fishmeal) can have a synergistic effect on fermentation in the rumen, so improving nutrient yield from silage and overall dry matter intake.

Well-made hay is less nutritious than silage, but extremely palatable. Soft meadow hay made from a range of grass species is particularly suitable for encouraging young calves to develop an appetite for roughages. Barley straw has little nutritive value (Table 3.12) but is a good source of fibre for cattle eating large quantities of cereals (e.g. barley-beef). The long, relatively indigestible fibre dilutes the cereal in the rumen and stimulates rumination and thereby increases salivation. Both these things help to reduce the risk of ruminal acidosis, which can occur through excessively rapid fermentation of starchy foods. It is possible to improve the nutritive value of straw by treatment with alkalis, sodium hydroxide or ammonia. These cause the cell-wall fibres to separate and release more fermentable cellulose from its bonds to unfermentable lignin. Alkali-treated straws have an ME value close to that of hay. Ammonia treatment also increases the nitrogen supply to the rumen microbes. The case for alkali treatment of straw in any year is strictly determined by economics: if the grass crop (hay and silage) is poor and expensive, it pays to treat the straw crop.

Maize silage

In the USA, the very high-yielding corn (maize) crop is normally ensiled at a more mature stage than in the UK, when most of the nutrients are contained in the seeds, or ears, and the remainder of the plant (the stover) is very fibrous and indigestible. Maize silage in the UK usually contains less grain and therefore less starch but more digestible fibre. It is, nevertheless, an excellent food (M/D = 11.4) for winter feeding to beef cattle in the warmer, drier parts of the UK where it can be effectively grown and harvested.

Cereals

The main nutrient in barley, oats and wheat is starch, which makes these cereals highly concentrated forms of ME for cattle. They are marginal or slightly deficient in protein, and low in most minerals, especially calcium. When they constitute the major part of the ration for a beef animal (e.g. barley-beef), they promote very rapid growth (over 1 kg/day), but may induce fattening which is too rapid in offspring of beef bulls like the Hereford and Aberdeen Angus. All cereals may dispose to ruminal acidosis through excessively rapid fermentation, unless diluted with long fibre. Oats is the safest cereal in this respect, being the most fibrous, but is more expensive than barley or wheat per unit of ME.

Root crops

The most popular root crops are fodder beet (in the south) and swedes or turnips (in the north). These have a dry matter content of less than 20 per cent, but are rich in sugars, and very palatable. However, they are deficient in protein and minerals. As indicated earlier, a small quantity of a root crop such as fodder beet (e.g. 5 kg fresh weight/day) can make an excellent supplement to grass silage. In the north east of England and Scotland, roots are often used as the main source of energy for wintering beef cattle, properly supplemented with protein, minerals and fat soluble vitamins.

By-products

Most pelleted compound rations for cattle consist largely of by-products from crops harvested primarily for direct consumption by man. The most common by-products are the protein-rich meals from plants such as soya and rape which are grown primarily for vegetable oils. Others include sugar beet pulp, which is a highly palatable source of digestible fibre, and fishmeal, an excellent source of high-quality protein, minerals and vitamins. The proper inclusion of by-products into rations for beef cattle depends on their precise nutritive value, and is beyond the scope of this chapter.

Environmental requirements

The environmental requirements of cattle are modest. The thermal environment should neither be excessively too hot nor too cold. The cattle should have access to a reasonably clean, dry, comfortable bed where they can rest undisturbed by other animals (including man). They should be able to get food and water without physical difficulty and without incurring conflict, but they should, wherever possible, be in social contact with other members of their own species. Last, but by no means least, management and environment should be designed so as to minimize the risk of injury and disease and ensure prompt diagnosis and treatment, should either occur.

Thermal comfort

Cattle are able to tolerate a wide range of thermal environments with little or no stress to their welfare, and at little cost to production. This is because they are a large, well-insulated species with an excellent ability to regulate body temperature. Figure 3.5 illustrates the heat exchanges of a pre-weaning calf. Within the range of 10–30°C (in still air), the calf is in the zone of thermal neutrality or thermal comfort.

Metabolic heat production (H_p) is unaffected by air temperature, and the calf maintains homeothermy by regulating evaporative heat loss (H_e) by sweating and thermal panting, i.e. increasing the rate but decreasing the depth of respiration so as to increase evaporation from the vascular heat exchanges in the nose. As air temperature falls, and sensible heat loss (H_n) by convection, conduction and radiation increases, the animal reaches the air temperature below

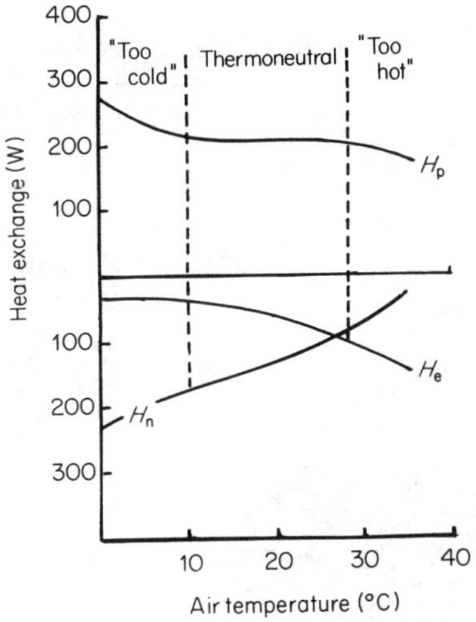

Fig. 3.5 Heat exchanges of a 90 kg calf at air temperatures from 0 to 40°C. Heat production (H_p) equals the sum of sensible (H_n) and evaporative heat loss (H_e) (from Webster, 1985).

which it can only maintain homeothermy by increasing H_p. This is the *lower critical temperature*. Increasing H_p at colder temperatures diverts food energy from production to maintenance, and so reduces productive efficiency. The *upper critical temperature* is reached when the calf cannot dissipate all metabolic heat by evaporative and sensible means and, therefore, it must reduce H_p in order to maintain homeothermy. This can only be achieved by reducing food intake, which also reduces productive efficiency in the growing animal. The more an animal eats per unit of body size, the greater its heat production. It follows, therefore, that the animal which eats the most (the fast growing beef steer or bull) is the least sensitive to cold and the most sensitive to heat. The animal that eats the least (the half-starved beef cow) is the most sensitive to cold. Table 3.13 gives approximate valucs for the lower critical temperature of different classes of beef cattle housed either in still air conditions (0.2 m/s) or in a moderate draught (2 m/s).

At birth, the calf in a well-bedded box and out of draughts begins to become affected by cold at about +10°C, but the stress is slight until the air temperature falls below −10°C, and dairy calves can be reared successfully out-of-doors in wooden hutches (Fig. 3.6) in the northern USA and Canada where temperatures may fall to −40°C! Farmers have discovered that cold stress while not ideal, is less of a killer than over-stocked, under-ventilated, polluted calf houses. By the time of weaning, the calf is unlikely to be stressed at all by low temperatures alone under UK conditions. Well-grown veal calves, by virtue of their enormous ME intakes of a high-fat liquid diet, are extremely tolerant to cold, but sensitive to heat. Yearling cattle and cows are unlikely to be cold-stressed unless thoroughly soaked by rain or snow. A cow standing in driving sleet at an air temperature of 2°C is colder than if she were in still air at −40°C.

A suitable bed

A reasonably clean, dry, comfortable bed is easier said than achieved. Cattle are undoubtedly very comfortable when lying in deep, clean

Table 3.13
Lower critical temperatures for beef cattle housed in still-air conditions (0.2 m/s) and in a draught (2 m/s)

	Lower critical temperature (°C)	
	Still air	Draughty
Newborn calf	+10	+18
Weaned calf, 6 weeks old	0	+10
Veal calf, 12 weeks old	−12	0
Yearling beef cattle		
rapid growth	−30	−15
store condition	−15	−5
Beef cow, maintenance	−15	−8

From Webster, in Swan and Broster (1981).

Fig. 3.6 A calf hutch in the north of the USA.

straw. However, straw is probably too expensive unless produced on the farm and, in some good beef country like the Orkney Isles, it is prohibitively so. The main problem is that cattle urinate and defaecate indiscriminately, which leads to inadequate amounts of wet filthy straw. Adult, non-lactating beef cows and growing cattle over 6 months of age may be housed on slatted concrete floors. This is far from ideal as a mattress to lie on, but it can be kept clean and dry if the gaps between the slats are 40 mm wide and the cattle are stocked densely enough to tread the dung through the slats. However, I feel young calves under 6 months of age should not be housed on slatted floors, because 40 mm gaps are too wide for their feet and the stocking density necessary to ensure that the slats stay clean carries a severe risk of provoking pneumonia in these young animals. It follows from this that, while pregnant beef cows may be housed on slatted floors it is no place for a beef cow to rear her young calf, still less give birth to it.

Beef cows and fattening heifers can be accommodated in cubicles. They are obviously unsuitable for males since these urinate in the middle. The dimensions of the cubicle are determined by the size of the largest animals that they are likely to accommodate, because the cow should be able to lie entirely within the cubicle without interference from its neighbours, stand up and lie down without difficulty yet, so far as possible, urinate and defaecate in the dunging passage. A cubicle for a typical Hereford × Friesian

Table 3.14
Space requirements for beef cattle

	Calves		Fattening cattle over 1 year	Beef cows
	0–6 weeks	6–12 weeks		
Floor space (m^2)				
straw bed	2	2	3–4	4
slatted floor	—	—	1.4–2.2	3.0–3.7
Air space (m^3)	6	10	12	12

beef cow should be 2 × 1 m. Fine tuning of position can be achieved by adjusting the neck rail.

Space requirements

These are listed in Table 3.14. When cows or fattening cattle are given restricted rations of concentrate food, it is essential that all animals can feed at the same time. Recommended trough spaces for cows and growing cattle are 0.6–0.8 m and 0.4–0.6 m respectively, depending on age and breed. A group of 10 yearling cattle on slats, with 0.4 m trough space per head and a floor area of 1.6 m^2 per head, would require a square pen 4 m long by 4 m deep. However, the recommended minimal air space is 12 m^3 per head. For pens 4 × 4 m either side of a 4 m wide feeding passage, the average height of the building should be 5 m. Examples of housing designs accommodating these basic environmental requirements will be given later (see p. 71 and Figs 3.8, 3.10 and 3.11).

Calf rearing

This section deals with the artificial rearing of beef calves born to dairy cows.

Birth to weaning

Unless the cow is to calve out-of-doors, she should be confined in a spotlessly clean and disinfected calving box and well bedded down with fresh straw to minimize the risk of infection to cow or calf at the time of parturition. Assuming a normal delivery, the stockman should ensure that the calf can breathe properly, then disinfect the navel with iodine or an antibiotic spray. The next essential is to ensure that the calf drinks an adequate amount of the cow's first milk (colostrum) in the first 12 h, because colostrum contains not only food but maternal antibodies that will protect the young calf against the common infections it is likely to encounter in early life. The abomasal secretions are not acid in the newborn calf, and the epithelium of the intestine permits the passage of the large protein antibodies. This state of affairs changes rapidly, so that by about 16 h of age the gut's capacity to absorb whole antibodies has practically ceased.

The teats of mature, high-yielding Friesian cows may hang far below the abdominal wall and be hard for the young calf to locate. If the calf is left with its mother, the stockman should ensure that it is feeding satisfactorily. Many stockmen like to remove the calf from the cow as soon as possible before the maternal bond has developed, so as to minimize distress for cow and calf. If so, the cow should be milked out and the calf offered three or four feeds of colostrum during the first day of life. For calves born to Friesian cows, each feed should be of 1.5 litres.

Having got the calf started, it is a matter of choice whether to rear it up to weaning in an individual pen on restricted amounts of milk replacer fed from a bucket twice- (or, later, once-) daily, or to rear it as one of a group with free access to milk replacer sucked through a teat. The feeding options are as follows (the amounts refer, once again, to the typical Friesian-crossbred calf).

Table 3.15
A comparison of calf feeding systems

Rearing method	12-week weight	Food intake (kg)		Food costs to 12 weeks	
		Milk powder	Starter	£	Pence/kg gain
Bucket feeding					
twice-daily, warm	100	14	115	35.08	60
once-daily, warm	100	15	115	35.85	62
once-daily, cold	90	18	112	37.54	78
Teat feeding					
dispenser, warm	105	30	100	44.62	71
mild acid, cold	105	29	105	44.86	71
Whey powders fed twice daily	95	16	110	33.67	64

Prices based on: High fat milk powder, £770 per tonne; dispenser and mild acid powder, £780 per tonne; whey powder, £650 per tonne; hay, £60 per tonne; early wean concentrates, £205 per tonne.
Figures provided by Liscombe Experimental Husbandry Farm (1987).

1. *Twice-daily bucket feeding*
   Week 1: 2 × 1.5 litres/day at 125 g powder/litre = 375 g powder/day fed at blood heat (40°C).
   During week 2: increase intake to 2.0–2.5 litres/feed at 125 g powder/litre = 500–625 g powder/day.
   Introduce a dry calf starter ration (based on cereals, etc.) and roughage in the form of hay (or possibly barley straw). Wean when intake of calf starter = 1 kg/day, which occurs at approximately 5 weeks of age.
2. *Once-daily bucket feeding*
   Weeks 1 and 2: feed twice daily as above. *Week 3:* feed 2.0–2.5 litres/day in a single feed containing 450–500 g powder; offer starter feed and hay as before. It is *essential* to provide water for these calves.
3. *Teat-feeding:* warm milk from dispensers. These are machines which dispense freshly mixed milk replacer at blood heat. They are rather expensive to buy, and calves inevitably drink more milk replacer than if their intake is regulated by bucket feeding (Table 3.15). Moreover, many calves may not be eating 1 kg/day of calf starter ration by 5 weeks of age, and will suffer a set-back if abruptly weaned at this time.
4. *Teat feeding:* cold (acidified) milk. Calves can be reared on cold milk provided they drink it slowly through a teat. (Fig. 3.7). The daily ration is mixed in advance and stored in plastic pails. The feeding system is completed very cheaply using a teat, a tube and a non-return valve. Milk powder for cold teat-feeding is often acidified, mainly to prevent the milk souring in storage. Milk acid powders are based on conventional milk replacer powders containing skim milk and added animal fats, and pH is reduced to about 5.7. Strong acid powders (pH 4.2) are sometimes called milkless powders, since they do not contain casein (the main milk protein), as this coagulates into curds and whey at a pH below 5.5. They are usually based on whey proteins (albumin and globulin) plus some vegetable proteins, specially treated to improve digestibility and reduce the risk of food allergies. In cold weather, the milk can be warmed to about 15°C using for example, an aquarium heater with a thermostat.

Calves drinking milk from a teat can be encouraged to eat more dry starter food after 4 weeks of age by restricting access to milk powder. The best way to achieve this is to clamp the milk line so that it becomes progressively smaller and calves get bored with sucking before they have filled their bellies.

Costs of these various systems, based on 1981 figures from the Liscombe Experimental Husbandry Farm, are given in Table 3.15. Twice-daily bucket feeding was most efficient in terms of cost per kg gain. Teat-fed calves were about 6 kg ahead by 12 weeks of age, having consumed twice as much milk powder. Once-daily bucket feeding of cold milk was not a success. The disease risks in these systems will be covered later.

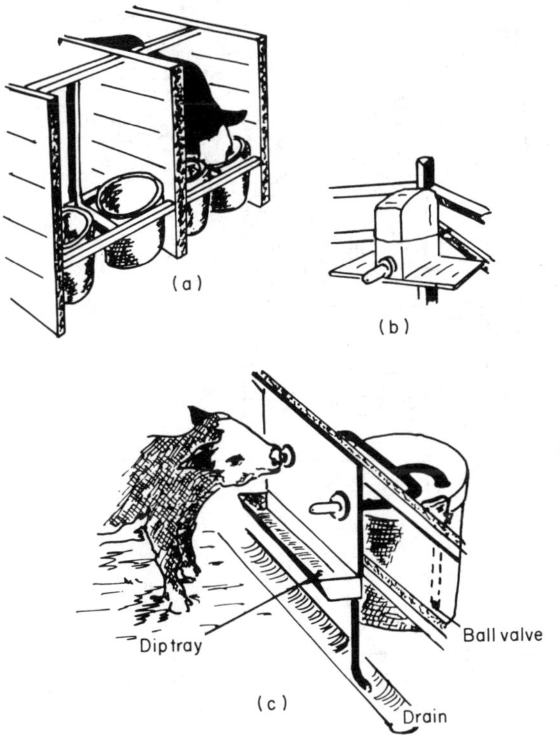

Fig. 3.7 Calf feeders. (a) Individual pens; (b) teat-feeding, starter; and (c) teat-feeding, cold milk.

The bought-in calf

The artificial rearing of calves on their farm of birth can usually be accomplished without mishap. Unfortunately, the majority of beef calves from the dairy herd are moved off their farm of origin at about 1 week of age, mixed through (usually) two or more markets, and transported on occasions the full length of the country. Such animals are deprived of normal food, water and physical comfort, and are confused, exhausted and exposed to a wide range of infectious organisms, of which the most important are the *Salmonella* bacteria. By the time they reach their rearing unit, they are infected, dehydrated and stressed, and need special care if they are to survive.

On arrival, bought-in calves should be rested in comfort in deep straw. In very cold weather, they may benefit from a little supplementary heating for the first 2–3 days. Water ought to be available, but they should be offered no milk replacer for at least 2 h after arrival. If they arrive in the evening, they can be left until the following morning. Some people feed 1.5 litres of milk powder at 125 g powder/litre for the first feed. I prefer to give two 1.5-litre feeds of a proprietary glucose/electrolyte solution (or one tablespoonful of glucose and one teaspoonful of common salt in 1.36 litres of water) to rehydrate the calves and provide a minimal supply of energy but keep the gut empty of nutrients until the calves have been able to eliminate most of the enteric bacteria acquired in transit. Thereafter, feeding can be as described above. An injection of fat-soluble vitamins A, D and E is also advisable, especially for calves purchased in the late winter.

Feeding after weaning

After weaning, calves are fed a concentrate ration plus hay, straw or silage *ad libitum.* The amount of concentrate fed and its protein concentration are governed by how fast the calf is expected to grow. If it is to be turned out to grass in the spring and reared for slaughter at 18–24 months, it should get no more than 3 kg/day of concentrate with a protein concentration of 180 g/kg plus hay or silage *ad libitum.* Calves in contract-rearing units, which need to achieve as much weight gain as possible by 12 weeks, and barley-beef calves, will probably be given unrestricted access to concentrates and consume more than 4 kg/day. In this case the initial protein concentration should be no more than 160 g/kg, and it may pay to feed roughage in the form of straw rather than hay or silage, to prevent overloading the gut with nutrients and provoking indigestion, abnormally soft faeces and impaired growth.

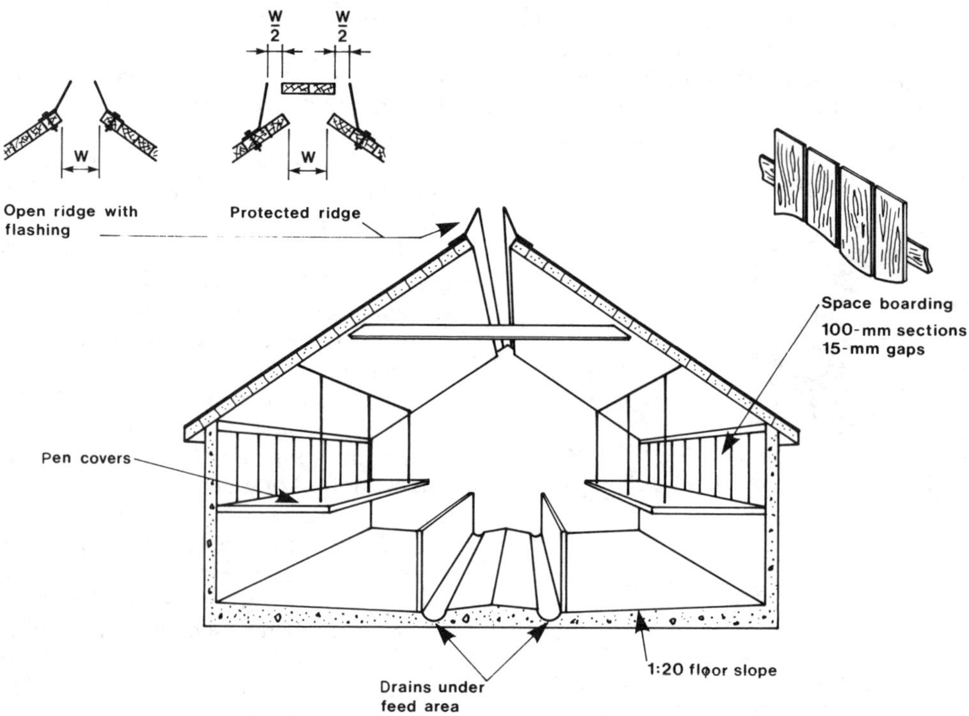

Fig. 3.8 Specifications for a naturally ventilated calf house (adapted from Mitchell, 1978).

Housing young calves

The greatest threats to the health and welfare of the young calf are infections which may cause septicaemia, enteritis (leading to diarrhoea or scours) and pneumonia. The organisms responsible for these conditions are widespread and young calves, especially those which are moved through markets, are practically certain to be exposed to some degree of infection. It does appear, however, that both the incidence and severity of these infections are determined by the numbers of pathogenic organisms and other irritant substances presented to vulnerable epithelial surfaces of the gut and lungs. The prime specification for a calf house is, therefore, that it should be as hygienic as possible, not only on the surfaces of the walls, floor and feeding utensils, but also in the air itself. It should be power-washed with hot water or steam and disinfected before the first calves of the season arrive, and this process should be repeated between successive batches. Whenever possible, the rearer of bought-in calves should practise an all-in, all-out policy, and avoid introducing new baby calves into an already infected building. Hygiene and humidity are also controlled by ensuring effective and appropriate drainage under the individual group pens (Fig 3.8).

Air hygiene is determined mainly by air space per calf (Table 3.14) and, to a much lesser extent, by ventilation. This is because the animals are the prime source of pathogenic organisms, but ventilation only removes a small proportion of these organisms from the air in the building; most die *in situ* (for further explanation, see Webster, 1985). Unweaned calves require 6 m^3 air space per calf to ensure reasonable air hygiene, and post-weaning calves 10 m^3 (Table 3.14). Under UK conditions, effective air movement (a minimum of four air changes per hour) can be achieved by natural ventilation through strategically placed inlets and outlets. Figure 3.8 illustrates one approach, which is to

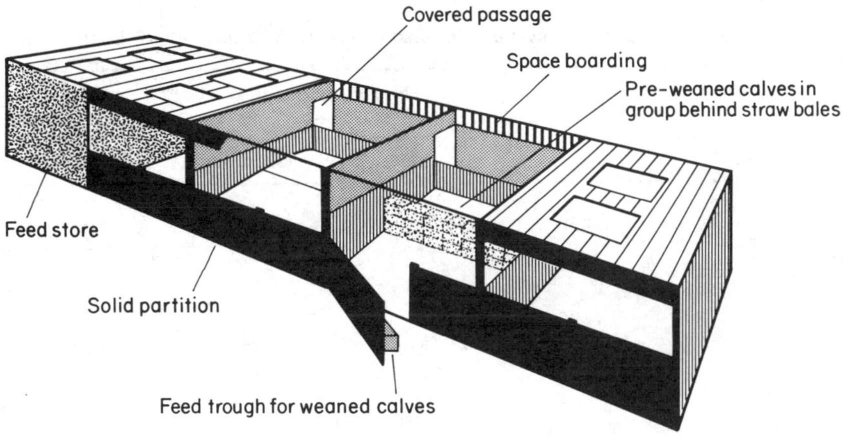

Fig. 3.9 Simple monopitch accommodation for rearing bought-in calves in groups. Calves stay in the same bay from arrival until turnout (from Webster, 1985).

place space boarding with 100 mm sections and 15 mm gaps below the eaves to provide 0.05 m^2 open area per calf and an open ridge providing 0.04 m^2 opening per calf. In still air, air enters the building below the eaves, is heated by the presence of the animals and exits through the open ridge. In windy conditions, eaves and roof openings can act as inlets or outlets, but the 15 mm gaps restrict the speed of the air entering the building and the pen covers prevent cold air descending to create cold draughts within the calf pens.

Whether reared in groups or individual pens prior to weaning, calves are normally grouped after weaning and reared in follow-on pens. The simultaneous stresses of mixing and weaning can increase the risk of disease at this time, especially pneumonia. One advantage of rearing batches of calves in groups and feeding them from a teat is that they can be kept in the same group and in the same accommodation from the time of arrival until turnout. Calves are initially restrained in the back of the pen by straw bales, which are taken down at weaning and used as bedding (Fig 3.9). The small increase in feed costs prior to weaning may be more than offset by reductions in the cost of housing and of disease.

Rearing calves for veal

Most calves reared for veal in Europe are confined for life in individual wooden crates and fed only a liquid milk replacer diet, deficient in iron, to ensure white meat.*

Conventional bucket feeding of crated calves

The calves are initially fed milk replacer in the same way as any bucket-fed calf, but given no access to solid food. The intake and concentration of the milk powder are increased, so that by 12 weeks they will consume 3 kg/day of powder at a concentration of 150 g/litre water at 40°C. Friesian/Holstein calves are slaughtered after they have consumed about 200 kg of powder and reached about 200 kg live-weight. This occurs after 14–16 weeks on the unit. The average iron concentration of the milk powder is 25 mg/kg or less. This system abuses the welfare of calves on several counts.

* *Editor's note. The Welfare of Calves Regulations, 1987*, which come into force on 1 January 1990, stipulate (para. 2) that: "No person shall keep, or knowingly cause or permit to be kept, a single calf in a pen or a stall on any agricultural land unless the following requirements are complied with—
(a) the width of the pen or stall is not less than the height of the calf at the withers;
(b) the calf is free to turn round without difficulty;
(c) the calf is fed a daily diet containing sufficient iron to maintain it in full health and vigour; and
(d) if the calf is more than 14 days old, it has access each day to food containing sufficient digestible fibre so as not to impair the development of its rumen."

Table 3.16
Comparative economics of different rearing systems for veal calves at the University of Bristol, 1981–1985

	Friesian/Holstein bulls		Hereford × Friesian heifers	
	Crates	Quantock	Quantock	Access
£s per calf purchase				
calf price	70	70	50	50
milk powder	125	153	132	110
dry food	—	—	—	5
straw, etc.	—	10	10	10
Total variable costs	195	233	192	175
Selling price	225	236	200	201
Carcass profit (ideal)	+30	+3	+8	+26
Less deaths/disease in group	30	30	30	3
	Nil	−27	−22	+23

From Webster *et al.* (1986).

1. The calves are malnourised, lacking dry, digestible food to encourage normal rumen development and sufficient dietary iron to prevent clinical anaemia.
2. They are enclosed in individual crates on slats. These crates make it difficult for the calf to stand when young, and impossible for them to adopt normal resting and sleeping postures when over 150 kg, nor groom themselves all over.
3. The calves are denied any exercise or direct social contact.
4. The incidence of both enteric and respiratory diseases is unacceptably high.

Alternative husbandry systems for veal calves

Several alternative husbandry systems for veal calves have been explored in an attempt to discover one that is more humane and also economically competitive with the conventional crates. The most common commercial practice has been to rear calves in groups on straw (if they are lucky) and allow them to suck milk replacer *ad libitum* from an automatic dispenser. In the UK this has become known as the Quantock system after its original proponents. The main drawback of this system is that the calves drink more milk than is necessary for optimal growth, or is indeed good for them, and this reduces food conversion efficiency. In brief, 1 tonne of milk replacer will rear five Friesian bull calves in crates but only four by the Quantock system. Moreover, the system still does not permit normal development of the digestive tract.

Trials at the University of Bristol (Table 3.16 and Webster *et al.*, 1986) have examined a variety of alternatives. Of these the most effective has proved to be ACCESS (A Computer-Controlled Eating Stall System) in which calves equipped with transponders that are recognized by the computer enter either of two computer-controlled feeding stalls, to receive controlled amounts of milk replacer or digestible dry food contributing respectively 90 and 10 per cent of nutrient intake. With this system we have been able to reduce feed costs below that for calves in crates (Table 3.16). Moreover, the health of the calves has been greatly improved by eliminating the worst abnormalities of feeding and housing. This has not only improved their welfare, but removed the economically crippling costs of death and disease associated with most veal systems. The profit margins for veal are very small and determined primarily by the price of calves at birth. Table 3.16 shows that they were largely non-existent, but a commercial veal producer would have paid less for his calves than we did.

Management of growing beef cattle

Summer grazing

The pastures grazed by young cattle during the summer need to be managed to guarantee that the grass is (1) sufficiently well grown to ensure maximum consumption but still sufficiently leafy to ensure optimum nutritive value, and (2) as free as possible from roundworms that can infest the gut and lungs. The most efficient way to utilize high-quality spring grass is to advance an electric fence across the pasture so as to graze a new strip every day. This encourages the animals to eat all the fresh grass exposed, and defaecate on areas grazed previously. This practice is more often associated with intensive dairy herds than with more extensive beef systems. The next most efficient approach is paddock grazing, whereby the herd is put on to a relatively small area of fresh grass at a high stocking density so that it grazes it down in, say, 3 weeks, before moving on to new pastures. The paddock is left to recover for about 6 weeks and then grazed once again by the young cattle or, preferably, older animals (or possibly sheep) who will not be affected by any build-up of parasitic roundworms.

However well a pasture is managed, the quality of the grass declines during the summer, due in part to natural seasonal changes in the pattern of growth and in part to the accumulation of dung pats which encourage the growth of rich, green patches of grass which the cattle will not eat.

The two most important species of parasitic roundworms for grazing cattle are *Ostertagia*, which causes gastritis, and *Dictyocaulus*, which causes pneumonia (see Chapter 2, p. 44). Both species overwinter on pasture and, therefore, can affect calves shortly after turnout. If, as is likely, the pasture was grazed by cattle the previous year, it pays to delay turning out young calves until after 7 May, by which time most of the overwintered *Ostertagia* larvae will have died. If calves do become infested and excrete eggs on to the pasture, there is a second build-up of larvae on the pasture by the middle of July. Ostertagiasis in cattle can be controlled effectively by pasture management, but many beef farmers consider it necessary to dose young calves in their first summer at grass regularly with anthelmintic drugs. Cows and yearling cattle who have previously been exposed to a low level of parasitic challenge develop an effective immunity.

The lungworm, *Dictyocaulus*, is a much more dangerous parasite which causes husk or hoose, a severe, incapacitating parasitic pneumonia. Affected animals have difficulty in breathing, develop a characteristic deep, husking cough and lose condition extremely fast. The most efficient way to control husk is to vaccinate the calves with two oral doses of irradiated (and thus attenuated) larvae before turnout.

Winter housing

When cattle are brought into yards, having been out at grass all summer, they are well adapted to a fibrous diet and reasonably immune to the viral and bacterial organisms that cause infectious pneumonia. Management, therefore, presents few problems. The objective is to feed the cattle so as to achieve pre-set targets for weight gain at the least possible cost. The most useful aid to management is, therefore, a good set of scales and an efficient arrangement for moving cattle in and out of their pens and over this weighbridge at intervals of about 4 weeks. Figure 3.10 illustrates such a system. Cattle may be driven fairly easily in groups into a collecting pen large enough to contain all the cattle from one pen in the beef rearing unit. They may then be forced down a race with solid sides and a curved approach to a weighbridge and a crush, where individuals may be restrained for routine treatment or preventive medicine. The high, solid sides to the race prevent the cattle from being distracted or startled by the presence of handlers or other alarming objects in front of them, and the curve in the race encourages them to follow their leaders until it is too late to try to escape.

Within the building itself, the cattle may be housed entirely on slats. Alternatively, and preferably, each pen may be divided into a well-strawed bedding area and a concrete area behind the feed fence at either side of the central feeding passage. Each pen is separated from the next by a gate which can be swung across daily to enclose the cattle in the bedding area and

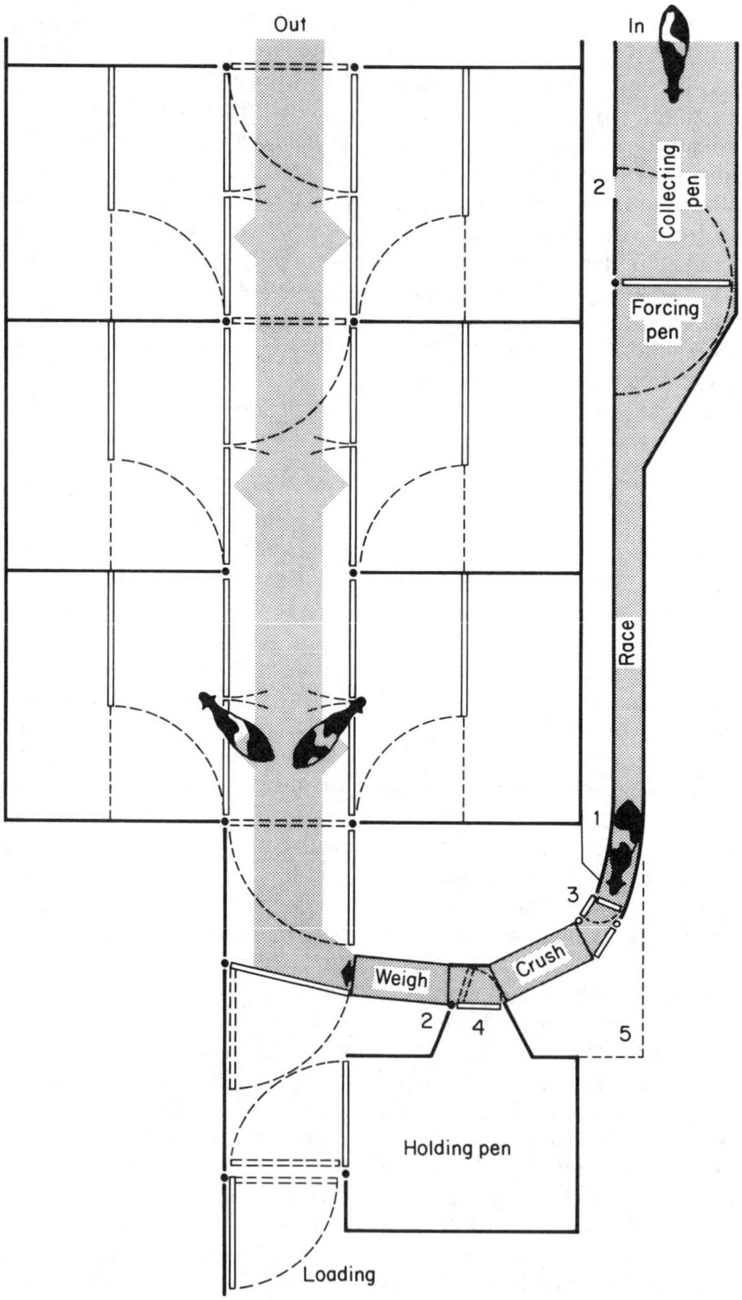

Fig. 3.10 Housing and handling facilities for in-wintered beef cattle (modified from MAFF/1984b Bulletin 2495). Crown Copyright. 1, Raised walkway; 2, squeeze gap; 3, Gates – access to rear of crush; 4, diversion gate; 5, line of roof over work area.

permit the concrete standing area behind the feed fence to be scraped down using a tractor (Fig. 3.10).

The building should be designed to permit the maximum amount of air movement without incurring draughts. Once again this is best achieved by installing space boarding to a depth of at least 1 and preferably 2 m below the eaves and an open ridge, with flashing upstands to either side (Fig. 3.8) over the central feed passage. The air, which has been warmed by the cattle, and which leaves the building by this open ridge, will prevent the entry of rain or snow in all but the most severe conditions. In addition, over the winter, the ridge will probably let out about 100 times more moisture (from the cattle) than it will let in.

Management of the suckler herd

I shall consider here only those suckler herds in which the animals are housed over winter.

Autumn calving

As indicated earlier, the main advantage of autumn calving is that the cows can give birth to their calves when they are well dispersed at pasture, which is hygienic. Moreover, the weather is usually mild and the cows will have eaten plenty of good grass, which ensures that they should have plenty of high-quality colostrum and, subsequently, milk to give the calves a good start. One possible drawback, is that calves by big bulls from cows in good condition may be very large at birth, and it is not exactly easy to give assistance to a half-wild beef cow having difficulty giving birth on open range.

If the cows calve between August and October they will require some concentrate feed, such as rolled barley, starting at about 0.8 kg/day and rising to 1.8 kg/day by November. This barley should be supplemented with magnesium to reduce the risk of deaths from acute hypomagnesaemia (see p. 77). The bulls should be introduced to the cows in November to try to get as many cows as possible in calf before they are brought into their winter accommodation. If beef cows are wintered in a deep straw yard (Fig. 3.11), mating indoors presents few problems but, if they are housed in cubicles or on slats, both cows and the bull can suffer injury due to slipping or falling on slimy concrete floors. To avoid any risk to the bull, cows can be observed for signs of oestrus and brought to the bull pen for service. Whether the cows are housed on straw, on slats, or in cubicles, it is essential to provide a creep area for the calves (Fig. 3.11) where they may go to lie on bedding in comfort and, more importantly, where they can receive good quality forage without competition from the cows, and probably a concentrate ration.

The amount of concentrate that is fed to the cows (0.5–1.5 kg/day) is, once again, determined by the quality of the forage. Cows that enter the winter in good condition (condition score 3–4, Fig. 3.12) may reasonably be expected to lose 80 kg over the winter and go out to grass the following spring at an average condition score of 2. This may appear harsh but is quite natural. Cattle evolved in environments in which the availability of food was seasonal, and are well-adapted to changes in body weight of up to 15 per cent.

From turnout to the middle of summer, cows and calves run together on plentiful grass to ensure not only that the calves grow fast but also that the cows regain the condition lost over the winter. Weaning occurs in mid-summer when the calves are about 10 months old. The very effective procedure adopted at the MAFF Experimental Husbandry Farm at Liscombe is to bring cows and calves into yards and separate them so that they can see each other but make no physical contact. Calves are fed hay for 2–3 days, then turned out once more. Cows are given a straw-only ration for 10 days to dry up their milk quickly, and an antibiotic is introduced into each quarter to minimize the risk of summer mastitis when they are returned to grass. There is no doubt that calves which are weaned and returned to grass for a period before being sold and/or housed for winter feeding are less likely to suffer problems such as shipping fever (pneumonic pasteurellosis) than those which experience the stresses of weaning, transport, mixing, winter housing and change of feed all at once.

Fig. 3.11 A cattle shed with a calf creep in the foreground.

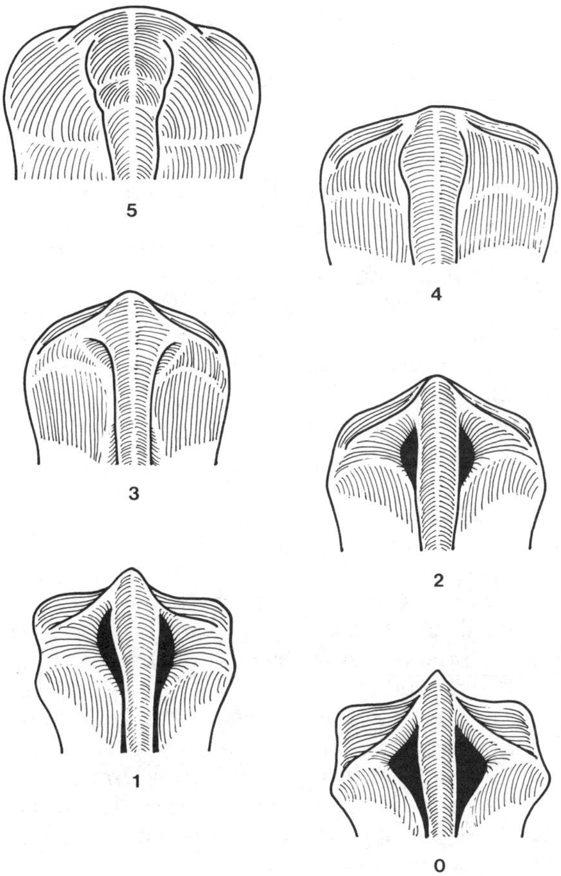

Fig. 3.12 Condition score of cows assessed for fat cover over tail/head and loin area. 5, Grossly fat; 4, fat; 3, good; 2, moderate; 1, poor; 0, very poor (from Webster, 1987).

Spring calving

The cows in a spring-calving herd will have probably had their calves removed in October and then been kept out at grass for as long as possible, entering their winter accommodation in November in Orkney, or as late as Christmas following a mild autumn on the moorlands of south-west England. These cows will also be expected to lose about 80–100 kg over the winter and calve down at an average condition score slightly greater than 2. Once again, the amount of concentrate required (if any) will depend on the quality of the silage. It is, however, important to ensure that all beef cows have adequate access to a mineral supplement, which can be sprinkled over the silage to ensure even distribution.

Ideally, spring-calving herds should be turned out to grass before calving but, because of the fickle nature of the spring and the shortness of the summer in classic beef country, this is usually impractical. This means indoor calving, which carries severe risks of infection.

The main problems of infectious disease for the newborn calf in a suckler herd are septicaemia, usually caused by *E. coli*, and enteritis caused by viruses such as *Rotavirus*. Both organisms are inevitable inhabitants of winter accommodation for adult cattle. Calves born into such an environment can be infected with *E. coli* at birth and the resultant septicaemia may (at worst) kill them within 2–3 days. *Rotavirus* and other enteroviruses usually induce diarrhoea beginning in the second week of life, which can also be fatal or severely stunt the calf's development.

The most effective means of controlling these two conditions is excellent hygiene at calving. There should be sufficient calving pens to allow each cow to deliver her calf on clean straw, stay with her calf for 2 days before returning to the herd, and enough time to permit the box to be cleaned out, disinfected and re-bedded with fresh straw before the next cow enters. This is expensive, but so are dead calves. Very recently vaccines against *E. coli* and *Rotavirus* have come on to the market. These are given to cows to boost the antibodies to these organisms in their colostrum and milk. The milk antibodies are not absorbed but can confer local protection on the gut surface. These vaccines appear promising but are no substitute for good husbandry and hygiene.

Turnout to grass takes place as soon as possible. The main risk for the cows at this time is hypomagnesaemia, and magnesium supplements are essential. If the cows are not getting a concentrate ration, these minerals can be incorporated into a block or added to the water supply. The bulls are introduced to the cows in June, when the weather is good and the cows are on a high plane of nutrition and gaining condition. In these circumstances, fertility should be high. The calves are weaned in October and usually moved directly into winter housing. As indicated above, simultaneous weaning and movement carries a high risk of provoking stress-related diseases like shipping fever.

Routine veterinary procedures and problems

A review of even the more common veterinary problems and procedures with beef cattle is beyond the scope of this section, which describes, very briefly, only the routine procedures in husbandry and preventive medicine common to most beef farms. The skills that enable the stockman or veterinary surgeon to carry out painful procedures like castration and dehorning with speed, safety and humanity are not to be learnt from textbooks but by direct, practical instruction from trained operators. The castration and dehorning of calves is controlled by the Protection of Animals (Anaesthetic) Acts 1954 and 1964 which deem the following to be operations performed without care and humanity unless carried out under anaesthetic:

1. The castration of a bull by a device that constricts the flow of blood to the scrotum (a rubber ring) unless it is applied within the first week of life.
2. The castration of a bull by any means after the age of 2 months (after this age it must be carried out by a veterinary surgeon).
3. The dehorning of adult cattle or disbudding of calves except by chemical cauterization in the first week of life.

Castration

Some operators apply a rubber ring to castrate male calves at about 1 day of age. This appears to cause less distress than if applied later in the first week, but it is undoubtedly a more painful process than surgical removal of the testes. The most usual time for surgical castration (and dehorning) is when the calves are about 7 weeks of age. An anaesthetic is not compulsory at this age and it is possible that the operator who removes both testes rapidly without anaesthetic causes less distress than one who restrains the calves, injects local anaesthetic, waits for it to take effect and then restrains them again for surgery. The alternative to surgical removal is bloodless castration, i.e. crushing the spermatic cord, blood and nerve supply to the testes. This causes less risk of infection but prolongs the period of pain and distress. Hygienic, surgical castration has to be the most humane approach.

Everybody who rears male calves for beef should ask himself the question, "Is castration necessary?" If the cattle can reach slaughter weight in 18 months or less, the answer is almost certainly no. Entire male cattle grow lean tissue faster than steers or heifers, and the production advantages of bulls have increased in importance since the ban on the use of anabolic steroids as growth promoters for beef cattle.

Dehorning

Disbudding calves less than 1 week old by chemical cauterization with caustic potash is seldom practised nowadays. It does not always work and produces pain that is prolonged long after the operation is complete. The most common method is to remove the horn bud under local anaesthetic using a hot iron when the calf is at least 3 weeks old and the horn tip can be clearly felt. An alternative approach which has been practised with success very recently is to destroy the horn tip with hot air. Both techniques should only be attempted after personal instruction from a competent operator.

Dehorning and castration are commonly done at the same time. This is acceptable practice but it is not a good idea to combine these two stresses with that of weaning because, all together, they may predispose to stress-related diseases such as pneumonia.

Preventive medicine

Septicaemia and gastroenteritis. To minimize the risk of septicaemia (*E. coli*) and viral gastroenteritis (*Rotavirus*, etc), calving accommodation should be clean and calves should receive ample colostrum. In suckler herds calving indoors, it may be necessary to vaccinate the cows against *E. coli* and *Rotavirus* to boost antibody concentrations in the colostrum. Calves with diarrhoea rapidly become dehydrated and sodium-deficient. When treating such calves it is even more important to replace lost water and electrolytes (up to 5 litres/day) than to attempt to treat the primary infection with antibiotics.

Salmonellosis. Infection with (especially) *Salmonella typhimurium* is a very common cause of enteritis and toxaemia in calves that have travelled through markets. Many strains of this organism are resistant to a wide range of antibiotics. The most effective preventive medicine appears to be to restrict food intake but otherwise minimize environmental stresses for the first few days after the calves arrive. Vaccination against salmonellosis *after* arrival can be dangerous.

Pneumonia. Infectious pneumonia of housed calves occurs most commonly at an age of 6–12 weeks. There is no vaccine that confers effective protection against all the infectious organisms that can damage the lung. The most effective approach to the control of the incidence and severity of pneumonia is to ensure good air hygiene through proper attention to stocking rate and ventilation, to avoid mixing calves as far as possible during their first winter and to minimize the stresses at and around weaning.

Ringworm. This is a common condition of housed calves in their first winter. The fungi are difficult to remove from porous surfaces like walls and woodwork and the disease is expensive to treat. However, it does not appear to cause the calves any great distress and normally disappears when they are turned out to grass. Most cases are therefore left to clear up spontaneously.

Lice and mange. These conditions are caused by external parasites and can cause hair loss, considerable irritation and loss of body condition in housed cattle at any age. They are easily controlled by local or systemic treatment with insecticides.

Parasitic bronchitis. Parasitic bronchitis or husk affects calves in their first summer at pasture. It is effectively controlled either by oral vaccination with attenuated *Dictyocaulus* larvae before turnout or by regular dosing with appropriate anthelmintic drugs.

Ostertagiasis (stomach worm infestation). This can also be controlled by regular dosing, or by pasture management.

Warble fly. The warble fly (*Hypoderma* sp.) lays its eggs on the hairs of the legs and abdomen of cattle. The larvae hatch out, burrow through the skin and migrate, eventually, to the back of the cattle from whence they emerge as maggots 25–30 mm in length. This painful and economically costly condition is now a notifiable disease and cattle farmers are obliged to treat all cattle with a systemic insecticide between 15 September and 30 November to kill the larvae after they have penetrated the skin but before they have grown and migrated to the stage where they can do damage.

Shipping fever. This disease, also known as fibrinous pneumonia and associated with *Pasteurella haemolytica*, commonly affects calves from the suckler herd when they are brought off open range and concentrated into buildings or open feedlots. The calves are normally carrying the organism in the upper respiratory tract on arrival. Pneumonia develops in response to stresses of mixing, disturbance, exhaustion or improper introduction to concentrate rations. Once again, the disease is better controlled by attention to management rather than vaccination.

Hypomagnesaemia in cows. Magnesium has a controlling, inhibitory effect on nerve and muscle activity. Cattle, especially adult cows at pasture, can teeter on the brink of magnesium deficiency for weeks before a sudden stress such as a rainstorm can push them over the brink into an acute crisis. The animal proceeds through nervousness to madness and may die in convulsions within 1 h of the first clinical signs. Because of the high risk of fatalities from hypomagnesaemia, it is important to make doubly sure that susceptible animals receive ample magnesium in the diet. Cows cannot be relied on to control their intake of high magnesium mineral licks, and these should be supplemented by adding magnesium to the concentrate ration or the water supply.

For details of other diseases, and metabolic disorders of adult stock, the reader is referred to Chapter 2, pp. 39–44.

Casualty and emergency slaughter

From time to time a farmer or veterinary surgeon is faced by an animal that is so sick or injured that it is clearly better to slaughter it as soon as possible rather than attempt treatment. In this event, the most humane procedure is for either the veterinary surgeon or a licensed slaughterman to kill the animal on the spot using an approved procedure such as the captive bolt. If, however, the animal can be transported to an abattoir, it is more likely to be killed and butchered in such a way as to render the meat fit for human consumption. The farmer is forced, therefore, to test the strength of his compassion against a cash loss that may amount to hundreds of pounds. Veterinary surgeons are required to write a casualty slaughter certificate for sick or injured animals, but this is only necessary to establish whether in their opinion the meat is likely to be fit for human consumption. The welfare of the animal is covered by the Transit of Animals (Road and Rail) Order 1975 which states "no animal which is unfit shall be permitted to be carried... by road or rail if, by reason of its unfitness the animal is likely to be subjected to unnecessary suffering". The only exception in law is that in an emergency a road vehicle can be used, in effect as an ambulance, to carry the animal "with all practicable speed... for veterinary treatment... or to the nearest available place where it can be slaughtered". In other words, an animal can be moved in an emergency if it is kinder to do so than, for example, to wait several hours for the arrival of a veterinary surgeon or a licensed slaughterman. Thus defined, the emergency transport of cattle is likely to be a very rare event.

Transport of weak, sick or injured cattle to a slaughter house will clearly cause unnecessary suffering if the animal is in severe pain which is exacerbated by movement; a broken leg is an obvious example. A calf that is too weak to move could reasonably be transported to a slaughter house if it is carried on to (and off) the lorry. The larger animal, such as the downer cow that collapses and develops paralysis post-calving, presents a more difficult problem. She may be reasonably comfortable where she lies and equally so if on deep bedding in a lorry. It would be reasonable to transport her if she could be rolled on to a board or sledge which was then winched carefully on to and off the lorry. Any procedure that involved dragging the live recumbent cow herself over the ground to enter and exit the lorry would constitute unnecessary suffering and the owner should arrange for the animal to be killed on the spot.

Stockmanship

The Preface to the *Codes of Recommendations for the Welfare of Livestock: Cattle* (MAFF, 1983) states:

> The basic requirements for the welfare of livestock are a husbandry system appropriate to the health and, so far as practicable, the behavioural needs of the animals and a high standard of stockmanship.

Husbandry systems differ in the extent to which they can ever meet the health and behavioural needs of the animals but, within any system, the quality of the stockman (or stockwoman, since, on average, women are demonstrably better calf rearers than men) is the most important determinant of welfare. The criteria for good stockmanship are (i) a proper understanding of the animals, (ii) a sense of caring and (iii) sufficient experience based on proper training. This can only be gleaned in part from books such as this. In the UK, training courses and proficiency tests for stockmen are run by the Agricultural Training Board, local authorities and the association of Young Farmers' Clubs. The young person with a liking for farm life and a sympathy for farm animals is not yet a stockman, merely a promising pupil who acquires stockmanship through experience and proper, formal training.

References and further reading

Allen, D. and Kilkenny, B. (1985). *Planned Beef Production*, 2nd edition. London: Collins.

Blowey, R. W. (1985). *A Veterinary Book for Dairy Farmers*. Ipswich: Farming Press.

Farm Animal Welfare Council (1986). *Report on the Welfare of Livestock at Markets*, RB 265. London: HMSO.

Grandin, T. (1980). Observations on cattle behaviour applied to the design of cattle handling facilities. *Applied Animal Ethology* **6**: 19.

Liscombe Experimental Husbandry Farm (1977–1981). *Beef Bulletin 1: Calf Rearing* (1981); *2: Beef Systems* (1977); *3: Suckled Calf Production* (1981). London: HMSO.

McDonald, P., Edwards, R. A. and Greenhalgh, J. F. (1981). *Animal Nutrition*, 3rd edition. Edinburgh: Oliver and Boyd.

Meat and Livestock Commission (1986). *Beef Yearbook*. Milton Keynes: MLC.

Milk Marketing Board (1986/87). *Report of the Breeding and Production Organization*. Thames Ditton: MMB.

Ministry of Agriculture, Fisheries and Food (1976). *Annual Reviews of Agriculture*. London: HMSO.

Ministry of Agriculture, Fisheries and Food (1983). *Code of Recommendations for the Welfare of Livestock: Cattle*, L 701. London: HMSO.

Ministry of Agriculture, Fisheries and Food (1984a). *Energy Allowances and Feeding Systems for Ruminants*, RB 433. London: HMSO.

Ministry of Agriculture, Fisheries and Food (1984b). *Cattle Handling*, B 2495. London: HMSO.

Ministry of Agriculture, Fisheries and Food (1986). *Annual Reviews of Agriculture*. London: HMSO.

Mitchell, C. D. (1978). *Calf Housing Handbook*. Aberdeen: Scottish Farm Buildings Investigation Unit.

More O'Ferrall, G. J. (ed.) (1982). *Beef Production from Different Dairy Breeds and Dairy Crosses*. The Hague: Martinus Nijhoff.

Signoret, J. P. (ed.) (1982). *Welfare and Husbandry of Calves*. The Hague: Martinus Nijhoff.

Swan, H. and Broster, W. H. (1981). *Principles of Cattle Production*. London: Butterworth.

Webster, A. J. F. (1985). *Calf Husbandry, Health and Welfare*, 2nd edition. London: Collins.

Webster, A. J. F. (1986). Health and welfare ot animals in modern husbandry systems. *In Practice* **8** (3) : 85 (Supplement to *Veterinary Record*).

Webster, A. J. F. (1987). *Understanding the Dairy Cow*. London: Blackwell Scientific.

Webster, A. J. F., Saville, C. and Welchman, D. de B. (1986). *Alternative Husbandry Systems for Veal Calves*. Bristol: University of Bristol Press.

4 Sheep

H. Ll. Williams

The industry in the UK

Structure

The UK has a very diverse sheep industry, largely because sheep production is undertaken in very contrasting environments, ranging from the lowland cultivated pastoral and arable areas to the more severe upland and hill areas dominated by uncultivated rough grazings (see Fig. 4.2 for classification of grazing types). In these areas of rough grazing which represent one-third of the total agricultural land in the UK, the combination of severe climate, difficult topography, low soil fertility and inaccessibility allows little choice of animal enterprises; apart from populations of feral ponies and goats and farmed wild deer, sheep are the only farm animal resource. It is not surprising that the majority of breeding ewes are found in upland and hill flocks kept above 250 m. In lowland areas there is more competition between species and between animal and crop enterprises, and sheep may, in some areas, play only a minor and complementary role to dairying. There is a high degree of interdependence between lowland flocks and hill and upland flocks (p. 83).

The MAFF Census of June 1987 for the UK shows a total sheep population of 38.8 million, the highest ever recorded. The increase in the sheep population during the last 10 years is largely due to the favourable economic climate for sheep production, and particularly since the introduction of the EEC sheep meat regime. The number of ewes and ewe lambs kept for breeding increased by 30 and 35 per cent respectively during the period 1976–1986 (Fig. 4.1). During that period, the proportion of these lambs which were put to the ram in their first year increased from 27.2 to 43.4 per cent.

The UK government and the EEC provide a wide range of financial support to the sheep industry. The industry is well supported by a vast range of educational and technical services which provides good opportunities for the training of skills, assistance with management, and advice on health and welfare. The Meat and Livestock Commission (MLC) has, since it was established in 1967, promoted greater efficiency in the livestock and livestock products industries. It has vigorously undertaken a wide range of activities: livestock improvement, carcass evaluation and classification, market information and intelligence, and meat promotion. The National Sheep Association represents both pedigree and commercial interests in its contact with national and international official bodies; it also undertakes breed promotion on behalf of breeders.

The UK sheep industry is geared to the production of meat, mainly from finished lambs. More attention is now given to fatness, partly because of public concern regarding diet and health. Overfatness is an aspect of carcass quality which has contributed significantly to the 10 per cent fall in the consumption of sheep meat in the UK during the last 10 years. The production of lamb meat has been increasing as a result of the rise in the breeding ewe population and the level of importation has been reduced, and consequently the level of self-sufficiency

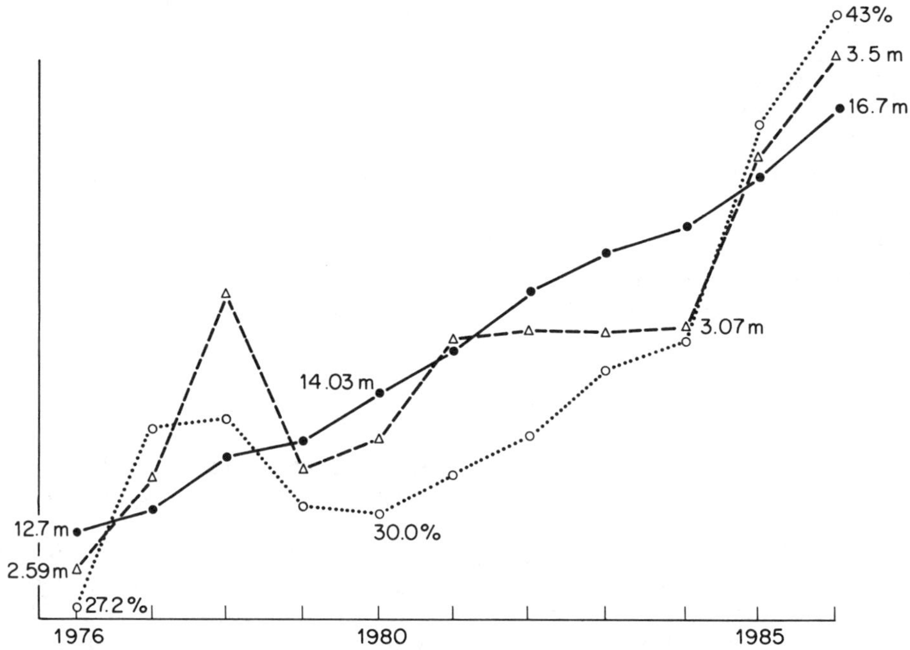

Fig. 4.1 Changes in populations of breeding females 1976–1986. ●, Total breeding ewes; ○, % ewe lambs tupped; △, total ewe lambs kept for breeding. *Source*: MAFF Dec. census.

increased from 55 per cent in 1970 to 77 per cent in 1985.

The UK remains the biggest wool-producing country in Europe, even though wool sales account for only 5 per cent of the gross financial return. Wool is marketed through the British Wool Marketing Board (BWMB), set up in 1950 to co-ordinate the grading and marketing of wool and to stabilize returns to its numerous registered producers (approximately 91 000). The Board is actively engaged in improving the general standard of the British clip by encouraging breeding for increased yield and better quality, and by setting high standards for shearing and presentation of fleeces.

Productivity

Comprehensive information is now available about the productivity of flocks kept in hill, upland and lowland areas. It is largely derived from the MLC's Flockplan Recording Scheme, which is mainly concerned with the analysis of physical and economic data and designed to encourage more efficient production. Of particular interest is the use of local data, during meetings of flock owners, to emphasize local problems and to demonstrate how management may be improved. The practice of publishing average performance, as well as that of the top one-third of producers, has encouraged the gradual improvement of productivity during the last 10 years. Target stocking rates per hectare for breeding ewes are as follows: upland, 9.0; lowland early fat lamb, 15.5; lowland fat lamb, 12.0.

It is clear from published data that levels of performance are well below the biological capacity of sheep. For example, in the case of fecundity, as measured by the number of lambs reared per ewe put to the ram, a flock average of 1.48 lambs per ewe is 25–30 per cent below what is possible under above-average management.

Breeds

One of the unique features of meat production in the UK is the well-established integration

Table 4.1
Classification of commercial sheep flocks in the UK

Category	Main characteristics and selection objectives	Breeds[a] and their mature live-weight (kg)[b]	Main function
1. Hill flocks	Hardiness; ability to rear 1 lamb; satisfactory quality and weight of fleece	Blackface (70) Welsh Mountain (45) Cheviot (64) Swaledale (64)	To make use of land, usually above 300 m, that would otherwise be unproductive
2. Flocks producing sires of cross-bred ewes	Prolificacy; milking capacity; growth rate	Border Leicester (100) Blue-faced Leicester (96) Teeswater (95) Wensleydale (90)	To sire crossbred breeding ewes for lowland flocks as a result of crossing with draft ewes from hill flocks
3. Flocks producing fat lamb sires	Growth rate; carcass quality	Oxford Down (110) Suffolk (90) Texel (87) Dorset Down (77) Southdown (60)	To sire fat lambs, mainly in crossbred lowland flocks. Sometimes referred to as terminal sires
4. Self-contained flocks (upland/lowland)	Prolificacy; milking capacity; growth rate; carcass quality; weight and quality of wool	Clun Forest (73) Romney Marsh (75) Devon and Cornwall Longwool (95) Dorset Horn (82)	To produce their own replacements, surplus breeding stock and fat lambs. The Dorset Horn is used for out-of-season fat lamb production

[a] The most numerous breeds in each category.
[b] Average of male and female sheep.

between regions, largely according to altitude. This is very evident in the sheep industry and is usually referred to as stratification. It uses land resources efficiently and exploits the wide range of genetic potential which exists within the variety of breeds found in the UK. The movement of stock between hill, upland and lowland is illustrated in Fig. 4.2. The hill and upland areas act as a reservoir of breeding stock and animals with fattening potential. Vast numbers are moved to more favourable environments in the late summer/early autumn. This is organized through regional sales of draft pure-bred ewes which have produced three crops of lambs on the hills, young cross-bred ewes (usually two-tooth) from the uplands, and store lambs (mostly wether lambs) from hill and upland flocks. The lowland farmer is, therefore, able to buy animals at an optimal stage of production, and he is also able to choose from a wide variety of breeds and crossbreds on the basis of body size, reproductive potential and milkiness.

There are approximately 50 breeds in the UK and a wide range of first crosses, many of them named. A full description of breeds and named crossbreds may be found in *British Sheep* (NSA, 1987). The breeds can be placed into four main categories (Table 4.1). The functions of each category are markedly different and form the basis of the planned cross-breeding associated with stratification.

Improvement schemes

It is evident from the criteria listed in Table 4.1 that there is great scope for combining desirable characteristics through cross-breeding and hybridization. Over the years, these methods have enabled the industry to adapt efficiently to changes in market demands and in systems of management. The fact that 55 per cent of all breeding ewes are crossbred emphasizes the importance of cross-breeding to the industry. The

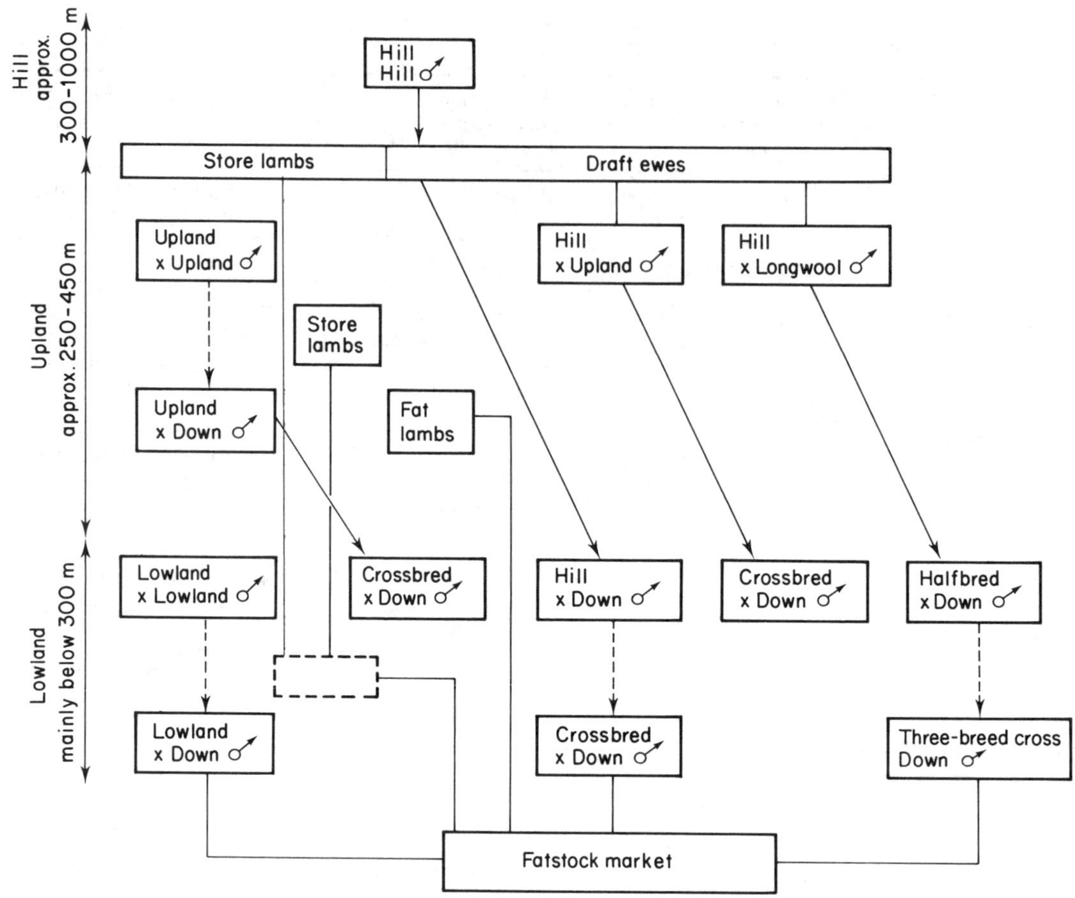

Fig. 4.2 Movement between regions and breeding plan.

majority of crossbreds are found in lowland flocks.

Improvement within a breed is more difficult to achieve, and methods which are acceptable to the UK industry are still being developed. In the past, attention has been given to visual appraisal and to breed characteristics such as colour, size, fleece, and presence or absence of horns. Purity of breed has been achieved by this method of selection, but little or no progress in terms of genetic potential can be claimed in respect of several important economic criteria. A better foundation to progress has now been established through better definition of objectives, more accurate identification of superior animals, and the introduction of schemes designed to utilize the superior genotypes advantageously. Performance testing, progeny testing, contemporary assessments, and the development of selected indices have all been developed and refined by he MLC and other organizations. Such developments provided the right background for the recent adoption in this country of Group Breeding Schemes as developed in New Zealand. These schemes require flock owners to pool the superior animals of their flocks to form a jointly owned nucleus flock, with the view to providing replacement rams, and sometimes ewes, for their members. The annual selection of superior animals is based on agreed criteria for commercially important characteristics.

Although artificial insemination (AI) of sheep is practised in many countries it is not yet offered as a commercial service in this country. Various centres in the UK, and particularly the MLC unit near Selby, Yorkshire, have greatly contributed to the development of procedures and technology of AI in sheep.

Outside pressures

Sheep have not followed the same pattern and scale of intensification as other farm livestock, and so there is a smaller degree of control over their social and physical environment. There has been less pressure from environmental and welfare lobbies than has been the case with some other species. Nevertheless, there are *Codes of Recommendations* (MAFF, 1977) which describe the basic requirements of management, feeding and husbandry of sheep, and which safeguard their welfare.

There is also widespread concern among consumers about the wide range of chemicals used in the production of food animals. Every flock owner must pay proper regard to the recommended withdrawal period for the wide range of chemicals used in the control of both internal and external parasites. The level of usage for these purposes is far higher in the sheep industry than the other food animals but, on the other hand, the use of antibiotics and growth promoters is far less. With sheep milk production gathering momentum, it will probably be necessary to place the sale of sheep milk and milk products under the same kind of legislation which applies to the dairy cattle industry.

Natural history

The sheep was domesticated about 10 000 years ago, probably in South-West Asia. It was one of the earliest species of domestic animal, largely because of its behaviour and size, and the fact that it was a ruminant and therefore did not compete with man for food. It was relatively easy for migratory early man to move the flock with him, and sheep provided a variety of resources: – milk, fibre, skin and meat.

It is thought that wild sheep had many behavioural traits which enhanced the process of domestication and these included a group pattern of social behaviour, readiness of the newborn lamb to imprint on substitutes for its dam, and adaptability to a variety of environmental conditions.

Sheep have remained an agricultural important species, particularly in adverse environments, and are to be found over a wide range of latitudes and altitudes, but they are less well adapted to moist tropical conditions than to other climatic areas.

There is a great range of live-weight (45–115 kg) among the British breeds and crossbreds (Table 4.1) and, therefore, considerable variation in the birth-weight of lambs. The mean birth-weight of individual single, twin and triplet lambs out of a 60 kg ewe would be approximately 4.35, 3.39 and 2.69 kg respectively. The survival rate of these lambs would be closely correlated with birth-weight, lambs weighing less than 2.0 kg being particularly vulnerable.

The productive life span of breeding ewes is greatly influenced by the prevailing policy regarding culling and this varies between farms and, sometimes, between years. Where the replacement rate is in the range of 12–20 per cent, the age at culling would be 5–8 years.

Normal behaviour patterns

There is more opportunity for sheep kept extensively to display innate patterns of social behaviour than those in intensive systems. A flock response to stimuli in the environment is a marked feature of sheep behaviour. It has been shown that during grazing some breeds disperse themselves more than others, but still maintain a flock pattern of movement. On unfenced hills or pastures, sheep will graze only within a prescribed territory, and subgroups may occupy sub-territories. The social interactions which govern flock behaviour are not well understood and, apart from reproductive behaviour, little attention has been given to communication between members of the flock. It is thought that leadership is not as precisely defined as in other species and is subject to frequent change. Subgroups of less than four sheep tend not to behave as a group and are difficult to manage. This has some bearing on the size of groups chosen for sheep dog trials.

There is no doubt that visual contact is important to sheep. A sheep which cannot clearly see its neighbour will raise its head frequently to try and establish contact. Sheep bleat when isolated. They alert other members of the flock by adopting a very upright posture with head held high and walking with tense steps, often audible to other members of the flock, and this may be accompanied by snorting. When disturbed, the grazing flock tends to regroup on higher ground and usually not very far from the cause of the disturbance. Where a predator or dog is pursuing, flight continues, often in an uncontrolled manner, and cases of ewes hurtling into ditches and streams are not uncommon.

Sheep alternate between periods of grazing and of rest. Total daily grazing time depends on the availability of food, but is usually around 8 h and rarely more than 10 h. During the rest phase short periods of sleep occur, amounting to 3.5–4 h per 24-h period. Most of the rest phase is spent ruminating. Sheep tend to have fixed eating patterns and, in some situations, are reluctant to accept new food when this is introduced abruptly (e.g. hill sheep given supplements during very bad weather). In the interest of their general welfare, and indeed survival, all sheep should be trained, during their first autumn, to accept cubed concentrate diets.

Trough-fed ewes exhibit dominance, the effects of which can be minimized by scattering troughs over a wide area. Lowland sheep quickly become conditioned to the signals of feeding time, such as the sound or sight of a bucket, or the appearance of the shepherd at the entrance to the building or at the gate. This can be used to advantage to move ewes between locations or into pens, or even into vehicles.

Ewes rarely show active aggression towards man and will accept interference around lambing time. On the other hand, they can be very aggressive towards dogs at this time, and this response may be used to heighten the ewe's maternal instincts during fostering. Housed ewes, especially of horned breeds, tend to fight more during housing than at pasture. Rams need to be carefully watched during the breeding season. Some rogue rams are aggressive at all times and should be culled.

Sheep are rarely seen grooming one another. Whilst in full fleece, they find it difficult to cope with local irritations and usually seek a post or rail. In the absence of a suitable object to rub against they will roll on to their backs. When in full fleece or heavily pregnant they may fail to get up and, if not seen, will die.

The reproductive behaviour of sheep is well documented (Hafez, 1969; Fraser, 1968). In the absence of the ram, ewes do not display the usual signs of oestrus, and therefore it is imperative to use a male to monitor the incidence of oestrus.

During oestrus, ewes stand quite firmly and may look backwards and also display tail fanning. Urination is usually associated with the rejection of the ram. Ewes in oestrus may indulge in ram-seeking, but most of the searching is done by the ram, sight and smell being the most important sensory cues. The initial approach may involve sniffing the perineal area, with or without lip curling and baring of teeth (flehmen). Depending on the degree of stimulation, it may be accompanied by a low-pitched gurgling sound, and nudging and pawing with the front foot. Alternatively, the ram may extend its neck and rotate its head slightly in an exaggerated forward movement alongside the ewe; some rams may show this patterm of behaviour some distance from the ewe. The ram may mount several times before thrusting, which is usually associated with ejaculation. After ejaculation, the ram dismounts and stands with head lowered alongside the ewe before continuing to search.

The understanding of the behaviour which occurs around lambing is probably the most crucial of the behavioural aspects which contribute to efficient stockmanship. Great dependence is placed on behavioural signs to indicate the imminence of lambing, the strength of the family bond, the well-being of the lamb, and the adequacy of the milk supply.

In a field situation, ewes will isolate themselves from the rest of the flock to establish a birth site. The ewe will stay around the selected site for a considerable period (up to 6 h has been recorded) and, if at all possible, should not be disturbed. The birth site becomes the focal point for the ewe. It is usually clearly visible as a result of her activity and it will have a recognizable smell. A period of conspicuous restlessness precedes the onset of full labour. When the ewe enters full labour, contractions become obvious and she will raise her head and point her nose to the sky, a signal which can be readily seen from some distance. Labour usually lasts for less than 1 h. The ewe is not easily disturbed but, if she is, she will resume contractions after a short interval.

Housed ewes become conditioned to the presence of the shepherd and to the noises associated with routine tasks. They should be allowed to lamb in their original pen, and then the family should be moved to an individual pen for 24–48 h. Irrespective of the system of management, every opportunity should be given for the ewe to lick her lamb(s) and to develop a bond between herself and her offspring. Both sound and smell play an important part in establishing this bond, and these, together with sight, contribute to identification of the lamb by the ewe during the early rearing phase.

The lamb attempts to get up within 20 min of being born and is attracted to the movement of its dam. This is an innate capacity; the newborn lamb will accept any moving object and thus it is relatively easy to imprint a substitute for its dam.

After a period of exploration along the underline of the ewe, the lamb locates a teat and starts sucking within the first hour. Sucking is usually accompanied by vigorous tail-wagging and occasional udder-butting. The ewe may help teat-seeking by manoeuvring herself into an advantageous position for the lamb and, as it approaches the teat, the ewe licks its hindquarters, particularly the perineal area, quite vigorously. The ewe suckles her lambs twice per hour to start with, and gradually less frequently as lactation advances.

Abnormal behaviour following lambing includes desertion, delayed grooming, mismothering and partial rejection of lambs.

Reproductive and breeding data

The reproductive capacity of the sheep may be described in terms of three main components: seasonality, fecundity (incorporating age at puberty, litter size, lambing interval and duration of breeding life), and fertility (the capacity to produce viable offspring). Some aspects of these are discussed below.

Seasonality

All British breeds of sheep except the Dorset Horn and Poll Dorset have a clearly defined breeding season commencing in the late summer/early autumn and continuing until mid- to late winter.

The main factor governing seasonality of breeding is the changing ratio of light to darkness. The photoperiodic response of the sheep is controlled by the pineal gland, which produces melatonin only during the hours of darkness. The changing duration and/or timing of melatonin secretion during a 24 h period acts as a biochemical signal to the higher centres of the brain, which control reproductive activity. This capacity of sheep to respond and to entrain to light rhythms played an important part in the establishment of sheep in higher latitudes, where the food supply is seasonal.

There are distinct differences between breeds in the onset of the breeding season (see Table 4.2). Individuals within a breed also show considerable variation. These aspects have to be given due consideration in the case of flocks in which early and compact lambing is sought. The extent of variation also influences decisions regarding the need for artificial control of breeding. The mean onset for a breed does not show much year-to-year variation, provided ram management remains unchanged. The range of body condition normally found in lowland flocks should have little effect on onset of breeding activity. Draft ewes from the hills show little change in onset during the autumn following their transfer to lowland flocks, despite a marked improvement in their condition.

With the exception of the Dorset Horn and Poll Dorset, British breeds of sheep are in seasonal anoestrum by the time they lamb in late winter/early spring and thus there is no opportunity for re-breeding until the subsequent autumn. The duration of lactation and the timing of weaning has little or no effect on their normal breeding season. In contrast, Dorsets have the capacity to re-breed, and those farmers who wish to lamb three times in 2 years aim for a mating period approximately 12 weeks after lambing. It is known that involution of the reproductive tract, lactational anoestrum and the presence of the lamb all contribute to

Table 4.2
Reproductive data: adult ewes

	Data	Comments
Seasonality		
Polyoestrous	7–12 oestrous periods/season	Variation between and within breeds. Variation according to latitude
Onset (mean dates)	Dorset Down (14 Aug.) Clun Forest (4 Sept.) Halfbreds (20 Sept.) Blackface (9 Oct.)	Significant between- and within-breed variation, high year-to-year repeatability. Ewes responsive to introduction of ram near to onset of season
Oestrous cycle		
oestrus	24–36 h	Only displayed in the presence of male
ovulation	Towards end of oestrus	First ovulation of season "silent" – not accompanied by oestrus
Inter-oestrus interval	16–17 days	
Fecundity		
Live lambs born/ewe lambed	Welsh Mountain – 1.23 Blackface – 1.40 Clun – 1.52 Mule – 1.78 Cambridge crosses – 2.40 Cambridge – 3.00	Markedly affected by body condition and season
Lambing interval	Annual	Restricted by seasonality
	Three lambings in 2 years for Dorset Horn and Poll Dorset	Aseasonal. With good management can achieve lambing interval of 240 days
Breeding life	Up to 15 years	Vast majority culled by 6 years of age
Fertility		
No. of ewes per ram		
hill	40–50	Vary according to size of flock, paddock size and terrain
lowland	50–60	
Synchronized flocks	10–15	
Percentage of ewes lambing	92–95	Lower in upland flocks. As low as 75% in harsh hill flocks
Pregnancy		
Duration	142–150 days	Variation due to breed and litter size

the *post partum* period of inactivity. Recent investigations have shown that, in Dorsets, ovulation occurs approximately 4 weeks, and oestrous cycles 6 weeks after lambing. There are no indications that it is an advantage to remove the lambs earlier than 5–6 weeks; thus artificial rearing should not be undertaken to accelerate the onset of breeding activity. Weaning at later than 5 weeks has the advantage of lambs requiring only dry food, and therefore their management is easier. It also provides ample opportunity for drying off to be completed and for body condition to be regained prior to tupping at 12 weeks post-lambing. There is, however,

Table 4.3
Reproductive data: ewe lambs

	Data	Comments
Age at puberty	7–9 months	Markedly photoperiodic. Must reach 50% of adult weight
No. of ewe lambs/ram	20–30	Use mature rams and small fields
No. of lambs born/100 to tup		
lowland flocks	56	Markedly affected by management, particularly nutrition
upland flocks	69	
>60% of mature weight	88	
<60% of mature weight	57	
Litter size/ewe lamb lambed	1.0–1.25	
Ewe lambs lambing/100 to tup	49–67	

considerable variation in performance in Dorsets and some would elect to reduce it through techniques of synchronization (see p. 97).

It is generally acknowledged that rams of all breeds are less seasonal than ewes in their sexual behaviour. Breed differences in mating activity during the summer have been demonstrated and seasonal fluctuations occur in semen quantity and quality. In practice, this may be counteracted by reducing the number of ewes allocated to each ram, by closer supervision of matings, and by confining the sheep in a small field or paddock.

Provided ram lambs have been born early, and thus have had every opportunity of growing well, it is a common practice to use them or offer them for sale as flock sires in their first autumn.

Age of puberty

There is an increasing interest in the early breeding of female lambs (see Fig. 4.1). Recent surveys have shown, however, that a much lower level of performance is to be expected, particularly with respect to fertility and litter size, than with two-tooth maiden ewes. Ewe lambs which have not achieved at least 50–60 per cent of their potential adult weight should not be bred from in their first autumn. Many of them will, in any case, not exhibit oestrus cycles until the following year. Data presented in Table 4.3 provide an indication of the reproductive performance of ewe lambs in different environments and at differing stages of development.

Ram lambs attain puberty at an earlier stage than ewe lambs, usually 40–50 per cent of their potential adult weight compared with 50–60 per cent for ewe lambs. The ram lamb should be capable of serving by 6.5–7 months, but with a smaller number of ewes than for adult animals.

Litter size

Litter size will be affected by the number of ova shed, the number of ova fertilized, and the number developing to viable lambs at birth. There is considerable genetic variation in ovulation rate, so levels of fertility are, to some extent, a matter of breed selection. They are also influenced by nutritional and management factors; age and stage of breeding season also have to be considered in the assessment of performance. These factors seem to exert their influence up to the early stages of pregnancy. Subsequent reduction in viable foetuses is usually associated with gross feeding mismanagement and pathological factors. An indication of variation between breeds and crossbreds in respect of litter size is presented in Table 4.2.

Control of reproduction

There is considerable scope for adopting methods such as hormone and/or light or melatonin treatments to control reproduction. Reasons for doing so may include the exploitation of superior genotypes, the induction of oestrus for out-of-season breeding, synchronization of activity to eliminate unacceptable variation within the flock, augmentation of fecundity in breeds/types with a low ovulation rate, or facilitating the use of AI (e.g. to control disease).

Current treatments for out-of-season breeding involve the use of intravaginal progestagen sponges coupled with pregnant mares' serum gonadotrophin (PMSG), or light treatment. Recent developmental work in this field includes the use of implants containing gonadotrophin-releasing hormone (GnRH) and melatonin given in the feed or as an implant in conjunction with photostimulation. Where AI is used, PMSG reduces the variation in interval to ovulation and allows insemination at a pre-set time and without the need to test ewes for oestrus. The use of prostaglandins may be considered where it is known that the ewes are displaying oestrous cycles; proof of its efficacy in sheep flocks requires more developmental work.

It has been known for many years that anoestrous ewes will respond to the unaccustomed presence of the ram (entire or vasectomized) during the transitional phase 2–3 weeks before the breeding season. This results in a high incidence of tupping on days 19 and 25 following the introduction of the male. The level of synchronization is low compared with that which occurs with progestagens, and the timing of the introduction of the ram is critical.

Further information on artificial insemination and the control of reproduction respectively may be found in Evans and Maxwell (1987) and Gordon (1983).

Nutritional requirements

Evaluation of foodstuffs

Being ruminants, sheep are well equipped anatomically and physiologically to graze and utilize a range of foodstuffs. Successful feeding management requires a sound understanding of the nutrient requirements, on the one hand, and a sensible assessment of the quality of the available foodstuffs on the other. The latter involves subjective judgement; it is rare for a flock owner to resort to sampling grass and other crops for the assessment of nutritional quality. Quality is usually assessed by noting stage of growth, density and type of sward, and the ratio of grass to clover. Hay, the most common winter feed for sheep, is assessed on the basis of colour, and absence of dust which often includes deleterious moulds. This provides a reasonable indication of the efficiency of curing and storage. The potential feeding quality of the crop at the time of mowing may be established by the botanical make-up of the hay and stage of growth. This also applies to silage and, in addition, colour, texture and smell will indicate how efficiently this was made and stored.

Details of nutritional values of a very wide range of foodstuffs can be found in MLC (1983) and MAFF (1986). It is essential to adopt some method of monitoring feeding, particularly during critical phases of the production year, so that corrective action can be taken. The most practicable are weighing and body condition scoring (see p. 113). In adult stock which have completed their skeletal development, condition scoring is a reliable indicator of their general well-being. In young stock, it is useful to monitor live-weight at particular ages, so that their growth and development can be assessed. This applies to both fatstock and breeding stock. The latter is expected to achieve a target weight at various stages prior to and during its early breeding life. The growth rate between weighings is a reflection of the adequacy of feeding, and the plane of nutrition may then be adjusted upwards or downwards as necessary.

Diet

The majority of sheep are maintained on grass alone during the grazing season, and hay or hay and silage during winter. Concentrates and/or supplements, if fed at all, are given during the last month or so of pregnancy and the very early stages of lactation, depending on availability of spring grass. Certain classes of sheep are fed for lengthy periods on arable crops and by-products, and early-weaned, autumn-born lambs may be reared and fattened exclusively on concentrates, usually given *ad lib.*, with only minimal roughage (100–200 g/day).

Grass

Grass supplies 90–95 per cent of the annual energy requirement of a high proportion of the national flock. The level of dependence on grass varies according to local resources and husbandry systems. Flocks managed extensively on low input–low output regimes (and these include many in hill and upland areas) rarely receive anything other than the available uncultivated rough grazing, with some hay during very severe weather conditions in late winter and early spring. In contrast, intensively managed land with high stocking densities provides grass only from the advent of good quality grazing in the spring to late summer or early autumn, depending on the season.

Forage crops

As stocking densities increase, the need for alternative forms of forage arises to bridge the gap between the end of the grazing season and the commencement of full winter feeding based on hay and/or silage. Forage and other foodstuffs used at this time of year for sheep feeding include green fodder crops (kale, forage rape, fodder radish), root crops (swedes, turnips, mangolds), and arable by-products (sugar beet tops, leafy brassica wastes). Green fodder and root crops are used during the autumn months and before the hard frosts of winter. The crops contain sufficient levels of energy and protein to meet the requirements of breeding animals and store lambs at this time of year.

The type of soil plays quite an important part in decisions concerning the provision of forage and root crops during autumn/winter. In high rainfall areas, and on heavy land, the use of tillage and root crops leads to a great deal of food wastage, heavy soiling of fleeces and a high incidence of foot ailments. The higher the clay content of the soil, the more difficult it becomes to prevent poaching and to provide early spring and late autumn grazing. High stocking rates are also difficult on such soils and winter housing is often necessary.

A major factor affecting consumption is the quality of the crops, particularly in terms of digestibility, and sowing is timed so that the crop is at the ideal stage of growth during the utilization period. It is customary to control the grazing of these crops using electric fencing or folds. The use of fall-back pasture or the feeding of hay during the introductory period helps to control digestive upsets. Over-dependence on the fodder crops can lead to loss of appetite and general unthriftiness due to the presence in these crops of brassica anaemia factors.

Root crops have the advantage that they may be lifted and stored for use later in the winter and around lambing time. Utilizing the crop *in situ* may be difficult for young breeding animals at the four-tooth stage (see p. 113). Roots are low in dry matter content, 10–12 per cent compared with 25 per cent in silage and 85 per cent in hay.

During the autumn, vast quantities of crop residues are available as sheep feed. These are used to finish store lambs, which are often bought in for the specific purpose of utilizing such by-products, and sold as finished lambs later in the season. Sugar beet tops need to be wilted for at least a fortnight, otherwise the oxalic acid content of the leaves will lead to digestive upsets. Breeding ewes fed tops for lengthy periods can easily become overfat by mid-pregnancy and therefore some means of control is required; if tops are fed through to late pregnancy, balanced concentrates should be used during the last 6–8 weeks. Fattening lambs may need a supplement of mineralized cereals (0.25 kg/lamb/day) during the last stages of fattening.

Waste vegetable material from packing centres may be used as sheep feed, but should be limited to 4–5 kg/sheep/day fed as a supplement to other foodstuffs. Vegetable trimmings deteriorate quickly and should be fed within 4 days of delivery. Vegetable waste is rich in protein and calcium.

Compound feeds and cereal supplements

The increase in nutrient requirements of breeding ewes during late pregnancy and early lactation, and of store lambs during the late stages of fattening, are usually provided for by a compound feed or by the use of cereal grain with added protein, mineral and vitamin supplements. Farm-produced compounds can be made up of a blend of cereal grains balanced with vegetable and animal proteins in the ratio 80 : 15 : 5 and with the final mix containing 2.5 per cent mineral/vitamin supplement. To avoid wastage, it is advisable to cube or pellet concentrates. Proprietary compounds cover a range of requirements.

Complete diets may be fed to housed sheep. These can range from cubed fortified dried grass to compounds that include ground roughage along with the ingredients normally included in such diets. Compound diets may also be fed as the sole provider of nutrients and with straw as roughage for winter housed ewes. Where this is practised, clean barley straw is fed *ad lib.* in racks or feed boxes, and ewes are expected to consume 1.5 kg/day. Attempts are being made to improve the feed value of straw by ammonia treatment, and investigation has indicated that this may lead to some saving in the use of compound feeds.

In recent years, considerable interest has been shown in the use of compressed feed blocks. The blocks weigh 25–30 kg, resist weathering and supply a good source of energy (usually based on cereals), combined with urea, minerals and, in some cases, anthelmintics. They are useful in hill and upland areas where transport is difficult.

Minerals, vitamins and other additives

The mineral/vitamin ingredients of sheep diets will, to some extent, reflect the known deficiencies of the region. Extra magnesium is required in most situations in the spring, and its inclusion in concentrates for breeding ewes is generally recommended. Where it is impractical to include magnesium in a feed, then the provision of mineral blocks with a high magnesium content, or the spraying of it in solution on to herbage, are alternative methods which may be considered.

Trace elements (such as copper and cobalt) may be given as a drench or by injection. This form of supplementation has the advantage of providing a direct and known dosage to every animal in the flock. In the hill and upland areas, where concentrate feeding is nil or minimal, it is the only reliable method. The frequency of drenching may be reduced by using slow-release boluses containing the element required, and it is expected that this form of supplementation will become widely practised. This approach is already used for the supply of cobalt and copper and to a lesser extent magnesium.

Vitamins are usually provided as an ingredient of compound feeds or supplementary concentrates. Cases of acute vitamin/mineral imbalances are normally corrected by injecting an appropriate proprietary product.

Growth-promoters and antibiotics are rarely used as feed additives in sheep production, although additives may be included to cure specific diseases, such as coccidiosis.

Water supply

Water should be available *ad lib.* in all situations. When sheep are housed, water may be provided from raised rectangular tanks, placed where there is no danger of contamination from foodstuffs. Buckets, frequently topped up, are usually adequate for small groups and those in individual pens. The water supply should be protected from frosts, but this is not easy under paddock and field conditions in severe winters. Fields without a natural source of water should be provided with tanks which are topped up periodically. Although under certain conditions grass and forage and root crops may satisfy the water requirement, it is inadvisable to rely totally on these foodstuffs to supply the needs of sheep.

Environmental requirements

Hill flocks

There is still little attempt at providing the same degree of protection for hill sheep against weather conditions as is practised in the lowlands, but there are one or two interesting developments.

The perpetuation of the fitness of a breed in any given hill area is ensured through a system of pure breeding and the use of only replacement female stock born on the farm. The practice of transferring part or all the flock to the new owner when the farm changes hands is also advantageous. These policies ensure that the native stock provide the lead, particularly with regard to the seeking of shelter during inclement weather, the source of water in dry seasons, and the flock or subflock grazing territories. Increasing use is being made of weather forecasts, and this may give the flock owner time to take action: to move the flock off exposed ground; to delay the return of the flock to the hill; or to delay shearing. In recent years much interest has been taken in the provision of shelter belts in hill and upland areas, largely through better communication and cooperation between the Forestry Commission and the farming community and their advisers.

Protection from the elements is also brought about through the continued use of traditional practices. In many areas, all breeding stock are gathered into the in-bye land or *ffridd* for a few weeks around lambing. The shelter provided by stone walls and banked hedges is very effective. A few flock owners use degradable plastic jackets for newborn lambs, which provide useful protection during the first 3 days or so. The transfer of ewe lambs to lowland farms over their first winter ensures that their growth and development continues unimpeded. This well-established practice is becoming more difficult to pursue in some areas, due to the cost and to the difficulty in finding lowland farms with surplus grazing. As a consequence, cheap housing is now being provided in the hill and upland areas for this class of stock, using feedstuffs transported from lower areas (see also p. 99).

Lowland flocks

Whereas hill flocks have the freedom to establish grazing territories, lowland flocks are under close management and may frequently change pasture, depending on the system of grassland utilization. The current systems – set stocking, rotational grazing and strip grazing – differ in the degree of freedom given to sheep and the time spent on any one site.

The availability of buildings and the proximity of the flock on lowland pastures make it relatively easy, compared with the hill situation, to undertake quick action and provide shelter when the need arises. There is also much more emphasis on the provision of shelter and it is usually routinely available during either part or the whole of the winter/early spring period. The longer the period for which it is required, the more permanent the facility needs to be.

Temporary or short-term shelter usually caters for the flock at lambing time, and provides the ewes with protection from the weather for 24–48 h after parturition. It could consist of individual pens using hurdles, straw bales and netting wire, or cheap boarding with some kind of temporary weather-proof cover over most of the pen, and with quick and easy access from the front, which should face away from the prevailing wind. Such shelters may form part of a small section of a paddock or field in which those expected to lamb may be held overnight. Strategically placed bales of straw can provide useful shelter in the main field for ewes and lambs newly released from such an area.

In recent years, long-term winter housing has become commonplace. This trend may be attributed to higher stocking rates, requiring land to be rested overwinter, the adoption of autumn and early winter lambing and, to some extent, the acknowledgement that well-trained staff deserve good facilities to demonstrate their competence. Some farmers provide purpose-built housing, but many use spare capacity in multipurpose buildings and fodder barns. In the interest of the general welfare of the flock, and irrespective of the type of building used, due regard must be given to the maintenance of good health, adequate feeding and watering arrangements, ample space for resting, good working conditions for stockmen, limitation of group size, easy flock inspection, and means of segregating individuals or small groups of sheep. Special arrangements are required for the newly lambed ewe, for fostering and for artificially reared lambs. The sick animal and the artificially reared lamb are the only classes of sheep requiring some local heat, usually in the form of infrared lamps placed well clear of the animal.

Fully covered buildings for sheep do not attempt to provide an artificial level of warmth; the general house temperature follows that of the local environment fairly closely. The design should allow free movement of air with freedom from draughts and the provision of dry conditions at all times. There is a very wide range of covered buildings in use (see p. 100).

Where sheep are kept in a controlled environment house (which is quite exceptional), an efficient ventilation system is of paramount importance. The ventilation system should allow uniform movement of air at the rate of 3 m^3/h/kg body weight. The relative humidity should not exceed 90 per cent, and good insulation (u value of 0.3) is required to minimize condensation. Larger houses of this kind should be subdivided and, ideally, no more than 80 mature sheep should be kept in one air space.

Choice of management system

The most appropriate system of management in a particular situation is largely dictated by geographical location of the farm, type of soil, type of crops and their seasonal availability, labour supply and marketing objectives. In general terms, sheep units in the west and in the uplands and hills are almost entirely grass-based. In the east, where rainfall is lower, arable farming prevails and crops other than grass are a feature of the feeding system at particular times of the year. Even in arable areas, leys (temporary grassland) maintain the flock for over half the year.

The timing of lambing will have a major impact on the seasonal pattern of management because the sequence of actions and assessments which has to be undertaken during the sheep year revolves around the following key events: *mating, lambing, shearing* and *weaning*.

The main choice is between spring and autumn lambing, and it can be readily seen from the example presented below that the timing of the main events occurs at contrasting times of the year, particularly in relation to the grazing season.

Spring lambing flock	Key events	Autumn lambing flock
Mid-October	Mating	Mid-July
Mid-March	Lambing	Early December
Winter or early summer	Shearing	Early summer
Early August	Weaning	Late January

Management of rams

In the past the care and preparation of rams for the breeding season has not received the same attention as that given to the ewe flock and, all too often, mature rams were simply taken out of the orchard paddock and immediately released with the flock for the tupping period. In recent years, there has been more awareness of the need to start the programme of preparation 6–8 weeks beforehand. The detection of any physical abnormality or disease which prevents or reduces the ability of the ram to search for ewes on heat, to serve, and to yield good quality semen is vitally important to the attainment of high levels of fecundity and fertility in flocks, particularly if only one or two rams are used, which is quite often the case. Ram lambs need to achieve 65–70 per cent of their mature weight by the date of the ram sale or breeding time (see Table 4.1). Those destined for the sales are usually selected from dams which lamb early, thus giving them a longer period for growth and development.

Mature rams will have been maintained on grass, often in a paddock close to the homestead, since the previous breeding season and with hay being fed during the worst of the winter months. Provided rams have access to reasonable grazing, and have been routinely drenched and vaccinated along with the ewe flock, they should be in good health and in good condition during the approach to the breeding season. Those in poor condition should be segregated and given a higher plane of nutrition than the rest. The aim should be to score all rams at $3\frac{1}{2}$ by the start of tupping.

All rams should be given some concentrates, partly to ensure an increased supply of energy, but also to provide minerals, vitamins and good quality protein – the type of diet normally given to ewes during early lactation. The level of feeding will depend to a certain extent on condition. Rams with a score in the range of $2\frac{1}{2}$–3 should start at about 200 kg/day, gradually increasing to 500–700 g/day. Training rams to respond to the sound of the feed bucket, and to accept bucket feeding, can be a great help during the 6 weeks or so of tupping. Rams trained in such a way can be preferentially fed, and are easier to catch for such things as harness adjustment, change of crayon and general examination for soundness. Rams tend to lose weight during the tupping period, largely because of the reduction in grazing time.

The examination of rams at the start of the preparatory period leading to the tupping season, or prior to purchase, should pay particular attention to general soundness and health and the capacity to work efficiently. All rams, irrespective of age, should be assessed in terms of breed type and soundness of limbs, particularly the conformation of the legs, pasterns and feet. The claws should be free of distortion, healthy and free of interdigital growths. The slightest hint of lameness or any other disease condition at any time before and during the tupping must be investigated promptly: anything which elevates body temperature can have a disastrous effect on semen quality.

During examination of the mouth, the age of young rams can be established and the alignment of the incisors checked. The eyes should be examined for signs of infection or scars from previous infections. There may be evidence of correction of inturned eyelids (entropion) and such rams should not be bought, or retained in the flock.

The general inspection of the body for vaccination abscesses, lesions and scars should give particular attention to the sternum and the skin leading up to the shoulder. This part of the body will carry the keeling harness and is rather prone to superficial injury especially if the harness has not been fitted properly or the straps have become hard.

The reproductive tract should be carefully examined, giving particular attention to the penis, appendage, prepuce, scrotum and testes. In the case of ram lambs, it is important to establish whether anatomical development has been completed.

The physical examination of rams gives no indication of libido or fertility. For this reason rams should be watched carefully to establish whether they go round teasing the ewes and serving those which appear on heat. Fertility cannot be assessed until the proportion of ewes returning is known. A return rate of over 15 per cent to first service and over 5 per cent to combined first and second services should be investigated. The effects of an infertile ram can be mitigated by circulating rams between groups at least weekly.

Some indication of libido and potential fertility may be established by subjecting rams to a serving capacity test, and by semen evaluation. A serving capacity test involves exposing the ram to a small group of ewes in which oestrus has been synchronized, in a pen for a fixed period. The number of mounts and services is recorded and this information is used to rank the team of rams. The test was developed in Australia and some concern has been expressed in this country about the welfare aspects of such a test. The welfare aspects of semen collection also have to be considered, for under farm conditions, there is no alternative to the use of electroejaculation. This technique is stressfull, and must be carried out by a veterinary surgeon (Veterinary Surgeons' Act 1966).

Management of breeding females

Preparatory phase

In lowland flocks, production objectives include a high lambing percentage (target of 170–190 per cent) and a compact lambing period. Attainment of these objectives is, initially, very dependent on feeding and management during the pre-service preparatory phase. Older breeding stock will have been assessed for soundness, and replacement stock selected or purchased.

Ewes in improving condition usually have a higher ovulation rate, and consequently a better lamb crop, and it has been traditional to flush them by providing a generous level of feeding a few weeks before service. Ideally, however, the preparatory phase should begin earlier in the season, at weaning.

Mature ewes should be in good body condition – score $2\frac{1}{2}$–3 (see p. 113) – 1 month from service. Those in poor body condition should be segregated and given generous grazing. An improvement in the score of $\frac{1}{2}$ during the flushing period would require approximately 2 kg live-weight gain per week in Scottish half-bred ewes. No ewes, even if considered overfat, should be allowed to lose condition during this period. Target live-weight, expressed as the percentage of mature body weight (see Table 4.1)

Table 4.4
Target live-weights of ewes at various ages and reproductive states, expressed as percentage of mature weight at conception

	State of pregnancy			Lambing		
	Conception	Day 90	Day 120	Pre-	Post-	Lactation (8th week)
Ewe lamb (bearing single)	60	65	70	75	64	62
Two-tooth ewe (bearing twins)	80	85	90	97	80	77
Mature ewe (bearing twins)	100 (score 3)	96	100	112	90	83

Calculated from data given in MLC (1983).

for the main age groups of lowland breeding ewes is shown in Table 4.4.

Ewe lambs and two-tooth ewes require preferential treatment in order that the demands of their continuing growth and development are met, and ewe lambs should remain segregated from the main flock. Ewe lambs are expected to achieve 60 per cent of their mature weight before they can be considered for breeding in their first autumn and therefore their progress during the post-weaning phase must be regularly checked, and forward projections made of their probable mating weight. Two-tooth maiden ewes should easily achieve 80 per cent of their mature weight; where they are reared elsewhere, early purchase is recommended so that the condition score can be checked and, if necessary, raised to the appropriate level.

The preparatory phase spans different periods of the late summer/autumn depending on the location and the breed or type of ewe. In the very early lambing flocks it would occur in July/August, for traditional grassland flocks in September/October, and in upland and hill areas in October/November. Normally, the preparatory phase coincides with the tail-end of the grazing season, when either the results of good grassland management or the use of arable by-products and forage crops should fulfil the nutrient requirements of all age groups within the flock. It is unusual for flockmasters to resort to the feeding of concentrates at this stage but, in the event of a crop failure or prolonged drought, the daily allowance per ewe would be 0.5 kg for 4 weeks before service.

During the preparatory phase, all ewes should be dagged so that soiled wool does not impair oestrous detection and service. Many flocks use vasectomized rams to initiate oestrous activity throughout the flock. They are normally turned in 4 weeks before the start of tupping and this ensures that the majority of ewes are tupped during their second or later oestrus when ovulation rate is high.

Tupping

Rams should be fitted with a keeling harness, designed to hold a coloured crayon firmly over the sternum, a few days before they are turned in with the ewes. Colour crayons are fitted, and any necessary adjustments made to the harness on the day the rams join the flock. The crayons leave a clear mark on the ewe's rump following each mounting, and the colour should be changed after not more than 17 days. Crayons are available in five colours and two textures: hard for summer/early autumn and soft for autumn/winter. Keel marks may be used to check the progress of tupping during the first 17 days and also the incidence of returns to service. More frequent changes of colour (e.g. weekly) is useful for grouping ewes in late pregnancy according to lambing date.

Ewe lambs should be tupped after the main flock, using adult rams. Young ram lambs should be used on adult ewes and in a separate paddock. Tupping efficiency of all age groups may be improved by avoiding very large fields (see Table 4.2 for ratio of ewes to ram).

Where synchronization of oestrus is required the following programme may be used:

Day 0 Insert intravaginal progestagen sponges.
Day 14 08.00 h: remove sponges. Inject 750 i.u. PMSG intramuscularly (500 i.u. for small breeds).
Day 16 a.m.: turn in one ram per 10–15 ewes, or artificially inseminate ewes, without testing for oestrus, from 16.00 h.
Day 28 (During natural breeding season only.) Introduce a harnessed fertile ram and check for returns to oestrus over a 2-week period.

The response to treatment varies according to the season. The incidence of oestrus is usually high at all times, and the litter size of ewes which lamb is also acceptably high. Unfortunately, the percentage of treated ewes lambing as a result of treatment in late spring/early summer is usually only in the range 40–60 per cent. If treated during late anoestrus, 50–70 per cent of ewes can be expected to lamb.

Pregnancy

Implantation of the embryos occurs during the third to fourth week of pregnancy. This is recognized as the period when most of the embryonic mortality occurs. The level of feeding should remain unchanged, and over-dependence on crops known to lower fertility, such as kale and rape, should be avoided.

The feeding management of the age groups changes during the second and third months. The feeding of ewe lambs and two-tooth ewes continues on a more generous scale than for mature ewes, and during this period they should show an increase of 5 per cent in body weight. In practice, the introduction of hay for the two younger age groups compensates for the decrease in pasture quality. Mature ewes, on the other hand, can tolerate more severe treatment and, if in good condition at mating, may be used to clear up pastures. A loss of 5 per cent in body weight at this stage is not detrimental to foetal growth and development. Over-fat ewes may benefit from this mild degree of undernourishment, because it reduces susceptibility to metabolic disorders and, in particular, to pregnancy toxaemia.

In mid-December, harnesssed vasectomized rams should be re-introduced to identify any empty cyclic ewes. These can then be segregated and their future decided.

From the beginning of the fourth month, there is an increase in the growth and development of the foetus and, consequently, in the nutrient requirements of the ewe. This phase coincides with winter weather and a limited range of available foodstuffs. Hay, and sometimes roots and silage, provide the basic feed at this time.

The main problem is to cater for the increased demands at a time when there is a fall in appetite, due partly to the physical effects of the foetal burden and partly to physiological changes. In practice, this can be achieved by feeding a basal diet of good-quality roughage and supplementing it with concentrates (10–12 MJ/kg). Provided the mature ewes are in the required body condition, further controlled depletion of body reserves, resulting in a loss of 5 per cent in basal body weight, is allowable. The live-weight of twin-bearing ewes should increase by approximately 16 per cent during the last 2 months. Failure to achieve this increase can result in reduced birth weights, less vigorous lambs, delayed onset of lactation, and poor development of mothering instincts. A widely practised feeding plan for this stage of pregnancy is presented in Table 4.5.

New regimens, such as housing, should be introduced before the last 6 weeks of pregnancy, and ewes penned according to their colour marks. Out-wintered ewes may be separated into lambing groups. The effect on food supplies of rapid changes in the weather (such as snowfalls) should be anticipated by conditioning ewes beforehand to alternative foodstuffs. Behaviour during feeding should be observed and a watch kept for shy feeders. Outwintered ewes should be range-grazed near to the proposed lambing site. Lambing pens (see p. 102) and aids required at lambing, and other equipment such as that required for identification, docking and castration, and for fostering and artificial rearing, should be prepared and ready for use.

Table 4.5
Feeding regimen (kg/day) for ewes in late pregnancy

	Ewe weight and no. of foetuses					
	50 kg			70 kg		
	Single	Twin	Triplets	Single	Twin	Triplets
Six weeks before lambing						
Hay[a]	0.83	0.83	0.83	1.00	1.00	1.00
or silage	2.60	2.60	2.60	3.50	3.50	3.50
plus concentrates	0.18	0.30	0.34	0.24	0.37	0.44
Four weeks before lambing						
Hay[a]	0.83	0.83	0.83	1.00	1.00	1.00
or silage	2.60	2.60	2.60	3.50	3.50	3.50
plus concentrates	0.28	0.45	0.51	0.36	0.56	0.66
Two weeks before lambing						
Hay[a]	0.83	0.83	0.83	1.00	1.00	1.00
or silage	2.60	2.60	2.60	3.50	3.50	3.50
plus concentrates	0.37	0.59	0.68	0.48	0.75	0.86

[a] Assuming ME concentration in the dry matter of 10.0 MJ for hay and 11.0 for silage and dry matter content for silage of 25%. Roots could be used to replace 75 and 50% of the concentrates on a dry matter basis at 6 and 2 weeks before lambing, respectively.
Source: MLC (1983).

Lambing

The flock should be inspected frequently, and particular attention should be given to changes in the behaviour of ewes during the approach to lambing, such as a long interval (more than 30 min) since first observation of the ewe's nose pointing to the sky, or appearance of the water bag. A high standard of hygiene should be maintained at all times and abnormal presentations corrected with great care. Appropriate medication is essential after all assisted lambings (see Eales and Small, 1986).

The behaviour of all lambed ewes, and particularly those lambing for the first time – ewe lambs and two-tooth ewes – should be kept under close observation. There is no doubt that an important contributory cause of lamb mortality is abnormal maternal behaviour, which may include desertion, delayed grooming or rejection of one or more members of the litter. The consequence for the new-born lambs can be debilitating or even disastrous. Timid ewes increase the teat-seeking time and thus the chances of the lamb imbibing pathogens lurking on the belly wool and other parts of the underline. Thorough grooming of the lamb, with a short interval to suckling, are necessary traits in ewes.

Lactation

The udder and teats should be checked soon after birth and the cleansings disposed of as soon as possible. After recording information regarding the ewe and her lambs, they are returned to a pen containing newly lambed ewes or, if appropriate, released to a paddock provided with shelters placed in strategic positions. A high level of surveillance must be maintained; early detection of ill-health or abnormal behaviour is a vital part of shepherding during the month following lambing.

Where late winter/early spring lambing occurs indoors, ewes and their lambs should be turned out of the sheep house within 1 week into nearby fields where surveillance can continue; delay can lead to cross-suckling and mismothering. The ewes are trimmed (dagged) before the start of the flush growth of grass, to reduce soiling and the risk of fly strike.

During early lactation, the encouragement of a rapid increase in appetite and an early attainment of peak of lactation is essential. This will be achieved more readily in ewes that have been properly managed during pregnancy. The feeding regimen is a continuation of that for late pregnancy, namely, the use of concentrates to supplement a low daily allocation of good quality hay. Diets with low digestibility are known to depress appetite and should be avoided. As far as the ewe is concerned, there is also the need to avoid undue loss of weight during the first 2–3 weeks. Provided peak appetite is achieved early, this loss should be approximately 5 per cent. Ewes suckling twins should be given 1–1.25 kg of concentrates (12 MJ/kg), which should meet the requirements of high milk yield in a 60 kg ewe. It has been estimated that an average weight gain of 250 g/day from twin lambs would be supported by a daily milk yield of 2.5 kg. Ewes suckling twins can be expected to produce 40 per cent more milk than ewes suckling singles, and they also tend to reach peak lactation earlier. They should always be segregated from the rest of the flock, and given preferential treatment to encourage high yield.

In the majority of lowland flocks, the early phase of lactation coincides with the onset of spring grazing. The change in the diet during the transition to total dependence on grass should be brought about gradually to prevent metabolic disorders and to maintain a high level of nutrient intake. Effective grassland management at this time of the year calls for sound judgement of the quality and quantity of grass available. The expected stocking rate under good lowland conditions is in the range of 14–16 ewes/ha; this range can only be supported where regular dressings of nitrogenous fertilizers are applied throughout the grazing season. Once the critical phase of early lactation and total dependence of the lamb on its dam's milk supply is over, any unforeseen crisis may be overcome by preferential feeding of the lamb or by separation of the ewes and lambs.

Winter housing

There are three main types of accommodation for sheep: open yards, partly covered yards and fully covered buildings.

Open yards

These are satisfactory in areas of low rainfall and are formed by using temporary walling of straw bales, linked, where convenient, to the sides of existing buildings. Where straw is used as litter, it prevents undue soiling, but where straw is in short supply a raised slatted floor should be considered. It is important to provide these topless pens with protection against wind and draughts.

Partly covered yards

These may be adapted cattle courts or open yards linked to limited covered space. In this system the sheep are exposed to weather for some of the time and this may lead to a very damp humid atmosphere in the covered area. It is essential that the open yard is well drained and that there is free movement of air in the covered area. With an apex roof, the ridge should have an opening 150–230 mm deep along most of its length. Monopitch buildings should have an opening 230–300 mm in depth along the back wall.

Fully covered buildings

There is a wide range of types in this category (Figs 4.3, 4.4, 4.5) and these include multipurpose climate houses, single row and face-to-face monopitch buildings, and short-life shelters clad with various types of plastics. Purpose-built, ridge-ventilated, wide-span buildings provide ample scope for an efficient layout of pens. Relatively cheap materials are used to clad the lower and upper halves of the sides and ends. Although, in some locations, a completely open space may be left between a low side wall and the eaves, some kind of baffle against wind, rain and snow is usually desirable. This may be provided by using traditional Yorkshire boarding in combination with fine wire mesh, or other materials which would give a comparable effect.

Internal features

The ease with which animals can be fed, watered, littered and supervised is very important and, at key times, such as the peak of

Fig. 4.3 Internal layout of sheep house at Nevie, Glenlivet. From Ferguson (1982). Courtesy of the Scottish Farm Buildings Investigation Unit.

lambing, the segregation of lambed ewes and the regrouping of ewes needs to be accomplished with the minimum of disturbance. The choice of floor is largely between slats and litter. Slats have the advantage of keeping the feet in good condition and providing a dry lying area with the fleece remaining clean. Less floor space is required per sheep (Table 4.6) and both the time required for littering and the problems associated with the rising level of litter are eliminated. The main disadvantages are the high capital cost of installation and the unsuitability of slats for ewes with young lambs at foot. Unsuitable materials and poor workmanship can lead to warping and splitting of slats, with consequent risk of injuries to legs and feet. Straw, on the other hand, may be readily available at low cost, and also be required as manure for the arable crops. It is more versatile than slats for all age groups, but regular littering is

Table 4.6
Recommended floor space (m^2) for sheep on slats and straw

Type of sheep	Slats	Straw
Large ewe (68–90 kg) in lamb	0.95–1.1	1.2–1.4
Large ewe (68–90 kg) with lambs	1.2–1.7	1.4–1.85
Small ewe (45–68 kg) in lamb	0.75–0.95	1.0–1.3
Small ewe (45–68 kg) with lambs	1.0–1.4	1.3–1.75
Hoggs (32 kg)	0.55–0.75	0.75–0.95
Hoggs (23 kg)	0.45–0.55	0.65–0.95
Lamb creeps (2 weeks)	—	0.15
Lamb creeps (6 weeks)	—	0.4

Fig. 4.4 Sheep house with "walk-through" troughs.

Fig. 4.5 Sheep in polypen housing.

laborious, and cleaning out has to be done annually. With built-up litter there is the danger that foot ailments, such as footrot, may spread through the group. Care and attention to the feet prior to housing is an important aspect of management.

For ease of handling and selection, the number of sheep in any one group should be kept low, preferably less than 30, to reduce the risk of injury and crushing under stress. The general layout of the pen is dictated by the feeding arrangements and trough space (Table 4.7) requirements. Feeding should be done from the outside; long narrow pens with a depth of not less than 3 m are the most suitable as they allow free movement of sheep during feeding time.

Table 4.7
Recommended trough and hayrack length (mm) for housed sheep

Type of sheep	Trough	Hayrack
Large ewes	475–500	200–225
Small ewes	375–425	175–200
Lambs (36–45 kg)	350–400	200
Lambs	300–350	150

"Walk-through" troughs (see Fig. 4.4) may be constructed for feeding from one or both sides and may also act as divisions between pens and along gangways. Where autumn-lambing ewes are housed, provision should be made to subdivide a part of the pen to form a creep area.

Segregation of lambing ewes may be done by providing portable hurdles for erection within pens or by moving ewes and lambs soon after lambing to more permanent mothering pens measuring 1.8 × 1.2 m.

The lamb

It is generally accepted that, in the majority of lowland flocks, preparations for lambing will include the provision of some kind of protection for the newly lambed ewe and her offspring. In inclement weather conditions, particularly combinations of high wind and rain, the body temperature of the wet or semi-dry newborn lamb can fall quickly and hypothermia sets in. Particularly at risk are lambs from ewes in poor condition, very old ewes, ewes showing abnormal behaviour, large litters, and difficult lambings. During the lambing period, the survival of the lambs should be the main preoccupation of all personnel. The first 24–48 h are the most critical, and due attention should be given to the need for resuscitation, disinfection of the navel cord, protection from the elements, early commencement of sucking and adequate intake of colostrum. In some situations, plastic coats for the newborn may be useful.

The newborn lamb requires an adequate intake of colostrum within about 5 h of birth, otherwise there is a high probability of it becoming hypoglycaemic and hypothermic. In addition to supplying energy for survival, colostrum is the sole source of antibodies which protect against the diseases prevalent in the environment. It is also a laxative and aids the discharge of the foetal gut contents (meconium) soon after birth. This reduces the risk of disease such as watery mouth.

In the case of a large litter, it is advisable to hand-draw the first colostrum, which usually contains the highest concentration of immunoglobulins, and to share it between members of the litter. This is a useful tactic in any situation where it is difficult to supervise progress during the lamb's first day. Lambs require approximately 50 ml of colostrum per kg live-weight four or five times daily; thus a large lamb would get 250 ml/feed amd a small lamb 150 ml/feed.

First-day colostrum can be taken from ewes with a surplus or which have lost their lambs and stored, deep frozen, for 1 year or more. Well fed crossbred ewes milked three times during the first day can produce 2.5 litres of colostrum. Colostrum for storage can also be obtained from cows vaccinated with sheep clostridial vaccine prior to calving. However, some lambs fed cow's colostrum have developed severe anaemia. This is an unusual occurrence but, if it happens, the remainder of the cow's colostrum should be discarded.

Hypothermia

If hypothermia is to be treated successfully,

early detection is essential. Any lambs which are seen to be reluctant to follow their dams, or which do not seem quite normal in appearance or behaviour, should have their body temperatures checked immediately. The following ranges are a useful guide: 39–40°C normal; 37–39°C at risk; below 37°C urgent action required. Basically, treatment consists of drying the lamb, keeping it warm and feeding it with colostrum, usually by means of a stomach tube. This consists of a rubber catheter, with rounded end and offset hole (usually about 30 cm long and 4.5 mm in diameter) which is used in conjunction with either a graduated container or a 50-ml syringe. Introduction of the stomach tube is relatively easy and, unless a lamb is excessively weak, it is unusual for the tube to go down the wrong way. The operator should be seated, with the lamb across his lap. The tube is first lubricated by drawing it across the lamb's tongue.

Coughing and rolling of the head are clear signs of incorrect insertion and, if these occur, the tube should be withdrawn and the procedure repeated. In large lambs, only about 5 cm of the tube should protrude. The tube is held, to prevent the lamb swallowing it, and the syringe or container connected to it. After the required amount of liquid (usually colostrum) has been administered, the tube is withdrawn while still attached to the syringe, and cleaned as soon as possible. In the interests of hygiene, it is advisable to have more than one unit, so that there is adequate time for soaking in disinfectant (hypochlorite).

The stomach tube is particularly useful for weakish lambs because it eliminates the risk of milk inhalation, which sometimes happens with bottle feeding. However, very weak lambs – those unable to raise the head off the ground – should not be tubed and, in their case, an intraperitoneal glucose injection is the most appropriate form of treatment (see Eales and Small, 1986 and MAFF, 1986 for full descriptions of treatments).

Fostering

Fostering of lambs may be necessary for many reasons, including death or abnormal behaviour of the dam, lack of milk or mastitis. For it to be successful, an understanding of maternal behaviour in the ewe is essential. There are several methods based on either modifying the smell of the lamb, or physical restraint of the ewe. Provided the recipient ewe has not had the opportunity of initiating a bond with its own lamb, the success rate can be quite high. Lambs born at the same time may be immediately transferred in the wet state, provided the ewe is still recumbent after lambing. If not, some means of imparting the smells of her own lamb on to the other lamb will be necessary.

The longer the interval to the first introduction of the foster lamb, the greater the likelihood of having to restrain the recipient ewe. Restraint may be achieved by using one of the following: yolked stall (available as single and multiple units), halter, or neck strap fitted with swivelling clip. Where multiple units are used, each stall must be self-contained, and the lambs prevented from straying between pens. Facilities for feeding and watering are an important aspect, and regular inspection for skin damage should be undertaken. Prolonged restraint is to be avoided. The reaction of the ewe towards the lamb(s) should be tested after the first day, and there is little point of prolonging the attempt to foster beyond the third day. Sometimes it has to be recognized that the ewe is temperamentally unsuited to be a foster mother.

The presence of a sheep dog, when ewes are first released from the fostering stall or when the orphan lamb is first presented, has long been regarded as an effective means of partially diverting attention and of intensifying the ewes' mothering instincts. Obviously the misuse of this tactic could impose unacceptable levels of stress on the ewe.

Artificial rearing

Acceptable ewe-milk substitutes, for use as liquid feeds during the first 6–8 weeks, are now available together with a wide range of feeding devices for either *ad lib.* or restricted feeding of groups of lambs. These can be used either to rear orphan and mismothered lambs which would otherwise have to be fostered, or as an aid to management, allowing lambs to be removed from their mothers at an early age.

Growth of the lambs can be readily manipulated by changing the frequency and level of feeding. The choice lies between *ad lib.* feeding on the one hand and restricted amounts at regularly spaced intervals on the other.

Liquid feeding is expensive in terms of both food and labour requirements, hence early introduction of palatable pellets is advantageous. During the first 3 weeks, intake of liquids has little effect on intake of dry food. The level of performance during this stage should meet three important requirements: maintenance of good health, adequate preparation for weaning, and economy of food use. The growth rate should be not less than 200 g/day. Since conversion efficiency is approximately 1 : 1, the total amount of dry milk substitute required per day would be 200 g.

Good quality and highly palatable roughage and concentrates should be offered during the third and fourth weeks to accelerate rumen development and to accustom the lambs to solid food. Artificial rearing provides every opportunity for adequate preparation for early weaning at 6–8 weeks.

Autumn lambing

Autumn lambing is markedly different from spring lambing in the range of management options which may be considered. The main features of autumn lambing are as follows:

1. It is restricted to Dorset Horn or Poll Dorset ewes, since all other types of sheep would require some kind of treatment to lamb on a flock basis in the autumn.
2. Outwintering can only be considered on very free draining land and it is usually linked to the availability of forage and root crops.
3. In many situations, housing is the preferred system of management. Lambing occurs during low labour demand for field work, and the system is therefore well suited for arable farms.
4. Ewes are expected to lactate for only 6–8 weeks, when lambs are early-weaned and transferred to high-energy diets fed *ad lib.*
5. The attainment of fatstock weight by 16 weeks of age is well within the growth capacity of lambs born in late autumn. Marketing coincides with the season of high market prices.
6. Many parasites are dormant during the winter. This, together with early weaning and dependence on grain-based diets, requires a different strategy for health control compared with spring-born lambs.
7. Ewes going out to grass in the spring without lambs at foot may be stocked at very high stocking rates.
8. The sale of lambs in the spring results in a rephased cash flow.

Management of hill sheep

Hill sheep are kept primarily to utilize uncultivated grasslands which would otherwise be unproductive (see also p. 92). Lack of cultivation, no fertilizer or lime treatment, herbage of low nutrient value, a short growing season, harsh climate and minimal control of the grazing animal, all contribute to low productivity. At this level of nutrition, even the small hardy hill breeds are subject to prolonged stress from early pregnancy to early lactation. This is mainly due to the seasonal nature of the nutrient supply and, in many areas, to the traditional reluctance of flockmasters to supplement the ewes' meagre diet during the crisis periods. The examination of the pattern of change in the basal body weight or "maternal tissues" (i.e. free of fleece, digestive tract and uterine contents) in relation to live-weight (Fig. 4.6) clearly shows that it reaches its lowest point during early lactation, despite the fact that this is a time of increasing grass supply. A high percentage of deaths in hill ewes occur around this time. High survival rates in hill flocks receiving little or no supplementary rations are dependent on good recovery of body conditions during summer/early autumn. This forms the basis of the reserves which are mobilized in late winter and early spring. Low body condition in early pregnancy, and severe winter weather, usually lead to high losses.

Moderate improvements in reproductive performance can be achieved without major changes of farming policy through the use of in-bye land (land near the farmstead, usually lower lying and of better quality) during the key periods around mating and lambing. Confinement of ewes during these periods also provides the opportunity to feed supplementary diets, including concentrates and minerals.

In many hill areas the away wintering of ewe lambs during their first autumn/winter is a well-established practice; some hill farmers use cheap housing as an alternative.

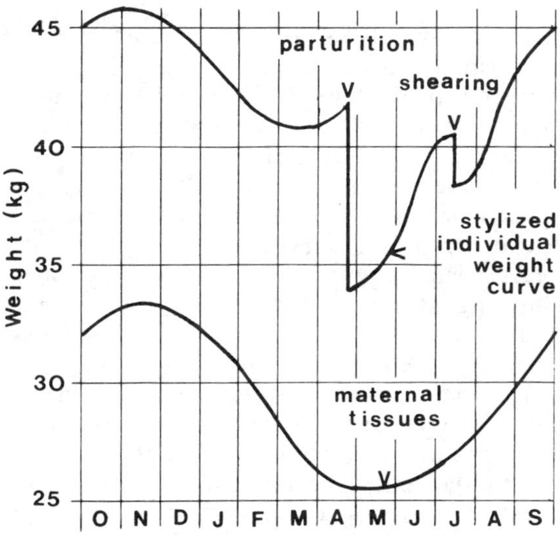

Fig. 4.6 Changes in live-weight and in the weight of maternal tissues of an "average" hill ewe throughout the year. From the Hill Farming Research Organization 4th Report (1964/67). N.B. In 1987 the HFRO was merged with the Macaulay Institute for Soil Research to form the Macaulay Land Use Research Institute.

Providing extra food for breeding ewes at key stages of the reproductive cycle, and a less severe environment for ewe lambs in their first winter, is generally regarded as only the first step in attaining a general improvement of productivity in hill areas. Improvement in terms of animal performance and land output requires an integrated and comprehensive approach. The improvement of favourable and accessible sites through drainage, reseeding and application of lime fertilizers, provides the basis for better nutrition. The provision of subdivision fencing for both the improved and adjacent area allows for an integrated system of grazing. Cleaner grazing and rest periods provide increased herbage of better feeding value. The higher potential of the improved area is reserved for use during the mating period and again in late pregnancy and lactation. Such a comprehensive approach results in a higher stocking rate, heavier ewes at mating, higher lambing and weaning percentages, heavier lambs at weaning and improved fleeces.

Meat production

Spring lambing is traditional in the UK, and most of the variation in timing will be related to the onset of grass growth. Farms will also differ in terms of stocking rate, marketing policy, type of crops available for autumn/winter use, and whether winter housing is adopted or not.

Approximately half the spring-born lambs are sold by the end of the grazing season, and the remainder in the autumn or as hoggets in the early months of the new year. Autumn-born lambs are sold during the late winter/spring.

The trade requirements for lamb carcasses fall into four weight ranges – 8–12.5 kg, 13–17 kg, 18–20.5 kg and 21–26.5 kg – the carcass weight being approximately half the live-weight. In the UK, the main demands are for carcasses weighing 15–19 kg.

Breeds and crossbreds show considerable variation in adult body weight (Table 4.1), a characteristic which affects the carcass weight of the lamb and thus the financial return per lamb sold. Choice of parents determines the potential carcass weight of lambs. As a general rule, lambs should be marketed at half their potential adult weight. This is the stage at which the carcass has a composition which is most acceptable to the British market. Delay in marketing, and the consequent increase in weight, leads to over-fatness, and down-grading or rejection by the grading authorities. The current MLC carcass classification scheme (MLC, 1986) gives particular emphasis to degree of fatness, muscle development, and carcass conformation and weight. Lambs of poor quality do not qualify for government subsidies. Graded lambs are ear-punched to prevent resubmission for grading.

Lambs sold off the ewe

It is within the growth capacity of single lambs reared on the ewe to reach fatstock marketing weight at 10 weeks of age. Such lambs will have grown at 400–500 g/day from birth, largely on milk and with a minimal intake of grass. This kind of performance can only be achieved when ewes sustain a high level of daily milk yield throughout the period. The lambs are selected for marketing on the basis of their live-weight and degree of finish, and the ewes are then dried off.

In the case of lambs marketed off the ewes at an older age, milk makes a smaller contribution to their growth and development during their last few weeks on the ewes. Their growth rate at this stage will largely depend on the quality of the grazing and freedom from internal parasites.

Conventionally weaned lambs

Weaning is usually based on the need to prepare the ewes for the next breeding season. In flocks with high stocking rates, it is not unusual for a high percentage of the lambs to be only at a store (below marketing weight and finish) stage. Weaned lambs are usually 16–20 weeks of age and will have been almost totally dependent on grass prior to weaning. Their post-weaning management will depend on the store market for which they are being prepared, but the majority will be fed on a system which allows a moderate gain in weight (200–250 g/day). This can usually be achieved on aftermath and late summer grazing. Those which remain throughout the autumn can be supported by grass/root crops or grass/arable by-products. Their growth rate may be boosted at any stage by providing concentrates as part of their diet.

Early-weaned lambs

It is advantageous, for a variety of reasons, to early-wean the following: artificially reared lambs; housed autumn/early winter-born lambs; members of a large litter; and lambs at risk of severe parasitism. The timing will depend on the opportunity to prepare the lambs for weaning. Where palatable, pelleted diets are provided from 3 weeks of age, weaning can take place at 6–8 weeks of age and without check to growth and development. Subsequent performance will depend on the plane of nutrition which is provided. There are two main options: rearing for either a high rate or a moderate rate of live-weight gain.

High rate of gain

This involves housing and a target level of performance approaching the genetic growth potential of the lambs. It requires high intakes of a diet of high digestibility and energy concentration. A growth rate of 400 g/day is a reasonable target for such lambs. Palatability and quality of both energy and protein components are of particular importance during the early phase. This is normally safeguarded by including ingredients such as flaked maize, ground nut, soya bean and fishmeal in the diet. These protein ingredients are readily utilizable at a stage when the supply of microbial protein produced in the rumen would not meet the protein demands of the young lamb. The physical form of the diet has a bearing on the incidence of digestive ailments such as rumenitis. The most suitable diet for lambs on an *ad lib.* regimen is a mixture of whole barley grain combined with a pelleted supplement of 4 mm diameter containing the main protein, mineral and vitamin source. No additional roughage need be provided.

Moderate rate of gain

This allows a much wider range of foodstuffs to be used after the early post-weaning phase. The diet can range from grass to coarse roughage supplemented with restricted amounts of medium quality concentrates. Many variables can influence the quality and quantity of the nutrients in these foodstuffs; it requires sound judgement to achieve the expected level of performance which, if unsatisfactory, can very easily place the lambs at a greater risk of disease.

Wool production

Wool yield and quality are influenced by genetic, seasonal and nutritional factors. The more important traits relating to wool quality are strongly inherited. The daylength has a direct effect on the rate of wool growth, which is at its lowest during the winter months. Under-feeding and ill-health can lead to reduced growth and thickness, which give rise to breaks in the fibre and may lead to partial shedding.

The assessment of wool quality takes into account fineness and length of the fibre, softness of handle, crimp or natural wave, and colour. The fleece should be free from discoloration from tars and paints, soiling due to dirt and manure, and contaminants such as straw, seeds, chaff and string.

Shearing

Shearing is usually carried out from late May in the more favoured areas of the lowlands, to mid-July in the harsher environment of the north west of Scotland. Only a small proportion of producers shear lambs, hence the majority of animals are shorn for the first time at 14–15 months old, and then annually.

When shearing is done early in the season, wet, windy, cold conditions can result in severe chilling, a high incidence of mastitis and, in some cases, death. Some protection may be given through using sheltered paddocks, or by using combs which prevent shearing too close to the skin. This practice is usually undertaken only in some hill areas.

Particular attention should be given to the weather forecast prior to shearing. During unsettled weather it is always advisable to house batches of ewes overnight so that work can proceed the next day; in wet and windy conditions, newly shorn sheep should be kept in sheltered paddocks or housed overnight.

Winter shearing has been widely adopted in some areas, particularly south west England, as a consequence of winter housing. Initially, the main benefit was the reduction in pen and trough space requirement. Since its introduction, several other features are now well established. Shearing should be restricted to ewes housed before early January and not turned out until mid-March or later. To minimize stress, housing and shearing should coincide, and it is thought that this will reduce the incidence of wool slip (partial baldness) during the weeks following shearing. The housing should have a solid wall (1.5 m high), using straw bales in some situations to reduce draughts and excessive shivering. Generous amounts of litter can also be a great help. Shearing before early January allows sufficient regrowth by turning-out time; deep combs should always be used for winter shearing. The respiratory rate of shorn ewes is always lower than unshorn ewes. If fed *ad lib.*, the roughage intake of shorn ewes is 10 per cent higher than in unshorn ewes. Shearing leads to easier surveillance at lambing, easier teat-seeking by lambs and a lower incidence of crushing, particularly in the post-lambing mothering pen. Lambs from shorn ewes have a higher birth-weight and a higher survival rate. However, the incidence of assisted lambings may be higher in shorn ewes. It has also been established that the gestation period is 1 day longer in shorn ewes.

Contract shearing is now well-established throughout the UK and such teams are usually well equipped, bringing with them their own equipment and shearing boards, and a person to roll the fleeces in the approved manner. The producer is expected to provide catching pens close to the shearing area and to keep a flow of sheep into these pens. Shearing is a highly skilled operation and requires a period of training, such as the courses organized by the Agricultural Training Board, combined with considerable practice under the guidance of a fully trained instructor. It is hard work and often carried out under less than perfect conditions. Only the larger flocks justify the expense of a fully equipped specialized shearing shed. Good control of the sheep during shearing minimizes stress. The shearer must work quickly and accurately without cutting the skin, and finish with the fleece intact and having been cut cleanly close to the skin. Double cuts depress the value of the fleece. Contract shearers clip 120–150 sheep/day.

A high standard of cleanliness should be maintained around the shearing area, and rolling should be carried out on a clean surface or special table. Trimmings, soiled bits and odd bits should be bagged separately. Rolled fleeces should be placed in woolsacks provided by the local grading centre and then stored in a dry place.

Milk production

The milking of sheep is widely practised in many parts of the world, and particularly in the north and east Mediterranean basin where nearly 50 per cent of world sheep milk production occurs. It has an important socio-economic role in many areas where the environment prohibits other animal enterprises. The practice was popular in the UK during the Middle Ages, but the use of sheep as dairy animals gradually declined and had disappeared by the eighteenth century. During the last 10 years there has been a renewal of interest, partly in the wake of the goat-milk industry, but also boosted by the need to establish alternative enterprises. It has also been encouraged by the availability of the Friesland breed in this country.

Sheep milk has many uses, but most of it is used for the production of cheese and yoghurt. The physical characteristics allow it to be frozen and stored prior to manufacture, sometimes at centres specializing in the preparation of products, but often soft cheese and yoghurt are prepared on the farm. It is imperative that producers establish an outlet for the milk and its products before embarking on this kind of enterprise.

Systems

There is a wide range of systems of sheep milk production. Most of the differences between them concern the management of the lamb and frequency of milking. The most intensive involves separation of the lambs from the ewes during the first week followed by artificial rearing. This requires a high standard of management and careful attention to hygiene, for it is on a much bigger scale than that described for the rearing of orphan lambs (see p. 103).

In contrast to the intensive system, the lambs may be weaned at 12–16 weeks of age, the stage when they are naturally dependent on grass, and the ewes are then milked until they dry off after a further 4–6 weeks of milking. This system may be useful where single lambs are marketed at an early age. There are two systems which fall between the two described above. The first allows lambs to remain with the ewes, but separates them for increasing periods from 3–4 weeks of age: this has the advantage of limiting the milking to once per day and not at all on some days. Where considerable capital outlay is involved in establishing the enterprise, a second system is more appropriate, involving weaning the lambs by 5–6 weeks of age, after which the ewes and lambs are managed quite separately. The ewes are milked twice daily until daily yield justifies milking only once daily.

Breeds

The meat potential of the lamb remains an important element in the productivity of all systems of managing dairy ewes, and therefore the fecundity of the dam and carcass quality of the sire are important criteria of selection. The availability of Friesland ewes has given a considerable impetus to sheep milk production in terms of high milk yield. Other breeds and crossbreds are also used as milk sheep, and the Dorset Horn is the only choice where a non-seasonal supply is required, as with this breed a proportion of the flock can be lambed in the autumn. In a pure-bred flock, particularly when the replacement rate is low, a proportion of the ewes may be mated to meat-type terminal sires, thus improving the meat potential of the lambs. As the sheep milk industry develops in this country, it is likely that performance and progeny testing will be adopted to improve production potential. Such testing may develop into sophisticated schemes such as that operated by the Roquefort Co-operative in France, and involving the use of tested rams from their own artificial insemination centre.

Milking units

Because of high labour costs, only mechanized methods of milking can be contemplated in the

UK. A great deal of development work has been undertaken in the design and operating efficiency of milking units and plants. As is the case in the dairy cow industry, there is a wide range of choice. Small flocks will require only a mobile bucket unit operated from a small wheeled power unit, whereas farmers with bigger flocks can choose between abreast or herringbone stalls in a specialized parlour, with milk storage, cleaning facilities and the machinery plant in the adjoining rooms. In the larger units in the Mediterranean region, and particularly in the Roquefort area, rotating carousel types of milking parlours are popular. These have the capacity to milk 500–600 ewes/h.

Although sheep milk production does not, as yet, come under the Milk and Dairies Regulations, most producers adopt the same standards of hygiene as are required on cattle dairy farms.

Lactation

The duration and yield of a lactation varies according to breed, frequency of milking, level of feeding and management of lambs during early lactation. The ewe may lactate for 6–7 months and a lactation yield of 130–150 litres is within the capacity of several British breeds and crossbreds. The yield of a Friesland may be as high as 300–400 litres with a peak daily yield of 2.4–4.0 litres. Milk quality changes during the course of lactation, and follows the same pattern as that found in cows, with fat content low in early lactation and high in late lactation, but markedly affected by diet. In terms of composition, sheep milk differs from other species in several respects, which can clearly be seen from Table 4.8.

The pattern of milk ejection differs from that in other dairy species in that "let down" may occur in two phases, with the second occurring 30–40 s after the clusters have been put on. It is thought that the first let down is cisternal and the second alveolar. The majority of breeds require machine stripping. This is mainly due to the placement of the teats to the side of the glands, and complete milking can only be accomplished by manipulation of the two glands.

Table 4.8
Average composition (%) of cow, goat and sheep milk

				Proteins	
Species	Dry matter	Fat	Lactose	Casein	Others
Cow	12.5	3.7	4.7	2.6	0.7
Goat	14.3	4.8	4.5	3.2	1.1
Sheep	19.0	6.9	4.9	5.0	1.6

Management

Ewes should be conditioned to using the parlour, at first for feeding, and then for milking. The majority of ewes accept the procedure within the first week. Those which find it stressful thereafter are normally taken out of the milking flock. During early lactation, milking takes 1.5–2 min. Great care should be taken to avoid leaving the clusters on too long and thus organization within the parlour is particularly important; automatic cluster removal may help in this respect. A high standard of hygiene, and efficient machine operation and maintenance are just as important in the control of mastitis in milk sheep as they are for dairy cows.

The general management of the milk flock is not significantly different from that of prolific lowland flocks. The same general strategy should be adopted, particularly during the approach to mating to achieve high fecundity, and during the approach to lambing to ensure the health of the ewe, high birth-weight of the lambs and a good supply of colostrum followed by a high-yielding lactation. The condition of the milking ewes should be closely monitored, particularly after the fourth week, and further loss of condition during the remaining phase of the lactation should be prevented. This requires particular attention to the quality of the grazing and the level and quality of the concentrates fed at milking time. The aim should be to supply approximately 30 MJ of energy per day for a 75 kg ewe during the first 4 weeks, with a gradual reduction during the rest of lactation.

From the welfare point of view, particular attention should be given to efficiency of milking, prevention of mastitis and general care of the udder. Lambs require particular attention following early weaning, and it is advisable to maintain them on a concentrate diet with some roughage until 12 weeks of age (see p. 106).

General care and handling

Stockmanship

The *Codes of Recommendations* (MAFF, 1977) for sheep draw particular attention to the importance of stockmanship in sheep husbandry. It is a term that is not easy to define; it encompasses stock sense and skill and stock tasks (for a full discussion, see UFAW, 1983). A wide range of tasks has to be undertaken during a sheep year and to complete these, with minimal stress to the sheep and least effort to the stockman, a great deal of skill is required. Fortunately, there are several sources of tuition available to the new entrant, such as the Agricultural Training Board and local agricultural colleges. The acquisition of stock sense is somewhat more difficult and it can only come from working with animals and alongside good stockmen, for it involves the use of powers of observation to recognize, at an early stage, abnormal behaviour and changes in the appearance of animals.

Gathering and driving

There are numerous occasions during the year when every sheep in the flock has to be handled, as either a group or a flock, for the prevention of disease, weighing and condition scoring, drafting and regrouping, and shearing.

The initial task of moving ewes from a field or gathering a flock off a hill will probably involve the use of a sheep dog. In fact, the management of sheep in upland and hill areas, where the flock is allowed free range, would be near impossible without a sheep dog. The outstanding harmony between the shepherd and his dog is without parallel in other forms of livestock husbandry, for it often involves communication between man and dog from a considerable distance. Although the collie has been selectively bred as a working dog, it is imperative that it is properly trained and able to undertake all the manoeuvres involved in moving a flock and segregating subgroups. Nothing can be more stressful and damaging to a flock than a dog which is not under proper control; in extreme cases it is as bad as the marauding dog. Those who wish to train a dog should either attend an ATB course or acquire the guide book produced by this organization.

A feature of sheep behaviour is the tendency to follow one another and to group when alarmed (see p. 85). During the early phase of the suckling period a lamb will follow its dam, and gradually this tendency develops into a group pattern of behaviour, such as a lamb play group, and subsequently the young sheep conform to the flock pattern of behaviour, particularly during gathering and driving. Sheep also become conditioned to the various signals, such as the sight of a bag or sound of a bucket, indicating feeding time. All these aspects of behaviour can be used usefully to minimize stress during the movement of sheep between fields, and from field to handling and loading areas. In the handling and loading areas, decoy sheep can be used to good effect during movement around the sheep pens and buildings and also to ease the task of loading.

Design and use of pens

The arrangement and design of pens and gates should at all times allow easy movement and good working conditions, and minimal stress. Interaction between sheep while going through the sheep pens is quite important. Unflustered older sheep which are accustomed to the various procedures will have a calming influence on younger sheep and sheep using the facility for the first time. The flow of sheep is greatly helped by pens which funnel towards narrow passageways and races, and by the absence of sharp corners and protrusions which slow up movement and may cause injury.

A sheep unit (e.g. Fig. 4.7) usually provides a large reception pen which leads through to smaller pens and races or narrow walkways, according to the task to be carried out, and has pens for subgrouping or regrouping in the area leading back to the pastures or loading bay. Where the unit is linked to some kind of housing, sheep can be held under cover in preparation for such activities as shearing, dagging and weighing. Where the unit is away from the farm buildings, consideration should be given to the prevention of pollution of water courses and, in particular, to the containment of chemicals used in quantity for spraying and dipping.

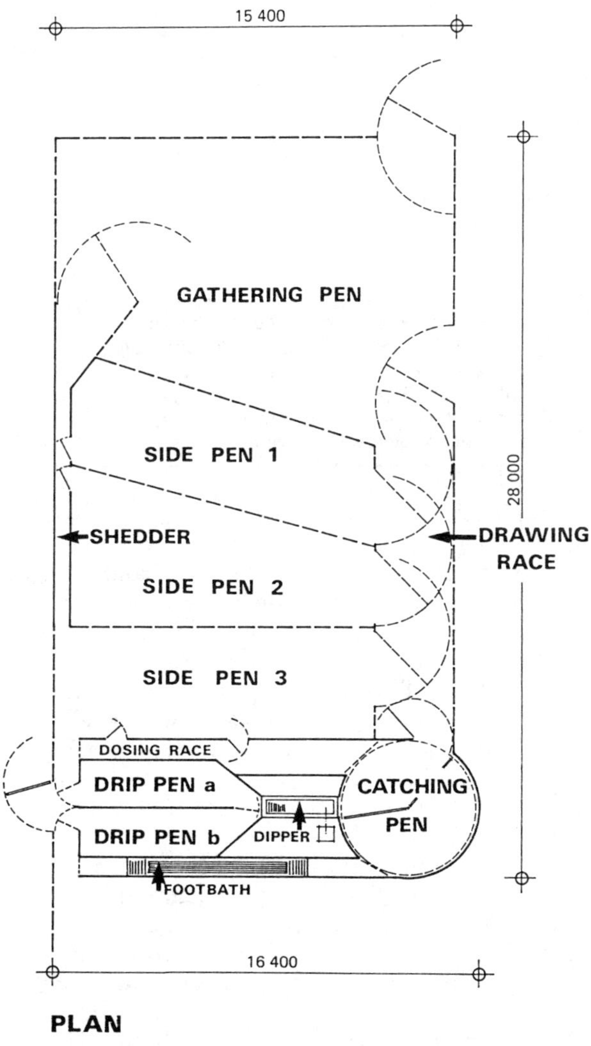

Fig. 4.7 Plan view of sheep handling pen. Reproduced from *Farm Building Cost Guide, 1988*, published by Scottish Farm Buildings Investigation Unit.

The type of facilities required to carry out the various tasks on the flock or groups are described in Table 4.9. Several of the tasks listed are carried out without any need to restrain or manipulate individual animals. Others involve contact and the use of restraint and are best carried out with the animals confined in a small pen or race.

Handling and restraint

In a sheep pen, movement is minimized by placing the open hand under the jaw and with the head held above the back line. The sheep may then be manoeuvred backwards or sideways towards a corner of the pen. Further restraint can be achieved by placing the other hand high on the neck behind the ears. For close examination of the head and mouth of adult horned breeds, both horns may be held with the handler astride the sheep.

When segregation is carried out away from the handling unit, in the corner of a field or at a gateway, the initial restraint can be done by using a crook to hook the sheep immediately above the hock, or by moving forward quickly to grasp a hind leg above the hock. No part of the fleece should be used to restrain sheep. The horns of young animals should not be used for restraint.

Casting

Full restraint often involves casting and this allows the examination and/or treatment of the feet and any part of the underline. This position is also used for the first stages of shearing, and for dagging and the removal of excessive wool around the udder.

There are several methods of casting; choice is usually determined by the size of the sheep. Light sheep are usually cast by the lift method, and heavy sheep by unbalancing the hindquarters and thus getting them into a sitting position. In both methods it is important to get the sheep standing quietly with the left hand under the jaw and with the handler standing close to the body, with the knees close together and behind the shoulder. In the lift method, the right hand is used to grasp a fold of skin and wool low down on the right flank region. This hand is then used to lift the hind leg off the ground, and at the same time a nudge of the knees prevents the sheep bracing on the left

Table 4.9
Restraint of sheep

Task	Restraint	Type of pen
Preventive treatments		
Dosing or vaccinating	Minimal	Small pen or race
Spraying	None	Spray race or shower pen with facility for recirculating solution
Dipping	Minimal	
Dipping	Minimal	Dip with draining pens
Foot bathing	None	Race with permanent or portable footbath
Foot trimming	Full	Small pen
Docking and dagging	Full	Small pen
Castration	Full	Small pen
Weighing and condition scoring		
Weighing	None	Near outlet from race
Condition scoring	Minimal	Race or small pen
Drafting and regrouping		
Tattooing, tagging, ear notching	Minimal	Small pen or race, depending on age group
Examination for soundness	Full	Race (teeth), or small pen (feet and udder)
Drafting and regrouping	None	Race with 2- or 3-way outlet or drafting gate
Marking for marketing	None	Race between weighbridge and outlet gates
Loading	None	Small pen leading to ramp or loading gate
Shearing	Full	Small catching pens adjacent to shearing floor, with sheep released to "counting" pens

hind. The sheep is gently turned so that it sits on its rump with the withers resting between the handler's knees (Fig. 4.8). For heavier sheep, the right hand is placed at the base of the tail and by bringing the head and tail towards each other, and, with the right fore-arm exerting some pressure on the rump, the hindquarters drop to a sitting position. The fore-end is then brought up so that the withers rest between the knees. The hindquarters can also be thrown off balance by pulling on the off-side hind leg with the hand placed just above the hock. There are several devices marketed as aids to this type of handling, such as cradles designed to hold sheep in the cast position. Mechanical casting and restraining devices usually hold sheep in the upside-down position. Sheep in late pregnancy should be held in the cast position for only a short time.

Routine procedures

Soundness, culling and selection

The examination of sheep for soundness is an annual event usually carried out after weaning the main flock and before the preparatory phase for tupping. It involves the close examination of all classes of stock of breeding age and may include an assessment of their productive performance (for ram assessment, see p. 95).

The first stage is a general appraisal which is a normal part of flock inspection at any time, when attention is given to signs of ill-health, abnormal behaviour, posture or gait, poor body condition and ragged fleece. The second stage involves the close examination of udder, teeth and feet.

When examining the udder, both teats should be carefully looked at for any evidence of injury during shearing or at the previous lactation, and both glands should be gently palpated for hard or abnormal tissue, which is usually associated with mastitis.

The examination of the teeth should give particular attention to alignment and to number, type and soundness. Correct alignment between the incisors and the upper pad is necessary for efficient grazing; attention should be given to this aspect in the selection of rams and breeding ewes during the transition from temporary to permanent incisors. The eruption of the first permanent incisors (the central pair)

Fig. 4.8 The lift method of casting. (A) Method of grasping sheep; (B) cast position.

occurs at approximately 1 year and 3 months, the adjacent pair at 1 year and 9 months, and the next two pairs at 2 years and 3 months and 2 years and 9 months. The number of permanent incisors (or broad teeth) has long been recognized as a means of ageing within the above range and is a part of sheep nomenclature. Two-tooth ewes are approaching their first year of full production potential and usually attract the best prices at auctions. Ewes showing four pairs of "permanents" are referred to as "full-mouthed" and this description remains until one or more teeth are lost, when they become "broken-mouthed". The age at which this occurs is variable. Broken-mouthed ewes tend to lose condition, particularly on hard grazings, and are usually culled.*

The feet should be closely examined for clinical symptoms of disease, distortions of the hooves and excessive growth of horn.

Body condition scoring

The method of body scoring involves five grades – 1 (very thin) to 5 (very fat). It is easily and quickly taught. It requires no equipment and may be carried out wherever it is convenient to pen the sheep, but preferably in a race, so that ewes can be regrouped according to their condition score. Those in poor condition may then be given extra feed to restore their condition to the required level, as outlined on p. 95.

The assessment involves handling immediately behind the last rib, the tips of the fingers being used to determine (a) the prominence of the two parts of the lumbar vertebrae, the spinous and transverse processes, (b) the amount of muscular and fatty tissue underneath the transverse process, and (c) the fullness of the eye muscle and its fat cover by pressing the fingers into the angle between the spinous and transverse processes. A full description of the five grades can be found in MLC (1983).

Weighing and selection of lambs

There are several types of walk-on weigh scales

* *Editor's note*: The Welfare of Livestock (Prohibited Operations) (Amendment) Regulations, 1987 *totally* prohibit tooth grinding in sheep, even by veterinary surgeons. There is no justification whatsoever for this procedure.

marketed for sheep. The scales may be sited at the outlet of the race and thus the task may be completed without the need for handling. Lambs quickly become accustomed to being weighed.

Lambs for the fatstock market should be selected at approximately 50 per cent of their potential adult weight. All lambs within 10 per cent of the desired weight should then be handled at the four points shown in Fig. 4.9 to assess level of fatness, which gives a guide to the potential fat class of the carcass. The two most important points are at A (the tail root, dock) and B (the vertebrae of the loin). The prominence of the vertebrae is determined by gentle palpation using the fingertips. The overall score of fatness is then determined from criteria which are described in detail in MLC (1986).

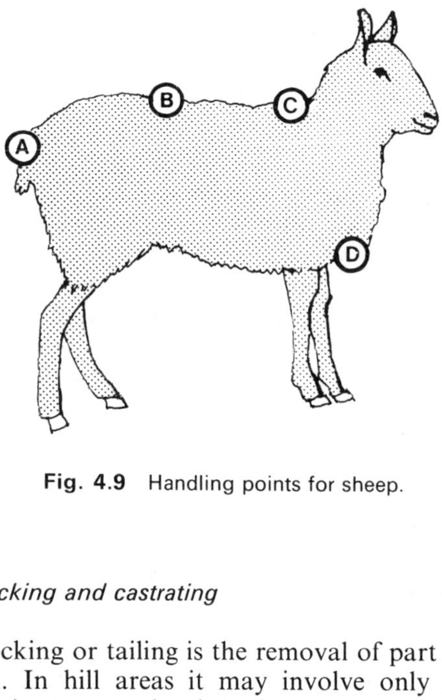

Fig. 4.9 Handling points for sheep.

Docking and castrating

Docking or tailing is the removal of part of the tail. In hill areas it may involve only a few centimetres at the tip, whereas under lowland conditions, the greater part of the tail is removed. In lowland areas, the main reason for docking is to reduce faecal contamination, which attracts blowflies.

Docking and castrating of lambs less than 1 week old are normally carried out with one operator holding the right legs in one hand and the left legs in the other, and with the lamb held against the chest, while a second operator carries out the actual procedure. When it has to be done single-handed, the lamb may be restrained by placing it between the knees and with the hindquarters facing the operator. Older lambs are usually done with the front legs being held and the rump resting on a rail or bale at a convenient height. There are several methods of docking: the use of a rubber ring to restrict the flow of blood to the tail; cutting with a knife; crushing with a Burdizzo bloodless castrator prior to cutting; or the use of a sharp-edged hot iron. In the first method, the ring is put on with an Elastrator applicator; it takes 7–14 days for the tail to drop off and normally there is no bleeding. Provided the ring is put on at the end of the wool-free skin on the underside of the tail, the remaining stump should cover the anus and vulva. This is a legal requirement under the Welfare of Livestock (Prohibited Operations) Order, 1982.

The rubber ring method is also used for castrating. The ring is placed at the neck of the scrotum and in front of the rudimemtary teats and with both testicles within the scrotum. It is an offence to use the rubber ring method to either dock or castrate lambs more than 7 days old.*

The other methods of castration are the use of a Burdizzo bloodless castrator to sever the spermatic cord (vas deferens), and the use of a knife to remove the testicles. Where a knife is used to dock or castrate it is important to maintain a high standard of hygiene and to treat the wound to prevent infection, particularly fly strike. Stockmen should not attempt these techniques without prior training. It is an offence to castrate male sheep by any means without the use of an anaesthetic after they have reached the age of 3 months and only a veterinarian may castrate a ram which has reached this age.†

Dagging

Dagging consists of the removal of excess wool around the tail and down the inside of both

* Protection of Animals (Anaesthetics) Acts, 1954 and 1964.

† Protection of Animals (Anaesthetics) Act, 1954 (Amendment, Order, 1982 and Veterinary Surgeons' Act, 1966 (Schedule 3 Amendment) Order, 1982.

thighs, to prevent soiling during early spring and prior to shearing. It reduces the risk of fly strike early in the season, and also the need to trim fleeces prior to rolling. It may also be done to unshorn lambs in the late summer. Whether mechanized or hand shears are used will depend on the amount of wool to be removed (shearing is described on p. 107).

Loading

The Meat and Livestock Commission have drawn attention to the need to reduce the incidence of injuries and bruising to fatstock which occur at various stages during their transfer to markets and slaughterhouses, and the consequent carcass damage. Attention to detail during the design and construction of handling pens can contribute a great deal to the reduction of stress and injury.

Loading at the farm usually takes place directly from the collecting pen, with the lorry's own ramp and side hurdles being used to get the sheep to a particular level within the lorry. A decoy or Judas animal greatly helps to start the flow of animals during loading. Particular attention should be given to the following recommendations:

1. Drive the sheep quietly.
2. Do not use dogs in confined spaces.
3. Never prod sheep with stocks.
4. Avoid dragging or lifting by the fleece.
5. Do not crush lambs as they go through shedding gates.

Where animals have to be lifted, the method adopted will depend on size. Lightweight lambs and fatstock may be lifted and carried by placing one forearm underneath the animal in the region of the flanks and the other arm firmly in front of the chest. Older and heavier animals should be lifted by firmly grasping the flank area and with the head and neck held firmly against the body. The handler's knee may be used as a lever during the lift and the body is slightly turned so that its weight is taken through the flank and neck.

Health and disease

Irrespective of the system of production in the UK, all flocks rely on hygiene, preventive treatments and early detection of ill-health, for the maintenance of a healthy flock. A good stockman will see the early symptoms of abnormality and will act upon it without delay. He relies on his vigilance and powers of observation to pick up subtle changes in behaviour, the early stages of lameness and other physical symptoms during regular inspections of his stock. The signs which indicate health include general alertness, good movement and absence of lameness, good uniform fleece and absence of rubbing, active feeding and rumination, and no visible wounds, abscesses or injuries.

The signs of ill-health in sheep include listlessness, abnormal posture and behaviour, persistent coughing or panting, absence of cudding, poor condition and, in some circumstances, separation from the flock.

Good hygiene and preventive treatments are vital aspects of sound flock management. In extensive systems there will be little scope to practise hygiene except at times of flock treatments, which may be minimal anyway. In contrast, lowland systems involving high stocking rates and winter housing will need high inputs of both hygienic measures and preventive treatments.

General hygiene

A high standard of cleanliness can be obtained in handling areas and houses by simply brushing and washing surfaces and by using clean dust-free litter. During housing, attention should also be given to ventilation.

Disinfection

General disinfection, after thorough cleaning, should be undertaken in buildings at the end of occupation, when all fittings and food and water containers should also be thoroughly disinfected. Similarly, collecting pens and handling yards should also be disinfected periodically. A high standard of cleanliness, coupled with thorough disinfection, is required during the lambing season of housed sheep, when particular attention should be given to the pens used for mothering, fostering, and artificial rearing of orphan lambs. Equipment used for routine procedures and treatment should be carefully cleaned, and disinfected or sterilized after use.

Preventive and other treatments

Preventive treatments are available for conditions caused by a variety of organisms including viruses, bacteria and other microorganisms, and external and internal parasites.

A programme of treatment is usually drawn up following consultation with the veterinary surgeon, and particular note is taken of the previous disease history of the flock and also the prevalent diseases of the locality. Treatments involve vaccination, dosing (drenching), spraying or dipping, foot bathing and foot treatment. Many are seasonal and some are related to the key events of the sheep year – mating, lambing, shearing and weaning.

An example of a plan of action for the main breeding flock is shown in Table 4.10. Similar plans may be drawn up for rams, ewe lambs, fatstock, and replacements purchased from markets or other sources.

Table 4.10
Relationship between key events and preventive treatment in a closed flock

Month	Key events	Type of preventive treatment
August		Spraying/dipping (all stock). Intensive foot care (all stock)
September		Vaccination[a] (all stock). Compulsory dipping (all stock)
October	Mating	Vaccination[a] (all stock)
November		
December		
January		
February		Vaccination[a] (all stock)
March	Lambing	Drenching (ewes before released to clean pasture)
April		
May	Shearing	Spraying/dipping (depending on season, ewes only)
June		Spraying/dipping (all stock)
July	Weaning	Drenching (depending on availability of clean grazing)

[a] 8 in 1 vaccine.

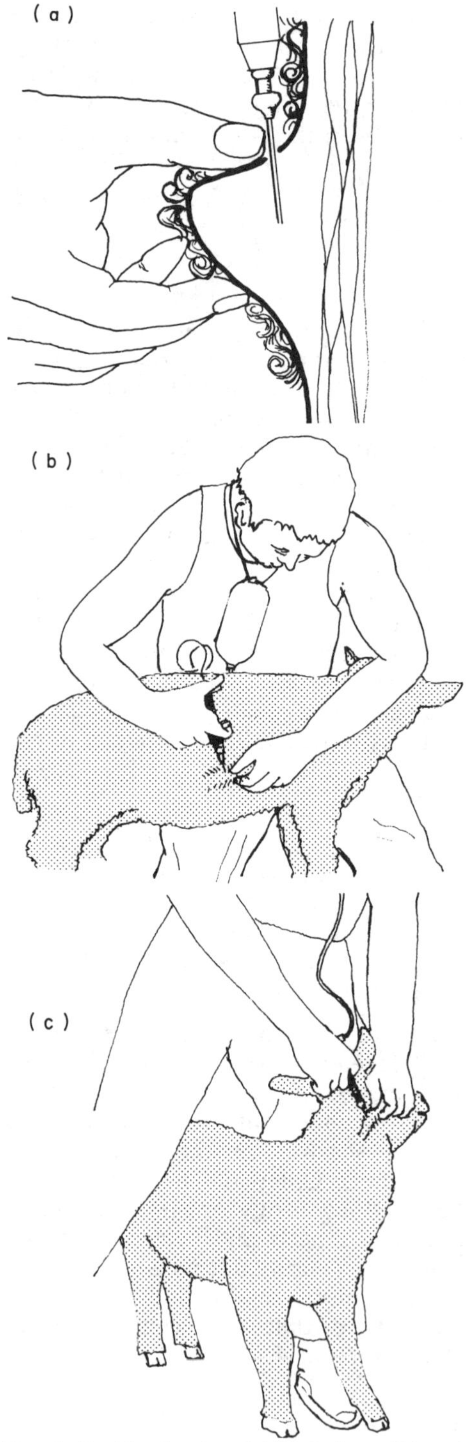

Fig. 4.10 Vaccination. (a) Technique; (b) and (c) sites.

Vaccination

Most vaccines are administered subcutaneously. The preferred injection sites are high on the side of the neck or over the lower rib cage. The needle should be inserted parallel to the body surface and into a cavity formed by raising a fold of skin (Fig. 4.10). If the needle is inserted at any other angle, it can easily penetrate the underlying muscle. Vaccination should be carried out as hygienically as possible. It is important to use the syringe according to the maker's instructions and to change the needle periodically. Incorrect or careless injection techniques produce abscesses which reduce the value of the carcass.

Some vaccines provide protection against many diseases; in the majority of flocks in this country a multi-vaccine (8 in 1) is used which provides protection against the following clostridial diseases: pulpy kidney; tetanus; lamb dysentery; struck; braxy; blackleg; post-parturient gangrene; black disease. Some of these vaccines also protect against *Pasteurella* pneumonia.

The multi-vaccines not only provide protection to the adult stock, but when given in late pregnancy, ensure the presence of antibodies in the colostrum to initiate passive immunity in young lambs. A vaccination programme for breeding stock involves two injections in the first year and an annual booster injection thereafter, usually during late pregnancy. Separate vaccination programmes are usually carried out as necessary against orf, louping-ill, enzootic abortion and foot-rot.

Dosing

Dosing may be undertaken to control internal parasites – stomach worms, nematodirus, tapeworms, liver fluke and coccidia. There is a wide range of drugs available; some are effective against more than one of the parasites listed above. Care should be taken in the choice of drug and in the timing of dosing. Consultation with the veterinary surgeon is advised. Farms with the opportunity to graze cattle and sheep alternately, and to include conservation in their programme of grassland management, are best placed to minimize the use of anthelmintics and the frequency of dosing.

The development of various kinds of plastic devices has made routine dosing a relatively easy task and not too time consuming, particularly if the flock is treated in a race.

For dosing, the head should be held at back level, or slightly higher, using an open hand under the jaw. The mouth-piece is then inserted at the corner of the mouth and with the tip resting on the back of the tongue.

Pills and boluses may also be used. Special devices may be used for administering these; it is recommended that this should be a two-man operation, with the handler concentrating on keeping the sheep still. The doser opens the mouth and positions the device so that the pill or bolus drops on to the back of the throat.

Dipping and spraying

These methods are used for the control of external parasites, such as ticks, lice, keds, mange mites and blowfly larvae, which can cause prolonged stress and discomfort and, if infestation remains undetected, can lead to serious losses. Effective control relies on dipping and spraying at particular times of the year. Dipping is the only approved method of controlling psoroptic mites; both options may be used for the control of other parasites. Both dipping and spraying involve some stress to the animal, and the procedures should be carried out with calm efficiency. Sheep should not be treated when they are hot, tired, wet, thirsty or full-fed. To reduce the risk of poisoning through absorption, dipping should not take place during very hot weather, and should preferably be carried out during the cooler time of day. Dipping and spraying may coincide with the breeding ewes having lambs at foot, and particular care is required when treating young lambs. Ewes and their lambs become separated during handling and the level of noise reflects the degree of general upset. It is advisable to check for mismothering after the flock returns to the field. Spraying is less upsetting than dipping, for it allows the ewe and her lambs to walk through together, and there is less risk of injury. Spraying is not as thorough, and consequently the period of protection is shorter than that which follows dipping.

Mites. Sheep scab is a notifiable disease caused by mites (*Psoroptes*). Compulsory dipping was reintroduced in 1976 under the Sheep Scab Order. In recent years there have been two specified dipping periods during the summer/autumn. The precise dates are announced annually. The local authority has to be notified in advance of the day and time of dipping, and only approved chemicals may be used. These are also effective against other parasites, and thus serve a multi-purpose function.

It is vitally important that the correct dipping procedure should be followed, as laid down in the manufacturer's instructions and recommendations. Sheep should be gently lowered into the dip; in some large units a slide arrangement eliminates handling. It is a requirement that the sheep be immersed for 1 min and during that time the head must be ducked at least once to ensure thorough wetting of the head and ears. The sheep must be plunged or revolved to remove air pockets in the fleece, so that the dip penetrates to the skin and complete saturation is achieved.

Handlers should always make sure that they wear suitable protective clothing and that containers are stored or safely disposed of so as not to be a danger to children and animals. Every precaution should be taken to ensure the safe disposal of the dip solution.

Blowflies. The blowfly season can extend, depending on location and weather conditions, from May to October and thus overlaps the sheep-scab dipping period. Blowflies lay their eggs in the fleece, usually around the hindquarters and shoulders. These hatch as maggots in about 12 h and invade the skin and flesh, causing intense irritation and discomfort. This results in characteristic changes in behaviour, and the sheep becomes very restless and endeavours to bite the affected area. Dagging helps to reduce the risk of strike, particularly before shearing and also late in the season. Detection of struck sheep is an important part of surveillance during the fly season. If it goes undetected, the sheep are "eaten alive" and usually die in 3–4 days.

Depending on the time of shearing, it may not be necessary to start preventive treatment until 2–3 weeks after shearing. The chemicals in current use remain effective for several weeks, the residues held in the fleece killing blowfly larvae as they hatch.

Ticks. Ticks are active mainly during the period March to August; they transmit a variety of diseases such as tickborne fever, louping-ill and tick pyaemia. They are mainly a problem of hill and upland farms, and flocks in these areas are treated during late pregnancy to reduce the tick infestation; this also reduces the risk to young lambs.

Keds and lice These parasites can only survive on the sheep and are not such a widespread problem as the parasites already mentioned. Heavy infestations do arise, particularly of lice in housed sheep, and this may lead to acute irritation and loss of wool as a result of rubbing. Dipping and spraying against other parasites are usually effective in controlling them.

Headflies. Headfly disease mainly occurs in horned sheep in the north of England and the Scottish Borders. Flies are attracted by skin secretions, particularly around the base of the horn. This causes intense irritation which results in sheep rubbing their heads against posts and fences with such vigour that wounds are caused which, in turn, attract more flies. The activity of the flies prevents healing, and often secondary infection sets in, which requires treatment with a broad-spectrum antibiotic. It is extremely difficult for wounds to heal during the fly season, and badly affected animals are usually housed. The recent development of pour-on sprays containing synthetic pyrethroids promises to be an effective means of control. Control can also be achieved by keeping sheep from pastures adjacent to woodland during the fly season.

Foot care and treatment

Lameness is probably the most widespread cause of pain and suffering in sheep and attention is rightly drawn to it in the *Codes of Recommendations* (MAFF, 1977), which require that stockmen should be experienced and competent in the prevention and treatment of foot-rot (MAFF, 1981b).

Although foot-rot is the main concern when it comes to foot inspection, care and treatment, lameness may be due to one of several other causes, including injuries to the sole of the foot from embedded stones, glass or thorns, and to the interdigital skin when walking through corn and brassica stubble. Interdigital growths can become infected, leading to painful lameness which can only be corrected by surgical removal of these growths under local anaesthesia. Congenital deformity of the claws can occur and this is best dealt with by paring them and then culling the animals as fatstock at the first opportunity. Sometimes acute laminitis occurs in all four feet after the introduction to a very high plane of nutrition, such as lush pasture, and this may immobilize the animal. Transfer to a poorer diet is the only effective course of action. Inspection and treatment is normally carried out in the handling pens leading up to the footbath. Individual lame sheep may be dealt with in the field. The essential items to be prepared beforehand are secateurs and a sharp knife for paring away infected tissue, a brush (if the sheep have muddy feet), a bucket with disinfectant and fresh solution in a footbath.

Feet which are extensively damaged are treated with an antibiotic aerosol or cream after paring and prior to bandaging: the animal is kept on dry straw in an isolation pen for 4 or 5 days and then re-inspected. Healing may take 2–3 weeks. Animals which are less seriously affected may simply be put through the footbath, or spot-treated with an aerosol antibiotic. Animals requiring treatment may be shed from the inspection pen and held until all sound sheep have gone through the footbath.

Sheep should be put through the footbath in batches and allowed to stand in it for the recommended time (usually 1–2 min). Whichever chemical is used, the manufacturer's instructions regarding safety and preparation must be followed. Currently the choice is between formalin and zinc sulphate. Formalin should be handled with great care to avoid workers being splashed in the face and eyes. If this occurs, they should wash immediately, using plenty of clean water. Particular note should be taken of the recommended concentration for routine preventive treatment and for the treatment of clinical cases. Excessively high concentration will be particularly painful to animals that have been pared. Of the two chemicals mentioned, zinc sulphate is now preferred to formalin in terms of penetration, effectiveness and welfare.

The prevalence of footrot infection in any flock will be governed by the sources of infection, both from within the flock and from purchased animals, and by factors which predispose the flock to infection. These include soil type and climatic conditions, drainage of the soil, stocking rate and, during winter housing, the quality of the litter and general housing conditions.

Careful inspection, foot paring and foot bathing are essential first steps to eradication. After foot bathing sheep should be held on dry concrete for about 1 h before returning to a fresh field. The causative organism cannot survive in the soil for more than 2 weeks. Since the incubation period is about 12 days, a repeat flock inspection should be made after this interval. Infected animals from the first inspection should have been segregated and subjected to more frequent inspection and treatment. Resistance to infection can be promoted by vaccination, but this should be considered only after the preliminary programme of inspection, treatment and pasture management has been completed. The maintenance of a "clean" flock will then greatly depend on stockmanship.

Nutritional and metabolic disorders

Pregnancy toxaemia (twin-lamb disease)

This disease of late pregnancy, usually found in ewes carrying more than one lamb, is typified by separation from the flock, listlessness, apparent blindness and unsteady gait; eventually the ewe is unable to rise and may lapse into a coma for several days before dying. The smell of acetone is usually detectable in the breath. Prevention is a very important factor. The condition is associated with a sudden shortage or change of food, which can arise as a result of winter storms, inadequate trough space, or inappetence, and therefore can be prevented by careful feeding and good management. The ewes should not be over-fat in late pregnancy and should be fed on a gradually rising plane of nutrition during the last 6 weeks (see Table 4.5).

Treatment involves intravenous injection of 40 per cent glucose solution, or oral glycerol and propylene glycol (55–85 g/day). The success rate of these treatments is very low.

Hypocalcaemia (lambing sickness)

The symptoms of this disease include listlessness, incoordination and recumbency on the chest with the head resting on the ground. If it is not identified and treated, it usually leads to death in 6–12 h. It is not directly related to calcium deficiency in the diet and therefore further supplementation is not advised. The condition occurs during the last month of pregnancy or in early lactation. It may be precipitated by over-excitement and enforced exercise, such as gathering for pre-lambing vaccination. Response to treatment is rapid, often within 30 min of receiving a subcutaneous injection of calcium borogluconate (see Hypomagnesaemia).

Hypomagnesaemia (staggers)

The main symptoms include excitability, nervous twitching, staggering and convulsions. It usually occurs after lambing, and is common in ewes grazing heavily fertilized improved pasture. Draft ewes from the hills are particularly prone to it during their first spring in the lowlands. The condition is a straightforward dietary deficiency which can be prevented by supplementation with magnesium, usually as calcined magnesite, in a mix giving 15 g/ewe/day over the first 3 weeks of lactation. Clinical cases quickly respond to subcutaneous injection of magnesium chloride. Due to the difficulty of distinguishing between "lambing sickness" and "staggers", and the importance of early treatment, a combination of calcium borogluconate and magnesium chloride is usually given to ewes in early lactation.

Copper deficiency (swayback)

Symptoms are usually seen in young lambs as loss of coordination in the hind limbs. The condition can also manifest itself as poor growth rate in lambs, and as dullness of the fleece in adults. Many areas in the UK are recognized as copper-deficient, but the condition can also arise as the result of excesses of other elements, such as molybdenum. Because of the danger of toxicity, copper should not be included in licks but added (at safe levels) to concentrates or given by injection. The copper levels in the milk of well-fed ewes provides protection to the lamb for a limited period.

Cobalt deficiency (pining)

Deficiency of this element leads to pining or unthriftiness in young and adult stock. It may also affect the viability of newborn lambs. Cobalt is required in minute quantities to enable rumen bacteria to function normally and to provide the essential vitamin B12, and it must be provided in such a way that it reaches the rumen. Top-dressing of pastures and the use of cobalt ruminal bullets are the most common forms of treatment.

Notifiable diseases

Notifiable diseases which may occur in sheep include anthrax, foot-and-mouth disease, sheep scab, maedi-visna, rabies and sheep pox (see Table 4.11). These diseases, even if only suspected, must be reported to the Divisional Veterinary Office, local authority or police. Suspected animals must be immediately isolated and all movement of livestock stopped. Comprehensive information regarding these diseases may be found in the recommended reading list.

Zoonoses

These are diseases which affect man following contact with infected sheep, or their tissues, fluids and faeces, or ingestion of infected meat, offal or meat products.

Reference has already been made to the danger of anthrax and rabies: other widespread zoonoses are enzootic abortion of ewes, toxoplasmosis, orf and hydatidosis.

Enzootic abortion of ewes

This is caused by *Chlamydia psittaci*. The infected animal usually aborts, and the evidence can be clearly seen in the cleansing. The cotyledons are heavily infected and the areas between them are thickened and have a leathery appearance. Fluids and membranes are highly infec-

Table 4.11
Main symptoms of the most common notifiable sheep diseases

Disease	Causal organism	Species affected	Symptoms
Anthrax	*Bacillus anthracis*	All animals including man	Sudden death; bleeding from nostrils, mouth and anus. Carcass should not be moved prior to veterinary inspection
Foot-and-mouth disease	Virus	All cloven-hooved animals	Profuse salivation; severe lameness
Sheep scab	Mite, *Psoroptes ovis*	Sheep	Intense skin irritation, with restlessness, and nibbling and biting of affected area. Wool loss
Maedi-visna	Slow virus	Sheep and occasionally goats	Panting, coughing and listlessness associated with a slowly developing pneumonia (maedi). Unsteadiness and paralysis of hind-quarters (visna) unlikely to be seen in UK

tious for at least 2 weeks. Personnel can develop flu-like symptoms. The most serious risk is to pregnant women, and several cases of human abortion have been attributed to close contact with infected sheep.

Toxoplasmosis

This is a widespread cause of abortion in sheep. It usually occurs during the last month of pregnancy but the lambs may go to term and may be born dead or weak. Infection is very common in humans, but there is only a low incidence of clinical disease, although pregnant women should be particularly careful to avoid infection. Cats are regarded as the main hosts for the protozooan parasite *Toxoplasma gondii*. Humans are infected by consuming raw or insufficiently cooked meat, by handling infected meat, foetal membranes and fluids, and material contaminated by cat faeces.

Orf

Sometimes known as contagious pustular dermatitis, this disease is caused by a virus which primarily affects the lips of lambs and the teats and udders of ewes. It may also affect the genitalia and the area adjoining the hooves in older animals. Protective gloves should be worn during treatment and vaccination.

Hydatidosis

The cause of this is the cystic stage of the dog tapeworm *Echinococcus grandosus*. Humans become infected by consuming infected sheep liver, which leads to the development of cysts in the lungs, liver and other organs; these usually require surgical removal.

On-farm slaughter

Irrespective of the method of killing, it should be accomplished quickly and humanely. The animal should not be frightened, excited or apprehensive. Shooting with a rifle or shotgun is usually the only on-farm method for killing sheep older than 2–3 weeks.

Stunning

Lambs younger than 2–3 weeks of age may be

killed by stunning. The lamb should be held by the back legs, and the back of head firmly struck using a heavy stick. The stick should have a shaped handle which can be firmly held. After stunning, and when the animal is unconscious, it should be bled by cutting a large blood vessel in the neck.

Shooting

A sheep may be killed with a rifle, revolver or shotgun. When using a shotgun (12-bore or 20-bore) with shot not smaller than No. 5 (A.A., B.B., No. 4 or No. 5) *the animal should be given some food and then shot behind the ear* from a distance of 1–2 m directing the aim so that the shot passes through the brain. A rifle or revolver may be used in the same way with the muzzle held 3–5 cm from the animal's head.

Disposal of carcasses

Foetal material and small carcasses should not be disposed of by putting them on the manure heap. It is essential that the method of disposal should be vermin-proof and that it should remove the possibility of humans or other animals becoming infected.

The most suitable method is to use a deep pit fitted with a fly-proof and smell-proof manhole cover, which is also child-proof. The pit should be approximately 3 m deep and 1.5 m in diameter. The sides of the pit should be lined, but the bottom left as earth to allow drainage (see MAFF, 1980). A well-drained site should be chosen, well away from a water course and above the water table. The site should be clearly marked and there should be some means of keeping heavy machinery away from it.

Large carcasses should be deeply buried or, if appropriate, sold to a kennels or a similar outlet.

The way ahead

Sheep have not followed the same pattern and scale of intensification as other farm livestock in terms of long-term control over their social and physical environment. There are, however, several trends in lowland sheep units which have changed the annual pattern of husbandry and also place more reliance on outside contract work. They involve increases in stocking rate and flock size, and housing for prolonged periods during winter. It is well-established that stocking rate is a major factor affecting the profitability of lowland sheep enterprises, and further increases are envisaged, thus making increasing demands on both management and stockmanship. Increasing the number of stock per person can lead to deterioration in the level of surveillance, hence the early recognition of changes in behaviour of the group or individuals may be missed. This aspect cannot be overemphasized, for frequent inspection is a requirement of good stockmanship.

Lameness is generally regarded as the greatest cause of pain and discomfort in lowland flocks. Increasing the stocking rate, and the consequential poaching of the land associated with it at certain times of the year, provide conditions conducive for the spread of foot ailments. This is also true of winter housing on damp litter. Intensification on these lines will require a high standard of foot care coupled, in some situations, with a vaccination programme.

Housing for lengthy periods during the winter will span the greater part of pregnancy. During this time ewes will be completely dependent on the diet fed to them for the nutrients required to maintain their own health, the fitness of the lambs at birth and the supply of colostrum and milk.

Since income from wool is low in comparison with other components of output, and labour costs are so high, it is not surprising that attempts are being made to develop "easy care" types of sheep. Current development concentrates on genotypes with the capacity to shed a rudimentary fleece, which dispenses with shearing and minimizes the need for dipping. Particular attention is also given to the ability to lamb unassisted and to soundness of teeth and feet.

The development of other breeds and crossbreds gives more emphasis to fecundity; these are more suited to management systems with a high level of husbandry. The recent identification of a British breed, the Cambridge, with a single gene affecting ovulation rate, will greatly improve the possibility of increasing ovulation rate in a range of genotypes. The use of new methods of artificial insemination and the technology associated with ova manipulation and transplant could significantly accelerate the availability of superior genetic material to sheep producers.

Consideration should be given to the suitability of genotypes for intensive systems which involve high stocking rates during the grazing season and housing during winter. Breeds show clear differences in behaviour, and in the ease with which they may be conditioned to accept the procedures and practices adopted in intensive systems.

In the hill and upland areas, sheep farming will remain as the most extensive system of livestock production practised in the UK, and there will probably be greater recognition of the need for supplementary feeding at critical stages of the reproductive cycle.

There is a great need for effective communication between farmers and the general public. In so far as the welfare of sheep is concerned, emphasis should be given to the consequences of disturbing and frightening groups of animals, the seriousness of the dog problem in some areas, and the need to respect fences and gates. These problems are not confined to the fringes of urban areas; there is no doubt that hills and moorlands will continue to attract more people and it is imperative that due regard is given to the countryside code and to the welfare of the animals which are such an important part of that environment.

References and further reading

Agriculture and Food Research Council (1984). *Grassland Research Today*. London: AFRC.

Agriculture and Food Research Council (1986). *Science and Quality Lamb Production*. London: AFRC.

Agricultural Training Board (1987). *Sheep Production: Foot Paring and the Use of a Footbath*. Booklet SICI. Beckenham: ATB.

Boundy, T. (1985). *Care and Examination of Rams* (Tape/Slide Programme). London: UVCE Royal Veterinary College.

Croston, D. and Pollott, G. (1985). *Planned Sheep Production*. London: Collins.

Eales, A. and Small, J. (1986). *Practical Lambing*. London: Longman.

Evans, G. and Maxwell, W. M. C. (1987). *Salamon's Artificial Insemination of Sheep and Goats*. Sydney: Butterworth.

Farm Building Information Centre (1983). *Housing Sheep*. Warwick: NAC.

Ferguson, I. B. (1982). Quality sheep housing. *Farm Building Progress* **69**: 5–6.

Fraser, A. F. (1968). *Reproductive Behaviour in Ungulates*. London: Academic Press.

Gordon, I. (1983). *Controlled Breeding in Farm Animals*. Oxford: Pergamon.

Hafez, E. S. E (ed.) (1969). *The Behaviour of Domestic Animals*, 2nd edition. London: Baillière Tindall and Cassell.

Haresign, W. (ed.) (1983). *Sheep Production*. London: Butterworth.

Kilgour, R. and Dalton, C. (1984). *Livestock Behaviour*. London: Granada.

MacCormack, J. A. D. and Sommer, M. (1988). *Farm Building Cost Guide 1988*. Aberdeen: Scottish Farm Buildings Investigation Unit.

Martin, W. B. (1983). *Diseases of Sheep*. Oxford: Blackwell Scientific.

Meat and Livestock Commission (1982). *Group Breeding Schemes for Sheep*. Milton Keynes: MLC.

Meat and Livestock Commission (1983). *Feeding the Ewe*. Milton Keynes: MLC.

Meat and Livestock Commission (1984). *Lamb Carcase Production*. Milton Keynes: MLC.

Meat and Livestock Commission (1986). *Sheep Yearbook*. Milton Keynes: MLC.

Ministry of Agriculture, Fisheries and Food (1977). *Codes of Recommendation for the Welfare of Livestock: Sheep*. L 705. London: HMSO.

Ministry of Agriculture, Fisheries and Food (1980). *A Simple Disposal Pit for Foetal Material and Small Carcasses*. L 648. London: HMSO.

Ministry of Agriculture, Fisheries and Food (1981a). *In-wintering the Lowland Flock*. B 2065. London: HMSO.

Ministry of Agriculture, Fisheries and Food (1981b). *Footrot in Sheep*. L 567. London: HMSO.

Ministry of Agriculture, Fisheries and Food (1982). *A System for Lowland Sheep*. B 2322. London: HMSO.

Ministry of Agriculture, Fisheries and Food (1983). *Grazing Management for Lowland Sheep*. B 2052. London: HMSO.

Ministry of Agriculture, Fisheries and Food (1984). *Store Lamb Finishing*. B 2394. London: HMSO.

Ministry of Agriculture, Fisheries and Food (1986). *Lamb Survival*. B 2525. London: HMSO.

Ministry of Agriculture, Fisheries and Food (1986). *Nutrient Allowances for Cattle and Sheep*. P 2087. London: HMSO.

National Sheep Association (1987). *British Sheep*, 7th edition. Ashford: NSA.

Nix, J. (published annually). *Farm Management Pocketbook*. Wye: Wye College.

Ryder, M. L. (1983). *Sheep and Man*. London: Duckworth.

Sainsbury, D. W. B. and Sainsbury, P. (1979). *Livestock Health and Housing*, 2nd edition. London: Baillière Tindall.

Speedy, A. W. (1980). *Sheep Production*. London: Longman.

UFAW (1983). *Stockmanship on the Farm*. Potters Bar: UFAW.

Wilkinson, J. M. (1984). *Milk and Meat from Grass*. London: Granada.

Williams, H. Ll. (1979). The sheep industry. In UFAW (ed.). *The Welfare of Food Animals*. Potters Bar: UFAW.

5 Goats

A. J. F. Russel and A. Mowlem

The industry in the UK

Goats are reputed to have come to Britain with Neolithic man some 5000 years ago and there have been many importations from the time of the Vikings to the present day. At one time, goats were more numerous in parts of Scotland than any other form of livestock but, with changing agricultural systems, numbers declined markedly in the nineteenth century.

There has been an interest in pedigree goat breeding in the UK for the last 100 years. This has recently been strengthened by a renewed awareness of goats' milk as a health food, by reactions against intensive animal production, and against food processing and the inclusion of additives in most food products, and, in the last few years, by an interest in goat fibre production.

Structure of the industry

Until recently, goats were not recognized as an agricultural species and were not included in the annual agricultural returns. The size of the UK goat population is not known, but has been variously estimated from about 75 000 (Locke, 1985) to 100 000 (Wilkinson and Stark, 1982).

The majority of goats are kept by smallholders for milk production and the average herd size has been estimated as approximately 9.5; only 0.6 per cent of herds have more than 100 goats (Locke, 1985). Larger herds of fibre-producing goats are currently being established.

Very few herds are maintained specifically for meat production. This is generally regarded as a by-product of both dairy and fibre-producing herds.

Productivity

Goat breeding has not been subjected to the degree of commercial pressure which has influenced the development of other livestock enterprises. However, selection for milk production has probably been associated with some loss of hardiness, and modern European dairy goat breeds require a high level of expert management to maintain productivity. The highest recorded milk yield is from a European breed, the Saanen, in Australia, where one animal produced 3500 kg (approximately 3500 litres) of milk in a single lactation. Most commercial dairy goat farmers aim to produce about 1000 kg (approximately 1000 litres) of milk per goat per annum.

Goat hair was used in the UK in the seventeenth and eighteenth centuries to make a wide variety of products ranging from ropes to legal wigs. In recent years there has been a renewal of interest in the commercial production of goat fibre, this time from two quite different types of animal, but both aimed at producing high value products – cashmere and mohair.

However, the goat fibre production industry in the UK is still in its infancy and it is premature at this stage to quote levels of production.

Breeds

The important breeds in the UK are the British

Saanen, the British Toggenburg and the British Alpine, all originating from Swiss breeds; the Anglo Nubian of Middle and Far East origin; the Angora, originally from Turkey; and the indigenous feral goat found mainly in western and northern Scotland.

The British Saanen is particularly suited to intensive farming for high levels of milk production, while the British Toggenburg and Anglo Nubian breeds are kept where milk quality is a primary consideration. Angora goats are farmed in many countries for the production of mohair fibre and have been imported in recent years from New Zealand, Tasmania and Canada. The potential importance of feral goats as producers of cashmere has only recently been appreciated and many of these animals are now being redomesticated.

Improvement schemes

Milk recording schemes and production classes at agricultural and specialist goat shows provide a stimulus to increase milk yields and improve composition through the choice of superior blood-lines.

Meat production, albeit as a by-product, is also receiving attention and consideration is being given to the possibility of mating the less productive dairy females to meat-type sires (Butler-Hogg and Mowlem, 1985), although the criteria by which conformation will be judged by the consumer remains to be established. It cannot be assumed that those characteristics which contribute to "good conformation" in other species, and particularly sheep, can necessarily be applied directly to goats.

The techniques of artificial insemination, using fresh and frozen semen, superovulation and embryo transfer have recently been applied to goats and are being used to increase the number of animals of certain breeds, such as the Angora, and to achieve more rapid genetic progress, particularly in fibre-producing goats and to a lesser extent in dairy breeds.

Influences on the industry

Goat milk production is not subject to the Milk and Dairies Regulations. However, in Scotland, there is a Department of Agriculture and Fisheries for Scotland (DAFS) Code of Practice in operation. A similar code is being drawn up to cover England and Wales.

As indicated above, one of the attractions of goats' milk and milk products is the non-intensive system of husbandry in which much of it is produced. Goats' milk has long been in demand from persons allergic to specific protein in cows' milk, and the relatively low fat content of goat meat may also stimulate a demand for this particular product from a more health-conscious consumer market.

The attainment within the UK and the EEC of self-sufficiency in some traditional forms of agricultural production and of over-supply in other commodities is currently stimulating interest in alternative enterprises. Mohair and cashmere production are particularly attractive in this respect as non-food enterprises, although these will, of course, generate meat as a by-product.

Goat production is more important in certain other European countries than it is in the UK. France, Greece and Spain have a combined goat population of over 8 million and there is a strong demand for goat meat in other EEC countries such as Italy. As the Common Agricultural Policy evolves, it is probable that there will be an increase in the legislation relating to goats and to the marketing of goat products.*

Natural history and normal behaviour patterns

Goats were among the first animals to be domesticated by man some 10 000 years ago and even now they exhibit an almost legendary independence which gives the impression that they tolerate domestication because it suits them.

The dairy breeds in Britain are prolific animals often producing twins or triplets, with individual birth weights of 2–5 kg. Given appropriate standards of nutrition, kids can grow at more than 200 g/day. Most dairy breeds have mature body weights of 60–70 kg for females

* *Codes of Recommendations for the Welfare of Livestock: Goats* (MAFF) are in preparation.

and 80–100 kg for males. Feral animals are smaller, females weighing about 35 kg and males 50 kg. Their kids are generally born as singles weighing some 2.5 kg and growing at less than 100 g/day.

In commercial herds, milking goats are usually kept for 6–8 years, although a recent survey established that the oldest goat in Britain was 19 years of age.

The developed dairy breeds have lost much of the hardiness of their ancestors and particularly dislike cold, wet weather. British feral goats have adapted by natural selection to withstand the extremes of climate, but even they do not like wet conditions and artificial shelter should be provided where natural shelter is not available.

Goats have a reputation for being inquisitive and their exploratory habits make them more difficult to contain within fences or houses than other livestock. Their ability to learn quickly can be used to advantage in modern milking systems using self-locking yokes.

Goats are naturally friendly animals and appear to benefit from the company of others. In the absence of other goats they will appreciate the presence of other species – even cats or chickens – and they always appear pleased to receive human attention. In any group of goats a hierarchy will develop, with specific animals becoming dominant over others. Bonds within family groups often last for life and, in adult herds, relatives such as sisters may often be observed to keep close company.

Although goats do not spend particularly long in grooming, they are naturally very clean animals. Their dry faeces are an advantage in this respect, and are an important factor in helping to maintain high standards of hygiene in the milking area.

They are noisy animals and will call out loudly when anything disturbs or interests them, such as the sight of their owner, a stranger, the feed bucket, or a dog. Females in oestrus are very vocal and this characteristic is useful as an aid to timely rebreeding.

Reproduction and breeding

Goats are seasonally polyoestrous, with females coming into heat at regular intervals during the breeding season which, in the northern hemisphere, extends from about August to February. However, it is not safe to assume that females will not exhibit oestrus or conceive at other times of year. Males are generally capable of breeding throughout the year, but libido and sperm production are likely to be poorer during the summer months.

Female kids become sexually mature and show oestrus from about 6 months of age, but are often not mated until approximately 18 months old, as yearlings. However, if they are well grown and have attained about 75 per cent of their expected mature weight, they can be mated earlier. Male kids are generally believed to attain sexual maturity at about 6 months, but there is ample evidence of 3-month-old kids being fertile. It is therefore important, where uncastrated kids are being kept for subsequent breeding, that a close watch is kept for signs of sexual activity, and that they are weaned and segregated from female stock before reaching sexual maturity.

The length of the oestrous cycle is about 21 days and oestrus generally lasts 1–2 days. The onset of cyclical activity in the autumn can be stimulated by the smell of the male. Oestrus is readily detected by behavioural changes including bleating, tail wagging and a swelling and reddening of the vulva. Where males are run with a herd of females, one male can be expected to serve 30–40 females. Where females are brought to the male for service, one male can serve 100–150 females or more in one season.

Gestation length varies from about 146–156 days, and averages some 150 days. Most females can be bred annually and there are reports of some animals kidding twice a year. This, however, is unusual. Dairy goats have an extended lactation, milking continuously for more than 18 months, and in some cases are mated only once every 2 years.

The number of kids produced varies according to breed, with feral stock in the UK seldom giving birth to more than one kid each year, while most dairy breeds commonly produce twins and some, such as the Anglo Nubian, have litters of up to four or more.

Pseudo-pregnancy, false pregnancy or "cloudburst", in which the female shows all the external and behavioural signs of being in kid but is, in fact, not pregnant, can occur. The type of ultrasonic scanning in which the uterus and its contents can be viewed on a TV-type screen, and which is now routinely used to determine foetal numbers in sheep, can be used to distinguish between true and false pregnancies as well as to count the number of kids carried.

Nutritional requirements

Goats are ruminants (for details of ruminant digestion see Chapters 2 and 3). Under natural conditions, the principal difference in the feeding behaviour of goats and sheep is that goats will readily browse on leaves and small branches of trees and shrubs, whereas sheep prefer to graze plant material nearer the ground. Given a wide choice of plant species, goats will tend to concentrate on those plants which are commonly regarded as weeds, such as willowherb, rushes, thistles and gorse, in preference to sown grass and clover. There is thus considerable scope for the complementary grazing of goats and sheep on certain types of pasture.

Despite their liking for roughage and plant material not readily grazed by other animals, goats are fastidious eaters and will refuse food which is not clean and palatable. They will not eat mouldy hay or food which has fallen from a rack or trough on to the ground. Neither can they be expected to produce well from diets comprising only low-quality roughages.

The goat's diet must be balanced in terms of energy, protein, vitamins and minerals. It should contain at least 40 per cent roughage and must be fed in relation to the animal's nutrient requirements, which in turn are related to its size and weight and to its level of production, whether that be of milk, live-weight gain, foetus or fibre.

For details of composition of diets and levels of feeding for different types of production the reader is referred to the sources given at the end of this chapter. It can only be stressed here that high levels of milk production from dairy goats require commensurately high levels of feeding. Goats kept for meat or fibre production have lower nutrient requirements than high yielding dairy goats, but adult females still require relatively high levels of feeding in late pregnancy and particularly during lactation when they are suckling kids. Similarly, kids destined for meat production will achieve high growth rates only when fed well.

Breeding males are too often left to look after themselves for much of the year, without any particular thought being given to their nutritional requirements. Special attention must be given to their feeding, particularly before and during the breeding season. They must not, however, be allowed to become too fat. It is particularly important that clean fresh water is available at all times to all types of stock.

Environmental requirements

Goats kept outdoors must have access to shelter at all times. Where herds are kept for meat and fibre production under extensive systems of management on rough hill country or with access to wooded areas, they may find adequate natural shelter. In most cases, however, it will be necessary to provide roofed shelters designed to afford protection from rain. All goats dislike wet weather and their coats provide less protection against wetting than does a sheep's fleece. Goats also seek shade on very hot days and will make use of shelters in all weathers. Unless the shelters can be moved frequently, their floors and surrounding areas become fouled and muddy very quickly. Where possible, shelters should be equipped with slatted floors to ensure that the animals have a clean and dry lying area.

The main environmental requirements for goats kept indoors, whether in purpose-built accommodation or adapted buildings, are that the housing should be clean, dry, well lit and well ventilated. Heating is not necessary, but the building must be free from draughts. Good draught-free ventilation can be provided by the use of spaced or Yorkshire boarding between a solid wall 1.5–2.0 m high at the eaves. In some locations, the space between such a wall and the eaves can be left completely open on the most sheltered side of the building, or covered with a light mesh or netting screen to keep out driving rain and snow. Any smell of ammonia in the atmosphere when the building is first entered in the morning is an indication of inadequate ventilation.

Goats must always have a dry floor to lie on. This can be bedded with clean straw, sawdust, wood-shavings, peat or some combination of these or other suitable materials. Alternatively, a raised solid or slatted wooden lying area is suitable and provides an object on which kids will play and obtain exercise.

Goats are sociable animals and, where individually penned, should be able to see readily other animals in adjoining pens. If no other goats are kept in the house they will accept animals of other species as companions. Where a goat has to be kept on its own it should be housed in a building from which it can see other animals or people.

Management of breeding males

Male goats are not difficult to manage and, if handled with care and respect, should not be aggressive. They are, however, extremely strong and thus require substantial housing. Block-built pens with metal gates are preferable. During the breeding season (August to February), they tend to go off their food and consequently lose condition. It is therefore important that they start the season in good condition. They have a repugnant smell, which is particularly noticeable in the breeding season.

Management of breeding females

Prior to the breeding season, female goats should be in good condition but must not be too fat. Fat goats tend to have lower fertility and lower milk yields. Young non-lactating goats have a tendency to become fat and care should therefore be taken when feeding concentrates.

The increasing possibilities for commercial goat farming have resulted in an interest in the ways of overcoming some of the natural limitations influencing reproduction. One major problem is the seasonality of breeding and the consequent seasonal lactation. Major retail outlets for milk require a year-round supply.

Out-of-season breeding is already practised by more than 5 per cent of commercial goat milk producers, and is attracting increasing interest. The two methods most commonly used are the use of intra-vaginal progesterone-impregnated pessaries (Table 5.1), and the use of artificially extended daylight during the winter. Ashbrook (1982) has recommended an extended light regime in which animals are exposed to 20 h of light per day for 60 consecutive days during the period January to March. Oestrus will occur approximately 10 weeks after return to ambient lighting.

Mating

Although the majority of goats are mated naturally, artificial insemination is becoming more common (Mowlem, 1983). In Britain most matings will be carefully selected, because many goats are pedigree animals. If a male is run with the females, a sheep raddle harness can be used to record mating dates. Artificial insemination using either fresh or frozen semen presents opportunities for extending the use of superior males. The female is restrained for insemination in a similar way to sheep, i.e. head down with the hind-quarters lifted up towards the inseminator.

Embryo transfer

The technique of transferring fertile embryos from donor to recipient females is now being practised in goat farming. Licences are not required for this technique, but it must be carried out by a veterinary surgeon. It is of particular value where a rapid increase in the numbers of a desirable breed or type is required. Up to 10 or more embryos may be removed from a donor female to be implanted, usually in pairs, into non-pedigree recipient females which act as surrogate mothers. This considerably increases the reproductive potential of one donor female. The technique requires the synchronization of the recipient and donor females' oestrous cycles using intra-vaginal pessaries and PMSG injections as described above.

Feeding

The requirements of non-lactating females can

Table 5.1
Alternative regimes for out-of-season induction of oestrus, using progesterone-impregnated intra-vaginal pessaries

	Geary (1985) (for good milker)	Corteel *et al.* (1982)
Progesterone pessaries (number of days inserted)	21	11
Prostaglandin injection	Not given	48 h before pessary removal
PMSG injection[a]	48 h before pessary removal	48 h before pessary removal
PMSG dose	600 i.u.	500 i.u.

[a] Pregnant mares' serum gonadotrophin.

be met from good quality forage. Milking animals will require concentrate supplements and, unless grossly overfed, are unlikely to become too fat.

During late pregnancy it is important to feed sufficient nutrients to meet the needs of the growing foetuses and to ensure a good subsequent milk yield. It is generally recommended that the energy and protein intakes should be increased during the last month of gestation (Oldham and Mowlem, 1981). Also, animals encouraged to consume a high level of forage during pregnancy will be conditioned to a high forage intake during lactation and this will benefit both milk yield and quality. An average dairy goat of around 65 kg live-weight should be offered a good quality forage *ad lib.* and, 2 months before parturition, 250 g of dairy concentrate/day rising to 450 g/day 1 month before parturition.

All goats require clean water which can usually be provided in a self-filling trough.

Housing

Related goats or long-established groups can be housed together. In fact, being companionable animals, they usually benefit from this. Much time will be spent lying together, particularly during late gestation. Care should be taken if some animals have horns, as injuries can occur, particularly during feeding. It is important that hooves are well trimmed at this time. Projections which could injure heavily pregnant animals should be avoided. (For further information on housing, see p. 133.)

Parturition and suckling

Ideally, when parturition is due, goats should be housed separately in a clean straw-bedded pen 2–2.5 m^2. The first sign that parturition is imminent is the increased engorgement of the udder. Twenty four hours before parturition the relaxation of the pelvis can be felt as hollows on either side of the tail. Sometimes a discharge will be seen and, providing this is clean and odourless, it is quite normal. The goat may become restless and may spend a lot of time pawing the ground and alternately lying down and standing up.

Most goats kid lying down and, once they begin to strain, regular checks should be made about every 15 min. If a goat has been observed to strain for more than 1 h without anything apparently happening it can be assumed that there is a problem. Unless the stockman is experienced at coping with difficult births, expert help from a shepherd, veterinarian or other such person should be sought. A great deal of damage can be done by inexperienced people attempting to assist an animal at this time.

Once the kids are born, it is important to ensure that the mother has the opportunity to lick them dry immediately. This provides essential stimulation for the kids and helps establish the mother–offsping bond. It is also important that the kids feed within 2–3 h of birth. Kids that are too weak to suck from their mothers should be fed by stomach tube (see Chapter 4, p. 103), after which they will usually gain enough strength to suck on their own. Some goats have such pendulous udders that it is

Fig. 5.1 Saanen kids with liquid feed dispenser (Lamb Bar Feeder available from Dalton Supplies Ltd). (Photograph: British Denkavit.)

impossible for the kids to feed. In these cases the kids should be artificially reared, but it is most important that they are fed colostrum for the first 24 h. It is useful to save some colostrum in a deep freeze for emergencies.

Rearing of kids

In commercial dairy units, most kids, whether intended for herd replacements or for meat, are taken from their mothers after receiving colostrum for 24 h. They can then be reared on a proprietary milk replacer. Many milk replacers formulated for calves and lambs, and a few for goats, are available – they are all based on either skimmed cow milk or cow milk and, for large-scale rearing, the less expensive calf milk replacers are often used. There is evidence, however, that the use of calf milk replacers can cause toxicity in Angora kids (Humphries *et al.*, 1987) and it is therefore recommended that a lamb milk-replacer be used for kids of this breed.

Kids can be weaned at 6–8 weeks (Mowlem, 1984). A suitable feeding regime is outlined in Table 5.2. For replacement female kids, it may be considered worthwhile to wean a little later, at 8–10 weeks, to ensure a good early growth rate. Some herd replacements may not go out until the following spring as kids grow less well when put out to grass early in life.

Table 5.2
Feeding regime for artificial rearing of goat kids (figures in parentheses show ages for a later weaning system)

Age	Milk feeds
Weeks 1–4 (1–6)	Supplied *ad libitum*
Weeks 5–7 (7)	½ amount consumed end of week 4 (6)
Week 6 (8)	½ amount consumed end of week 5 (7)

For *ad lib.* feeding it is best to offer the milk in a "lamb bar" type feeder where several kids can suck from several teats in one milk container (see Fig. 5.1). Feeders should be fitted with lids to prevent the milk becoming fouled by bedding, faeces, etc. Automatic calf-rearing machines can also be used. Milk replacers can be fed either warm or cold, but the latter usually gives the better results. A rearing concentrate containing 16–18 per cent crude protein, with good quality hay and clean water, should be offered from 2 to 3 weeks of age. Any dirty or fouled food should be regularly replaced. At weaning, kids should be eating 400–500 g/day of the concentrate feed, if a growth check is to be avoided.

Castration of males not required for breeding, and disbudding should be carried out during the first week of life (see p. 138).

Goats kept for fibre production may be run as a suckler herd, with the kids remaining with their mothers until weaned at around 12–14 weeks of age. Any male kids left entire should be weaned before attaining puberty and segregated from all female stock before they are 12 weeks old.

Kids enjoy the company of others, and should never be reared in single pens. Housing need not be elaborate – pens that are free from draughts and are provided with plenty of clean straw are all that is required (see Fig. 5.2). If the building is not particularly cold, kids penned in groups of 12 should not require supplementary heating.

Kids are very adept at finding their way out of pens, and therefore solid partitions should be used and pen sides should be about 1.2 m high. Approximately 0.5 m^2 of floor space should be allowed per kid up to 2 months of age, increasing to approximately 1.5 m^2 at 6 months of age. Care should be taken to ensure that there are no projections in the pens, or gaps where legs or feet might become trapped. Hay nets are not recommended as kids may jump up on them, catching and breaking their legs.

Infection is likely to spread through groups of intensively reared kids, and it is important to observe a high standard of hygiene and husbandry. Kids should be checked constantly for signs of ill health, and appropriate action taken as soon as disease or infection is suspected.

Milk production

The choice of milking system depends largely on the scale of the enterprise. Those who keep just a few goats will probably milk by hand, while

Fig. 5.2 Angora kids in pen. (Photograph: British Denkavit.)

those with large herds will use a milking machine. The choice of milking machine system is as varied as for dairy cows, ranging from simple small-scale bucket units to large sophisticated parlours (Fig. 5.3). Milking machines not only make milking large numbers easier but, if well-managed, may be instrumental in producing better quality milk containing fewer spoiling microorganisms.

Feeding systems

While it is not necessary as part of the milking routine, it is convenient to feed concentrates to goats when they are being milked, and it also allows more control over individual feed intakes. Automatic, in-parlour feeding systems are available for dispensing concentrates, but on economic grounds are probably not justified in other than the largest herds. Forage is fed in the yards, pens or paddocks.

Whether a small bucket system or a full parlour is operated, it is likely that a yoking system will be used to restrain the goats during milking and feed troughs or buckets can be attached to this.

Water is also an important requirement of milking goats and they should be provided with clean water at all times.

Housing

It is most economical to house goats in groups. This saves space, labour and bedding.

For loose housing, approximately 2 m^2 should be allowed per goat. If the animals are familiar with each other, there appears to be no upper limit to the number that can be housed together. New herds made up of unrelated or unfamiliar animals should be allowed more space until they have settled down, otherwise the resulting stress may affect their health and

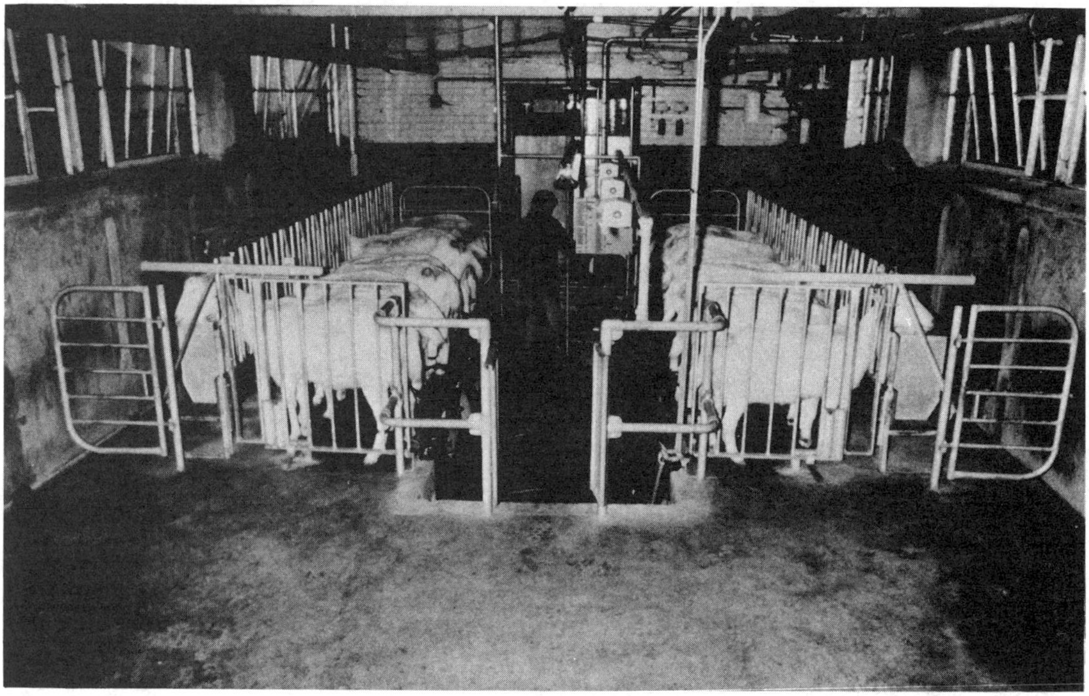

Fig. 5.3 Milking parlour.

almost certainly their milk yields.

Deep-litter straw bedding is suitable, but other materials, such as wood chips, can be used, except at parturition. Such materials stick to newborn kids and may dissuade the mother from cleaning them.

When designing goat buildings or modifying existing ones it should be borne in mind that goats can be destructive. They will chew soft material such as wood, rubber or plastic, and may reach up on their hind legs to a height of more than 2 m. Timber parts of buildings such as doors and door frames are best clad in sheet metal to avoid such damage. Feed troughs should be constructed of metal, and hay racks of heavy weld mesh or wooden slats. Routes in and out of buildings, and particularly the milking area, should be simple and straightforward for efficient operation.

Milking systems

Anyone contemplating setting up a dairy goat enterprise is advised to assume that, even if they begin with only a few milking animals, the herd will increase in size. At its simplest, a milking area can comprise a clean space away from the bedded pens, where the goat can be tethered and milked. Some goat-keepers construct a raised milking platform with a feed bucket at one end and a hole at the other to take the milking bucket. For machine milking, a raised platform on which a number of goats will stand is required. As goats readily climb ramps or steps it is not necessary to have an operator's pit, although, if adapting a cow parlour, one may already be available. Steps leading up to the standings should have treads 540 mm deep with 140-mm risers and those leading down should have 800-mm deep treads with 410-mm risers. Care should be taken to ensure the steps and standings do not become slippery when wet.

Milking machines for goats are in principle similar to those for cows. However, the mechanical settings are different (see Table 5.3), and it is important that this is noted if adapting cow-milking equipment. Goats milk quickly, and one operator cannot cope with more than six milking units. Little work has been done with automatic cluster removers with goats, and it is possible that the use of these could increase the number of units that one person could manage. Without automatic cluster removal, one person can milk approximately 40 goats/h in a 6-point, side-by-side milking parlour (see Fig. 5.3).

Table 5.3
Mechanical settings for milking machines for goats

Vacuum	Pulsation rate	Pulsation ratio
37 kPa(11″Hg)	70–90 p.p.m.	50 : 50

Like cows, goats become used to particular routines. Changes in these, or continual changes in milking personnel, cause stress which may in turn cause a fall in yield or, in extreme cases, death. Production often suffers when new goats are introduced into larger herds and, because of this, some farmers have adopted a policy of herd replacement from home-bred animals only. To prevent contamination of the milk, or infection of the udder, it is advisable, immediately before milking, to wash the udder with an approved disinfectant solution and to dry it with a disposable paper wipe. It has been argued that this procedure is unnecessary with clean goats coming straight from pasture. Certainly washing with plain water, not drying after washing, or drying all udders with the same cloth are likely to cause more problems with contamination of the milk than if the udder is not washed at all. Another sensible precaution is for the operators to wear protective overalls and hats reserved solely for working in the milking area.

In some countries, the milking equipment is often in the same area as the bedded pens. This is not recommended, because bedding and bedding dust will be a constant source of bacterial contamination. The milking parlour and milk-handling room should be sited some distance away from the goat pens.

Immediately after a goat has finished being milked, its teats should be dipped into a suitable disinfectant solution as a precaution against mastitis. A 4 per cent solution of sodium hypochlorite or a proprietary iodophor-based teat dip will be suitable for this purpose. A useful check for early signs of mastitis is to milk a stream of milk from each teat at the onset of milking into a strip cup which contains a black rubber disc which will show any clots in the milk. Clots are usually, but not always, the first signs of udder infection (see Chapter 2, p. 42).

Milk handling

It is important to transfer milk from each animal into a cooling system as quickly as possible. Ideally this should be a refrigerated tank, but in-churn coolers operated by running mains water are also quite effective. If the milk is not quickly cooled, the lipolytic enzymes present will break down the fats to produce off-flavoured fatty acids, giving the milk what is, to some consumers, an undesirable strong goaty taint. This lipolysis can also be brought about by excessive mechanical agitation when raw milk is pumped through elaborate systems of pipes.

All surfaces which come into contact with the milk, including milking machines, churns, buckets, tanks and pipe lines, must be cleaned thoroughly to prevent build up of bacteria and subsequent spoilage of the milk (Cousins and McKinnon, 1979). Proprietary dairy solutions are available for this purpose and the manufacturers should be consulted for the most appropriate cleaning routine. MAFF and DAFS Dairy Husbandry Advisers will also give advice on milking and dairy routines.

Fibre production

Most goats have two coats, a coarse, hairy outer or guard coat and a fine, soft undercoat which is cashmere. The quantities of cashmere carried by most modern breeds of dairy goats are negligible, but the feral goats found in many of the mountainous parts of the country have significant quantities of very fine cashmere. Levels of production, averaging about 100 g/animal, are too low for commercial production but are being improved by selective breeding and by crossing with imported stock. The Angora, on the other hand, has only a single coat which is

Fig. 5.4 Young Angora goats in polythene-covered shelter with open sides.

mohair. This fibre is coarser than cashmere but is produced in greater quantities, averaging about 4 kg/goat/year. In contrast to wool, which is generally much coarser than either cashmere or mohair and has a marked crimp or tight curl, goat fibres are straight or only slightly waved. That, and their different structure, enables them to be used in the manufacture of fine, light-weight garments which are much in demand and command high prices.

Mohair

Although the importation of Angora goats to the UK in 1981 was not the first attempt to establish the breed in this country, it appears at this stage to be the most successful, and numbers have expanded very rapidly. There has not yet been time, however, for systems of mohair production to have been developed. It would, none the less, appear likely that Angora goats kept for fibre production, as distinct from the breeding of pedigree stock, will be run in substantially sized herds and managed in similar ways to sheep flocks. The principal difference will be the requirement for shelter, if not housing, because the Angora's fleece does not afford the same protection against wet conditions as the sheep's more greasy wool (see Fig. 5.4).

Mohair is a continuously growing fibre which is harvested by shearing every 6 months. In some herds the age structure will probably be similar to that of a sheep flock with the fibre being harvested from the adult breeding stock and yearlings, and with perhaps one shearing being taken from those kids not required as herd replacements, before they are slaughtered for meat. In other situations the herd may contain a high proportion of adult castrates kept primarily for their fibre and sold for meat at an older age. (For rearing of kids, see p. 132.)

Fig. 5.5 Imported Tasmanian cashmere bucks with Scottish feral does.

Cashmere

The feral goats in the UK probably originate from stock which were among the first animals to be domesticated in the country, but their potential as producers of cashmere has only recently been appreciated (Russel *et al.*, 1986). The development of a cashmere production industry, based on feral goats crossed with imported stock (Fig. 5.5), is at an even earlier stage than that of mohair production, and again clearly defined systems of management have not yet been established.

The majority of cashmere-producing goats will almost certainly be run in similar systems to sheep and in many cases will be integrated with established sheep enterprises on upland and hill farms. Although these goats live naturally in some of the harshest conditions, farmed herds will require to be provided with some form of shelter from rain, unless the terrain is exceptionally well endowed with adequate topographic shelter. These artificial shelters need not be elaborate, but, unless on a very freely-draining site, they should have slatted floors to prevent fouling and poaching.

Cashmere is the very fine undercoat found beneath the coarse outer hair and, unlike mohair, it grows seasonally and is shed in the late winter and early spring. The fibre must therefore be harvested at that time of year, otherwise it will fall out and be lost. In the recently developed cashmere production industries in Australia and New Zealand the fibre is harvested by shearing, but in the UK this would necessitate the animals being housed for extended periods with consequent increases in costs. The female breeding stock are also more likely to be pregnant at the time of harvesting and the stress of removing the entire fleece might cause abortion. Present evidence suggests that cashmere harvesting in the conditions in which these goats are likely to be kept will most probably be by combing (Russel *et al.*, 1986). It is also possible that lightweight coats of a woven synthetic material which allows the passage of water vapour, could be put on the animals prior to shedding of the undercoat and used to contain the fibre and make combing at a later date an easier and less time-consuming operation.

Cashmere will be harvested from adult breeding stock, and most probably also from the previous season's kids, at about 10–12 months of age. Those young stock not required for breeding may be sold for slaughter at that point or kept for a further 6 months, being used, for example, to control or eliminate weeds from sown pasture (Russel *et al.*, 1983, Grant *et al.*, 1984), and to make further live-weight gains from relatively inexpensive grazing. It is also probable that, in some situations, herds of castrates may be maintained for purposes of grazing management and the production of cashmere and meat.

Meat production

Goat meat is always lean and, at a time when there is discrimination against animal fat, it is being consumed in increasing amounts. Most of the goat meat presently produced comes from kids, usually males, which are surplus to dairy herd requirements. Although there is a limited specialist market for entire males, their meat is liable to carry the strong characteristic taint of such animals, which most people find unpleasant. Castrated males are more generally acceptable for meat production, and are also more easily managed. (For rearing, see p. 132.)

General care and handling

Goats require similar environments, feeding and husbandry to other types of ruminant livestock. Their independent nature and behaviour, however, is different from that of other livestock and it is only after much experience that the understanding necessary for successful management is acquired. Anyone contemplating starting a goat unit, however small, is advised to seek out another goat farmer who would allow them to gain some experience by working with their goats even for a few days.

Handling and restraint

Dairy goats are usually naturally inquisitive and will crowd round their handler. It is thus relatively easy to catch individual animals. Once caught they can be restrained by holding firmly but carefully around the neck just behind the head. It is advisable to attach a collar and lead if an animal is to be restrained for some time or is to be moved.

Unlike sheep, groups of dairy goats cannot be driven, and are most easily moved by being led. It is, however, advisable to have a second person walk behind the group to move along those stopping to investigate points of interest. Feral goats which have been taken from their natural habitat are more difficult to handle, but they too learn very quickly and can be worked like sheep by dogs. Feral kids reared as pets or in small groups are as tractable as dairy goats. Redomesticated feral goats kept in substantial numbers for cashmere production can be managed as would a sheep flock, but the sides of pens and handling yards require to be higher and more secure than those designed solely for sheep.

Routine procedures

During the first week of life, males not required for breeding can be castrated using the rubber ring method. Care must be taken to ensure that both testes have descended into the scrotum before the rubber ring is applied.

It is also desirable to disbud kids within 1 week of birth. Goat kids' horns grow more vigorously than those of calves and more care is needed. The kids are anaesthetized, ideally with a general anaesthetic, and the horn bud is cauterized using an iron with a flat head approximately 2.2 cm in diameter that has been heated to become red hot (Buttle *et al.*, 1986). In Britain, goats of any age may be disbudded or dehorned only by a qualified veterinarian.

Goats' feet, and particularly those of dairy breeds, require to be trimmed regularly to prevent lameness due to overgrowth, especially with heavily pregnant animals. Foot rot is less of a problem in goats than in sheep. The technique for hoof trimming is the same as for sheep, using foot-trimming shears and/or a sharp knife. Unlike sheep, goats should not cast, but should be tethered and have the feet trimmed in a "horse shoeing" position.

Identification marking of goats, particularly white ones, can be a problem, because ear-tags and other identification aids tend to be chewed. Ear tattooing is the only permanent method and is a requirement of the British Goat Society for registered pedigree stock. Coloured animals can be freeze-branded.

Oral drenching is carried out in the same manner as for sheep and using the same type of drenching gun. Vaccinations are relatively straightforward because of the thin covering of hair and lack of subcutaneous fat, and are generally given in the loose skin around the neck.

The jugular vein is easy to locate in the goat, making the collection of blood samples also relatively simple.

Goats are able to reach almost all parts of their body with their mouths and it is therefore difficult to apply dressings in such a way that they will not be pulled off. The only completely satisfactory method is to restrain the goat so that it cannot turn round. This can be achieved by using a halter or by temporarily confining it in a narrow pen which restricts movement.

Transport

If goats are to be moved long distances, a suitable vehicle must be used. Goats generally travel well, being able to maintain their balance easily, provided they do not slip on the floor. The ideal covering for a truck or van floor is a rubber mat covered with a thin layer of straw. Space should be adequate but not excessive, otherwise the animal may fall if the vehicle stops or turns suddenly. In the interests of safety, a partition must be fitted between the driving cab and the back of the truck or van.

Health and disease

The maintenance of a healthy herd demands a high standard of stockmanship and the ability to recognize diseases and disorders in their early stages so that proper attention and treatment can be given promptly. All animals must be observed at least daily by an experienced person who can recognize signs of ill-health. It is important not only to recognize the obvious signs such as diarrhoea, lameness and laboured breathing, but also to know when animals are showing atypical behaviour or signs of discomfort. The stockman must be able to judge what conditions are amenable to treatment by non-veterinarians, and when it is essential to call on the services of the veterinary surgeon.

Proper feeding, the provision of clean, dry and well ventilated accommodation, and strict adherence to a preventive medicine programme of vaccination and dosing, designed, in consultation with the local veterinary practice, to suit the particular herd requirements, will minimize disease problems and make a positive contribution to herd health.

Bacterial and viral diseases

All adult and young stock should be vaccinated regularly to provide protection against clostridial diseases, such as enterotoxaemia, pulpy-kidney and tetanus. Where pneumonia is a persistent problem, tests should be carried out to determine whether the causal organism is one for which there is a vaccine.

Foot rot should be controlled through good management including regular foot examination, hoof trimming, timely treatment of affected animals and the frequent use of footbaths. Where these measures fail to control the problem, vaccination should be considered.

Louping ill is a disease caused by a virus transmitted by ticks. In dairy goats there is a potential danger to human health, because the virus can be secreted in the milk of apparently healthy animals. The disease can be prevented by vaccination, and advice on the occurrence of louping ill in a particular area should be sought from the local veterinary practice.

Mastitis can be caused by any one of a number of organisms which gain access to the udder through poor hygiene, bad milking technique or injury. In severe mastitis the udder is destroyed and the animal may die. If diagnosed in the early stages, most cases respond to antibiotic treatment applied directly into the teat, although sometimes this may require to be supplemented by intramuscular injection.

Johne's disease generally occurs in adult

goats and is characterized by a general ill-thrift and weight loss progressing to emaciation and death. There is no treatment for the disease and, if it is suspected, it is imperative that tests be carried out so that all reactors can be culled immediately.

Parasites

Internal and external parasites are a common cause of ill-health and poor performance in goats and, in some cases, can lead to death. There is some evidence to indicate that goats do not develop as good resistance to internal parasites as do cattle and sheep. Virtually all stock harbour various species of parasitic worms, ingested from the pasture as larvae, in the stomach and intestines, and all adults and kids should be dosed regularly with a suitable anthelmintic. The more intensive the stocking, the more frequent will be the requirement for dosing, and in some cases it may be necessary to dose kids every 3–4 weeks during the summer months. Zero grazing is sometimes practised where these parasites are a particular problem.

Other internal parasites which should be considered when designing a routine dosing programme include lungworm, coccidia and, particularly in wetter areas, liver fluke.

Of the external parasites, lice are particularly common on goats but are readily controlled by dipping, dusting with an appropriate powder or treatment with "pour-on" preparations. Mange is rarer, but much more difficult to control, and requires frequent treatment at short intervals over a period of months.

Where milk is produced for human consumption, particular attention must be paid to the need to withhold milk from sale for varying periods after the administration of certain preparations used to control parasites. Veterinary advice should be sought as to the most suitable preparations to use.

Metabolic disorders

Metabolic disorders such as pregnancy toxaemia, hypocalcaemia (milk fever) and hypomagnesaemia (grass tetany) are the result of dietary inadequacies, although this does not necessarily imply that the level of feeding *per se* has been unduly low. Pregnancy toxaemia is most likely to occur in the final weeks before kidding, and is most common in overfat females carrying multiple foetuses. To prevent excessive mobilization of body fat reserves, energy intakes should be increased as kidding approaches, using, for example, good quality concentrates. To prevent hypocalcaemia, a *low* calcium diet should be fed prior to kidding to stimulate mobilization of the calcium reserves in the skeleton; normal feeding with a calcium-supplemented diet should be resumed after kidding. Hypomagnesaemia is most likely to occur in goats grazing lush pastures in the spring, and can be prevented by supplying additional magnesium, either in a magnesium-fortified concentrate or by addition to the drinking water. Feeding some hay or straw can also be effective in the prevention of this condition.

On-farm slaughter

For those who do not wish to rear kids, particularly males, a suitable method of euthanasia will be required. Kids can be killed satisfactorily using an overdose of barbiturates or a humane killer. These may not be available to many goatkeepers, in which case it will be necessary to seek the help of a veterinary surgeon. It should be remembered that if barbiturates are used, the carcass must not be fed to dogs or other animals.

If an animal is to be slaughtered to provide meat for home consumption, or has to be destroyed because it is seriously ill or injured, this should be carried out either by a veterinary surgeon or by a person from an abattoir which provides an emergency service.

The way ahead

There is a small but growing market in the UK for goat dairy products produced under as near as possible "natural" conditions, and the future for specialist, small-scale dairy-goat enterprises appears to be reasonably secure. As farmers seek alternatives to conventional agricultural commodities, there is likely to be an increase in larger-scale, more intensive enterprises concentrating solely on production and relying on others to undertake the processing and marketing of their product. There appears to be a sufficient demand for goat dairy products for both types of enterprise to co-exist, so that in the future the range of sizes of dairy-goat enterprises is likely to increase.

At present there is a potential annual supply of some 2000 tonnes of goat meat in the UK, but many of the animals which could contribute to this are destroyed at birth because goat farmers either cannot identify a market, supply a market on a regular basis, or obtain an economic price for this product. At the same time, there is a considerable unsatisfied demand for goat meat. There would thus appear to be a need for specialist goat-rearing enterprises which would bridge the gap between supply and demand. These would pay producers a fair price for surplus kids and would be of a sufficient size to be able to supply markets on a regular basis and thus command realistic and reasonable end-product prices.

The rapidly expanding interest in mohair and cashmere production is likely to result in a further substantial increase in goat numbers.

All these factors combine to point to a significant increase in the goat population of the UK over at least the next decade. Such an expansion will bring in its train both problems and opportunities, and it is important that goat husbandry and welfare be incorporated into the curricula of colleges and universities, and included in the teaching not only of those who will be farming goats, but also those who will be expected to provide specialist research, advisory and veterinary support.

References

Ashbrook, P. F. (1982). Year-round breeding for uniform milk production. In *Proceedings of 3rd International Conference on Goat Production*, pp. 153–154. Scottsdale, Arizona: Dairy Goat Journal.

Butler-Hogg, B. W. and Mowlem, A. (1985). Carcase quality in British Saanen goats. *Animal Production* **40**: 752 (Abstract).

Buttle, H., Mowlem, A. and Mews, A. (1986). Disbudding and dehorning goats. *In Practice* **8**: 63–65.

Corteel, J. M., Gonzalez, C. and Nunes, J. F. (1982). Research and development in the control of reproduction. In *Proceedings of 3rd International Conference on Goat Production*. pp. 584–601. Scottsdale, Arizona: Dairy Goat Journal.

Cousins, C. M. and McKinnon, C. H. (1979). Cleaning and disinfection in milk production. In Thiel, C. C. and Dodd, F. H. (eds), *Machine Milking*, Technical Bulletin No. 1. Reading: National Institute for Research in Dairying.

Geary, M. R. (1985). *Use of Chronogest Sponges and PMSG (Folligon) in Goats*. Cambridge: Intervet Laboratories.

Grant, S. A., Bolton, G. R. and Russel, A. J. F. (1984). The utilisation of sown and indigenous plant species by sheep and goats grazing hill pastures. *Grass and Forage Science* **39**: 739–751.

Humphries, W. R., Morrice, P. C. and Mitchell, A. N. (1987). Copper poisoning in Angora goats. *Veterinary Record* **121**: 231.

Locke, D. G. F. (1985). *Goat Husbandry Survey Report*. Argyll: Islay and Jura Goat Society.

Mowlem, A. (1983). Development of goat artificial insemination in the UK. *British Goat Society Yearbook*, pp. 4–6.

Mowlem, A. (1984). Artificial rearing of kids. *Goat Veterinary Society Journal* **5** (2): 25–30.

Oldham, J. D. and Mowlem, A. (1981). Feeding goats for milk production. *Goat Veterinary Society Journal* **2**(1): 13–19.

Russel, A. J. F., Maxwell, T. J. Bolton, G. R., Currie, D. C. and White, I. R. (1983). A note on the possible use of goats in hill sheep grazing systems. *Animal Production* **36**: 313–316.

Russel, A. J. F., Lippert, M., Ryder, M. L. and Grant, S. A. (1986). Goat production in the hills and uplands. *Hill Farming Research Organisation Biennial Report 1984–85*: 135–141.

Wilkinson, J. M. and Stark, B. A. (1982). *Some Possibilities for Goat Production in the UK*. London: Ministry of Agriculture, Fisheries and Food.

Wood, J. D. (1984). Composition and eating quality of goats meat. In Wilkinson, J. M. and Mowlem, A. (eds). *Proceedings of the Inaugural Conference of the Goat Producers' Association* pp. 7–8. Reading: Animal and Grassland Research Institute.

General reading

Books

Devendra, C. and Burns, M. (1983). *Goat Production in the Tropics*. Farnham: Commonwealth Agricultural Bureaux.

Dunn, P. (1982). *The Goatkeeper's Veterinary Book.* Ipswich: Farming Press.

Gall, C. (ed.) (1981). *Goat Production.* London: Academic Press.

Guss, S. (1977). *Management and Disease of Dairy Goats.* Arizona: Dairy Goat Publishing Corporation.

Hetherington, L. (1979). *All About Goats.* Ipswich: Farming Press.

MacKenzie, D. (1980). *Goat Husbandry*, 4th edition. London: Faber and Faber.

Ministry of Agriculture, Fisheries and Food (in prep.). *Codes of Recommendations for the Welfare of Livestock: Goats.*

Salmon, J. (1981). *Goat Keeper's Guide.* Newton Abbot: David and Charles.

Thear, K. (1985). *Commercial Goat Keeping.* Saffron Walden: Broad Leys.

Wilkinson, J. M. and Stark, B. A. (1987). *Commercial Goat Production.* Oxford: BSP Professional Books.

Conference Proceedings

Proceedings of the 3rd International Conference on Goat Production, College of Agriculture, University of Arizona, Tucson, January 1982. Scottsdale, Arizona: Dairy Goat Journal.

Development in Goat Production 1984: Proceedings of the Inaugural Conference of the Goat Producers' Association. Reading: Animal and Grassland Research Instutute.

Development in Goat Production 1985: Proceedings of the 2nd Conference of the Goat Producers' Association. Reading: Animal and Grassland Research Instutute.

Nutrition and Systems of Goat Feeding: Proceedings of Symposium, May 1981. Paris: ITOVIC (149 Rue de Bercy).

Journals

Goat Veterinary Society Journal. Department of Clinical Medicine, Madingley Road, Cambridge.

Monthly Journal of the British Goat Society. Lion House, Rougham, Bury St. Edmunds.

Magazines

Home Farm. Saffron Walden: Broad Leys.

Smallholder & Goatkeeper. Stowmarket: Charter Publications.

6 Pigs

I. J. Lean

The industry in the UK

Structure of the industry

Modern pig production is a blend of business ability, good husbandry and the rapid adoption of new technology. Efficient pig production aims to convert feed to lean meat in such a way as to make maximum profit for the producer. There is considerable variation between producers and between their products, and efficient producers maintain a margin on their products even when national production statistics suggest that the market is grey and uncertain.

The pig industry has, since the Second World War, been characterized by a decrease in the number of individual pig producers, coupled with an almost four-fold increase in the national herd size. Units have become specialized, using relatively small acreages of land and relatively sophisticated intensive housing. This increase in herd size and specialization was fuelled in the late 1940s by price incentives backed by Exchequer support, followed in the 1950s by a move to more economic output which took advantage of technological change (Fig. 6.1).

The industry has always reacted quickly to fluctuations in market price. This has led to a pig cycle of production – erratic swings between over-supply with low prices and under-supply with high prices. This has been accentuated by the relative ease with which pigs intended for slaughter can be kept on as breeding animals, making it easy for producers to enter the industry or to increase production, and by the rapid and prolific breeding cycle of the pig.

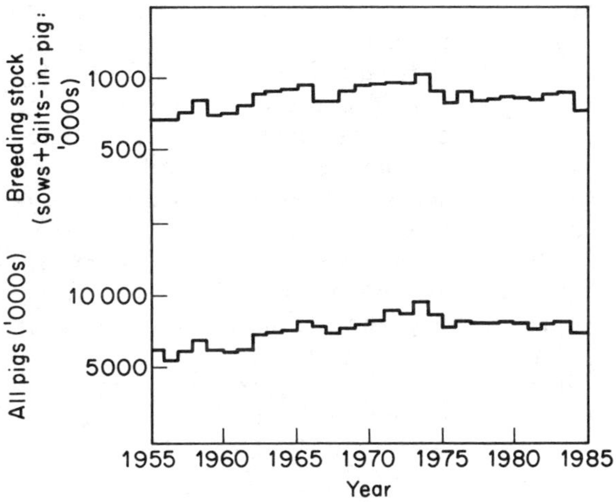

Fig. 6.1 UK pig populations 1955 to 1985.

Stabilization of prices was attempted in the 1960s and early 1970s by use of a flexible guarantee scheme, but this was abandoned following EEC entry in 1973, its smoothing effects were lost and, once again, pig populations fluctuate in relation to market conditions. One consequence of the pig cycle, coupled with the increasingly stringent requirements of the pig meat market, has been to speed the decline of the small non-specialist pig producer, because short-term fluctuations have greatest effect on small herds.

About 50 per cent of all the meat eaten in the world comes from pigs. In Europe about one-third of the production is eaten fresh as pork, one-third as cured bacon or ham, and the remainder is processed into pies, sausages and other manufactured products.

Pigs are sold for slaughter between 55 and 125 kg live-weight. Consumer demand is for lean meat, hence genetic, feeding and environmental aspects are all of importance in producing, in the optimum time, a carcass with a low quantity of subcutaneous fat. The efficient producer has to manipulate these factors continuously if his animals are going to be slaughtered in the most financially attractive markets.

In the UK, slaughter pigs fall into four main categories: the pork pig (55–62 kg) for small joints on the bone, sold as fresh meat; the cutter (64–82 kg) trimmed of fat and skin, prepared for multiple retail shops and supermarkets, with some of the carcass used for manufacturing purposes; the bacon pig (90–100 kg) traditionally cured; and the heavy hog (100–125 kg) often trimmed of skin and fat and most frequently used for manufacturing into sausages, pies and processed meats. Culled adult stock – both sows and boars – are also used for manufacturing purposes.

The aim, in all cases, is to produce a lean carcass. Most of the fat in a pig is found subcutaneously, and this has led to the development of comprehensive payment schemes for carcasses based on subcutaneous fat measurements, weight and, in the case of bacon pigs, a minimum length (see MLC, 1987 for details of carcass classification).

Breeds

There are three principal white breeds of pig in the UK: Large White, Landrace, and Welsh. Breed differences are not very obvious other than in relation to the position of the ears, which are pricked up in the Large White and pointing forward in the other two breeds. The Large White is usually considered to be slightly more prolific and faster growing, with meat of a better quality. British Landrace stock are derived from Scandinavian importations, but Landrace from other European countries have also been imported for breeding. A number of other British breeds, such as the British Saddleback, Tamworth, Gloucester Old Spot and Large Black, have a more limited following. These are often kept on outdoor extensive systems.

Exotic breeds have also been imported in recent years. These are generally coloured breeds, characterized by shorter length, greater eye muscles and larger hams, and include Pietrain, Hampshire, Duroc, Lacombe and Poland-China. These have been crossed into British breeds and the performance of the progeny, in terms of reproduction potential, growth rate and carcass quality, has been evaluated. The useful genetic strains resulting from these programmes have subsequently been made available to commercial producers.

About 75 per cent of UK pigs are crossbred, and these animals usually perform better than the pure-bred lines. Traditionally, commercial producers crossed only two breeds of pig, for example the Large White and Landrace, to produce high-quality piglets which showed the beneficial effects of hybrid vigour. Pig breeding companies have improved on this technique by selecting high performance lines from various pure breeds and combining these to produce first or second cross-gilts for selling on to commercial producers. Meat-type male lines are also selected to mate with the gilts. Stock bred in this way is generally known as hybrid stock. The resources of a large breeding company allow considerable control of all aspects of selection, and guarantee the producer high performing animals at all times.

Improvement schemes

In the 1950s, recording and testing of superior stock, particularly boars, was undertaken by the government-sponsored National Pig Industry Progeny Testing Board, the activities of which were taken over in 1959 by the Pig Development Authority (PIDA). This body developed and improved the traditional structure of the industry, and encouraged research into all aspects of pig production, including nutrition, reproduction and artificial insemination, carcass quality, disease and marketing. PIDA also amassed data on all production aspects of the industry through a series of recording schemes. These recording schemes were taken over in 1968 by the Meat and Livestock Commission (MLC), which also maintained some of the research interests of PIDA.

The traditional breeding structure of an animal industry is that of a pyramid. The apex is composed of those breeders considered most advanced in the field (nucleus breeders); they make up a small percentage of producers who, in turn, supply multipliers with male and female animals, and the commercial producers in the lower stratum of the pyramid with some male animals (Fig. 6.2). Thus, over a period, improved genetic material percolates through the industry, and overall improvement is obtained.

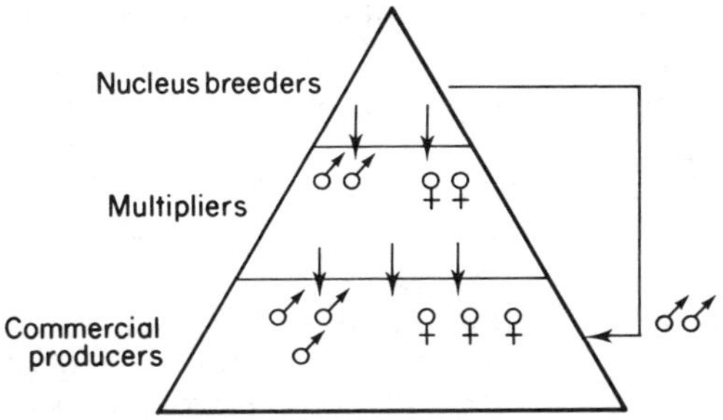

Fig. 6.2 Structure of the pig breeding industry.

In the immediate post-war period, pedigree pig breeding had a long and successful record of improvement. In recent years, the upper strata of the breeding pyramid for pigs have become increasingly filled with commercial breeding companies who have maintained and improved top quality pure-bred animals by selection under rigorous conditions, as well as combining various genetic lines in order to take full advantage of hybrid vigour. These companies now supply a very large proportion of the gilts needed by commercial producers.

The usual method of selecting pigs of high genetic merit is to feed them a known diet under controlled environmental conditions. These may be group-housed pigs fed *ad libitum* or individually penned animals fed on a fixed time-based ration scale. After a fixed period, such pigs are assessed on live-weight gain and on back fat thickness as measured with an ultrasonic scanner. The top performing band of tested animals will then go forward into a continuing breeding programme, and progeny will eventually reach the commercial breeder and rearer, who will purchase the type of stock which suits his production methods and chosen market.

In the breeding herd, the number of piglets successfully reared per sow per year is a major indicator of efficiency of production. In the feeding herd, efficiency of lean meat production is of main importance. Both of these parameters are limited by many biological and managemental factors.

The genetic improvements achieved in the pig industry have been largely achieved by natural mating of superior stock. Deficiencies in artificial insemination technology, together with problems of handling and storing semen on the pig unit, has meant that AI has played only a minor role in breed improvement. Recent reports suggest that this situation may be changing.

Standards of production achieved throughout the industry are shown in Table 6.1. Despite the best efforts of producers and research workers to consistently improve these standards, the figures have been relatively unchanged for some years.

Outside pressures

In common with other branches of intensive animal husbandry, pig production has been the target of considerable criticism from the welfare lobby. Public pressure on pig producers to alter intensive systems – to allow, for example, more freedom of movement for pregnant and lactating sows – is persistent, but not yet loud enough to encourage a legislative approach across the whole of the European Community. Producers themselves are aware of it and show considerable interest in alternative systems (e.g. the housing of dry sows) which allow greater freedom of movement while still controlling feed intake. Detailed advice on the welfare needs of pigs is contained in the *Codes of Recommendations for the Welfare of Livestock: Pigs* (MAFF, 1983), which reflects the changing attitudes to these needs.

Public concern is also expressed on the possible effects on human health of eating pig meat which contains large quantities of fats, and perhaps drug residues from products used for maintaining pig health in intensive fattening

Table 6.1
Suggested production standards

	Average	Best
Breeding stock		
Live pigs born per litter	10.5	12.0
Weaners per litter	9.0	10.5
Weaning at less than 20 days		
litters per sow a year	2.4	2.6
weaners per sow a year	22.0	25.3
sow feed used per weaner	53 kg	42 kg
Weaning at 20–29 days		
litters per sow a year	2.25	2.5
weaners per sow a year	21.0	24.0
sow feed used per weaner	58 kg	45 kg
Weaning at 30–39 days		
litters per sow a year	2.15	2.35
weaners per sow a year	20.0	22.5
sow feed used per weaner	63 kg	50 kg
Weaning at 40 days and over		
litters per sow a year	1.9	2.1
weaners per sow a year	18.0	21.0
sow feed used per weaner	68 kg	60 kg
Feeding Stock from 8 weeks of age (18 kg) to slaughter		
Mortality rate	2.5%	0%
Feed conversion rate[a]		
porkers (69 kg live-weight)	2.8	2.2
cutters (85 kg live-weight)	3.05	2.5
baconers (93 kg live-weight)	3.15	2.6
Quantity of feed per pig (kg)		
porkers	140.0	110.0
cutters	205.0	170.0
baconers	235.0	195.0

[a] Quality of feed will influence the conversion rate and quantity per pig.
Figures taken from Ridgeon (1984).

systems. Drug safety is checked by licensing authorities, but there is room for misuse, abuse or carelessness of use. Producers are encouraged to keep accurate records of drugs used and to follow instructions regarding withdrawal periods and slaughter.* There is a small but growing market for pig meat products which can be identified as coming from a humane, drug-free system. A number of producers are attempting to fulfil this requirement, and one presumes their numbers will increase.

Despite these pressures, the consumption of pig meat continues to rise, and its consumption is encouraged in various promotional campaigns. Producers have aimed at the requirements of the British market and produced mainly porkers, cutters and baconers. To expand sales opportunities in export markets, there is a need for carcass weights of 70–80 kg with low backfat measurements. This will necessitate some modifications in production, because growing animals will have to be reared to a higher weight than commonly occurs at the moment.

A final point of public concern which is of great importance to the pig industry is that of environmental pollution. As with other intensive systems, the pollution potential from pig units is great, particularly in those units where effluent is disposed of in the form of slurry. There is a considerable risk of contaminating the water table when slurry is spread continuously on relatively small acreages of land, and there are also odour problems at spreading time. The industry is well aware of this problem, and strenuous efforts are being made to find improved methods of incorporating slurry into soil (since it has a manurial value), as well as to evaluate other possible uses, such as methane production. Planning regulations concerning the establishment of new pig units vary from one local authority to another, but there is overall legislation which can control the pollution of water tables and water courses. Control will presumably become more important as increasing pollution risks are identified.

Natural history and normal behaviour patterns

The pig belongs to the non-ruminant section of the artiodactyls. The wild pig is a forest animal feeding on a wide range of plants, small animals and insects. It consumes some herbage and has a reputation for despoilation of crops in its soil-turning search for feedstuffs. It has been claimed that domestication began as an effort to

* The National Office of Animal Health (NOAH) has included a Code of Practice for the safe use of veterinary medicines on farms in their *Animal Medicine Record Book* (1987).

stop the destruction of man's early efforts in cultivation.

Domestication and genetic change, coupled with husbandry practice, has produced marked changes in the behaviour patterns of the present-day pig, and many husbandry methods affect the way in which pigs react to their farmed environment.

Wild and feral pigs have a social unit composed of a sow and her litter. The wild pig farrows once annually and has a litter of between one and four piglets. At weaning, which usually occurs at about 12 weeks, two or three sows and litters will join together. There is often a matriarchal hierarchy of sows involved in the grouping, which holds until the onset of the breeding season. In European wild pigs, this is protracted and often extends from autumn through to the following June. Old boars, which generally live alone, join the groups at this time, driving off the young boars and serving the sows. Young boars will live together during this period. Breeding females and sub-adult animals usually regroup after breeding.

Pigs have poor eyesight, but highly developed senses of smell, taste and touch. The tactile, highly sensitive snout is essential for rooting behaviour. Outdoor pigs maintain this habit, and it is common practice to ring the upper edge of the snout so as to limit the damage done by rooting. The pattern is maintained in intensively housed pigs, which will persist in nosing the surface of concrete pens. Outdoor pigs can spend many hours per day exploring their environment, food-seeking and eating. Indoor pigs spend a relatively short time eating and have little opportunity for exploratory behaviour. Competition for feed is pronounced in intensive situations, appetite and intake being affected both by feeding methods and by the total availability of food and water, Pigs which have only limited access to feed will often spend the time drinking. Hence adequate feeding space and watering points are essential.

Weaned, growing and finishing pigs live in a barren environment. The physical dimensions of their living space, their companions, the stocking rate and the feed type are all decided by man. They are commonly put together in groups of the same age, sex and weight. In this situation they have to learn to recognize feed, water and dunging sites and to fit into a social hierarchy. Various behaviours are shown as this hierarchy develops. Newly mixed pigs will fight to establish a linear hierarchy. Fighting may be intense for up to 48 h after mixing and will continue intermittently thereafter. The social rank which develops as a result of fighting usually reflects the weight of individual pigs within it – the heaviest being most dominant, the smallest least dominant. Management techniques, such as the provision of adequate feed and living space, can ease this settling period after mixing. Aberrant behaviours, such as tail-biting and ear-biting, may occur, suggesting that managment is at fault in being unable to satisfy the behavioural needs of the pigs. Considerably more work needs to be undertaken in this area to identify and measure behaviours which affect production and which might be manipulated to the advantage of both pigs and producers.

Management can exploit the inquisitive and learning behaviour of pigs in a number of ways. Perhaps the most common is to take advantage of the pig's demarcation of dunging and urinating places. It is common practice to drive pigs into a pen via the dunging area and hold them there for a short time, because young animals, when moved, tend to dung and/or urinate, and the pigs rapidly associate with these functions the area they first fouled; the rest of the pen is then kept clean. Overcrowding or inadequate control of temperature leads to a breakdown in this system of demarcation, and soiling of both pigs and pen then becomes general. Once soiling has occurred, pigs seldom return to their previous clean state.

Conditions for the manifestation of sexual behaviour in breeding animals need to be manipulated if optimum reproductive performance is to be achieved. Young boars should not be reared in isolation; they benefit from the stimulus of being in the company of others. They generally become sexually active between 5 and 8 months of age. Gilts reach puberty at the same mean age as boars and should react positively to the sudden introduction of a boar into their close environment by showing oestral behaviour within 7 days. (For management aspects of oestral behaviour, see p. 156.)

Under intensive conditions, the sow has little opportunity to exhibit the behaviour patterns which occur in more natural situations before and during parturition. The extreme restlessness commented on by many workers has led to a re-evaluation of the type of accommodation provided for this time, but has not led so far to any changes in common commercial practice. It has been suggested, for example, that an increase in farrowing space might lead, under certain conditions, to a reduction in piglet mortality, but producers are wary of change because of the costs involved in providing efficient and humane farrowing accommodation.

Suckling piglets respond to the milk let-down call of the sow by nuzzling the part of the udder they identify as their own. A change in note is associated with milk release, and suckling piglets prefer the anterior teats of the sow, which offer more protection and possibly have a greater milk supply. Teat order establishes quickly after birth. As in other competitive situations, the greater the litter size, the more varied is the weight of the individual piglet, both at birth – because of intra-uterine competition – and at weaning – because the sow has a limited number of teats (usually 14), which vary in the quantity of milk they secrete. Fostering of piglets can take place soon after birth with little disruptive effect and units practising synchronized farrowing (p. 160) use this technique for balancing up litters.

An awareness of the normal behaviour pattern of animals is essential for good stockmanship and management. The Preface to the *Codes of Recommendations* (MAFF, 1983) supports the view that pigs should be allowed to show as many behavioural patterns as acceptable to efficient management. Commercial conditions do not necessarily always allow a full behavioural repertoire to be shown.

One area of behavioural interaction which has not yet been adequately studied is the relationship between pig and stockman. One of the main objectives of good stockmanship is to maintain a contented relationship between pig and handler. Pigs should not be stressed by the presence of stockmen, and the use of adverse human behaviours such as hitting, kicking or goading must be avoided. This difficult area of interaction needs close analysis if we are to improve the welfare of our livestock.

Reproductive and breeding data

These are summarized in Table 6.2 Further details are given below, under "Management of Boars" and "Management of Breeding Females".

Nutrition and feeding

The pig is a monogastric omnivorous mammal which resembles man in its nutritional requirements, there being many similarities between the anatomy and physiology of the digestive systems of the two species. To satisfy both the nutritional needs of the animal and the desire of the producer to offer feeds of acceptable cost in relation to the level of production demanded requires the manipulation of a range of feedstuffs. The major components of most pig diets are cereals, animal and vegetable proteins, and cereal by-products. A wide range of vegetable products may be included, depending on the management system involved. Adult stock may, for example, be fed root crops or other arable by-products, while a more traditional and closely controlled ration may be fed to slaughter pigs.

Rationing

The allowance of food given to a pig per day is a major determinant of performance and profitability. A very restricted feed supply leads, in growing pigs, to a barely edible carcass and, in breeding sows, to an emaciated animal incapable of adequate production. Over-availability of feed results in obese animals which are not wanted by a discerning, meat-eating public and breeding sows or boars which have high maintenance cost and poor reproductive performance. Feeding scales must be pitched at such a level that animals grow as fast as possible or, in the case of breeding animals, maintain sufficient body reserves, without in either case laying down too much fat. The pig's appetite is large and the provision of feed *ad libitum* throughout the growing period will usually lead to overfat carcasses.

To arrive at diets and feed scales which suit production patterns, the producer needs to have

Table 6.2
Reproductive data

Sow	Mean	Range
Age at puberty	200 days	135–250 days
Age when first bred	221 days	156–271 days
Oestrous cycle length	21 days	18–23 days
Duration of oestrus	53 h	12–72 h
Time of ovulation		38–42 h after the onset of oestrus
Rate of ovulation	16 ova	10–25 ova
Gestation length	114 days	110–120 days
First normal ovulation after parturition	5 days post-weaning	4–12 days
Cycle type	Polyoestrous all year	
Fecundity	Gilts average 9 live-born Sows average 11 live-born	
Breeding life	Approximately 2 litters/sow/year. 3–5 litters in modern systems with longer day lengths	

Boar	Mean	Range
Age at puberty	200 days	150–250 days
Age when first bred	*c.* 245 days	
Seasonality	Long days and high temperatures reduce sperm production and viability but are not considered of practical significance in temperate conditions	
Breeding life	Usually culled between 3–4 years of age	
Sow/boar ratio	25 sows per boar is a practical ratio for "natural mating"	

a knowledge of the energy and protein concentrations of the feedstuffs available,* the biological value of the protein, and the concentration and the nature of the fibre content, because digestion of fibre by pigs occurs to only a limited extent. Giving diets rich in fibre and thus low in digestible energy concentration will limit feed intake and growth rate.

Energy is needed for maintenance functions, and for lean tissue growth. Once these needs are satisfied the remainder will be used to lay down fat.

Energy is quantified as digestible energy (DE) expressed as megajoules per kilogram of food (MJ/kg), and protein as crude protein (CP), or digestible crude protein, expressed as a percentage (per kg). Nutrient density can easily be altered. The usual values are medium density 12.5 MJ DE/kg, high density 13.5 MJ DE/kg, and very high density 14.5 MJ DE/kg. The inclusion of cassava and/or fat will increase the energy concentration. Table 6.3 gives recommendations for DE and CP allowances for different classes of stock.

Typical food conversion ratios (FCR) (kg feed/kg gain) for a medium density diet are: weaner production, 1.5; porkers (20–63 kg), 2.8; baconers (20–90 kg), 3.1, and heavy hogs, 3.6. Using a diet of higher DE concentration will produce better ratios.

* MAFF (1980a) contains tables of feed composition, as do many other texts.

Table 6.3
Some typical diets

Class of stock	DE (MJ/kg)	CP (g/kg)
Young growers[a] (5–20 kg live-weight)	14–16	200–220
Growers (20–60 kg live-weight)	13–14	150
Finishers (60–100 kg live-weight)	12–14	135–150
Pregnant sows	12.5	140

[a] The most concentrated diet would be necessary for piglets weaned at 2–4 weeks of age and would probably include milk products, such as dried skim milk; nutrient density could be reduced when 10 kg live-weight is reached.

The nature and quality of the protein in the diet requires careful consideration because it must meet the pig's requirement for amino acids. Some 22 amino acids are commonly found in pig diets and, of these, some nine are essential, in that they cannot be synthesized by the pig and must be included in the diet. A shortage of these will limit the production of lean meat and so have an adverse effect on growth. For example, lysine, histidine and methionine are needed in greater amounts than valine and isoleucine. Unfortunately, the aforementioned are less abundant in common feedstuffs than the latter. Cereals supply about 66 per cent of the total protein in the diet of growing pigs, but they provide less than 50 per cent of the requirement for lysine. Protein concentrates, such as fishmeal, which are rich in lysine, must be included. In diets for growing pigs (15–90 kg live-weight), lysine concentrations should be about 7 g/100 g crude protein. The ideal protein in a diet for a pig should be one which best satisfies its needs. Since amino acids are being used for meat protein production in growing pigs, and body maintenance and milk production in sows, the ideal protein in feedstuffs should bear a resemblance to those in meat and milk.

Minerals and vitamins should be included routinely in all diets at or above the requirement levels suggested by experimentation. Tables of these requirements may be found in many textbooks (see references). The literature suggests that much has still to be learned about both minerals and vitamins, particularly vitamin deficiencies, and the interrelationships which occur between various minerals.

The energy-to-protein ratio is an important factor in diet formulation. Small piglets and growing pigs have lower maintenance needs (energy) and higher protein needs (lean tissue growth) than a pregnant sow, for example. The ratio runs between 1:14 (DE:CP) for a young growing animal down to 1:11 for a pregnant breeding animal or a growing pig approaching slaughter at 100 kg live-weight (see Table 6.3).

Gilts are often kept on the same feed or ration scales as bacon pigs until 14 days before the second or third oestrus. At this time their live-weight will be about 90–115 kg, and feed containing 14 per cent crude protein is then offered *ad lib*. After service, feed is often restricted to about 23 MJ DE/day for a period, but extra nutrients for the growth of the foetuses and associated tissues will be required during the latter part of pregnancy, especially the last 30 days. Litter size, birth weight and sow body-weight can be affected by feeding during pregnancy. It is often suggested that gilts and sows should be fed to gain 20, 15 and 10 kg in net body-weight per parity for the initial three parities, but the amount should be adjusted to the weight gain and condition of the individual animal.

Sows draw on their body tissues during lactation, and a 10-kg weight loss over this period is usually acceptable, but excessive weight loss should be avoided. A lactating sow will need about 50 MJ DE/day for up to five piglets and 5 MJ DE/piglet above that, while the crude protein concentration of the ration should be 15 per cent for 5-week weaning, and 17 per cent for earlier weaning.

Feed allowances and diet calculations, as described above, should only be looked upon as guides, and it may be necessary to modify either or both in accordance with the response of the pigs. Diets may be either bought-in compounds or home-mixed: quality control is as important for the home-mixed as for the factory product. Indications are that compounded pelleted feeds are preferred for younger stock – piglets and weaners – both from the point of view of convenience and from the need to maintain a small, regular supply of a fresh, palatable feedstuff.

Feed ingredients

The major ingredient in all pig diets is cereal. In the UK, barley has been the preferred ingredient for many years, though this may in some part reflect the vast quantities grown. Barley is a medium-energy ingredient and can be fed at high levels to growing and breeding stock. It contains a good level of lysine in its protein content. Wheat and oats are less commonly used, probably because they are less readily available for use as animal feed. Wheat has a higher and oats a lower digestible energy content than barley. The higher fibre level in oats limits the use of this cereal to older classes of stock, where one is not necessarily looking for maximum feed conversion efficiency. Other cereals, such as maize, rye, sorghum and rice, may all be used in pig rations depending on availability and price, and this emphasizes that the British pig may be fed on rations of a cosmopolitan origin.

The animal protein sources which have been traditionally used in pig foods are fishmeal, meatmeal, meat-and-bonemeal, milk by-products, bloodmeals and feathermeals. These products are all dried and ground, care being taken to avoid over-heating, which renders the lysine content unavailable by causing denaturation of the protein.

Vegetable protein includes oilseeds, pulses and microbial protein. The usefulness of these products is dependent on their amino acid composition and the presence or absence of toxic or antinutritional factors (plant products which may inhibit enzyme activity, or act as deterrents to consumption by the pig). Their crude protein content ranges from 72 per cent for microbial protein down to perhaps 9 per cent for sunflower meal.

The methods of milling cereals for human use have resulted in a wide range of cereal by-products which are frequently included in pig rations. These products often have a high fibre content, being the outer husks of the grain. Their protein, energy and oil levels reflect the degree of milling. Any one of them may be used in pig rations, as long as thought is given to their effects on productivity.

Root crops are a useful source of energy but are low in dry matter and protein, and are more commonly used in rations for adult stock. They can be macerated and put through pipeline feed systems. Potatoes, particularly if cooked so as to make the starch more digestible, are used as efficiently as cereals by pigs.

Other feedstuffs, such as manioc, dried sugar beet pulp, molasses, dried green crops and distillery by-products, may all be used in pig rations at relatively low levels. The development of very high density diets has resulted in the increased use of oils and tallows as energy sources in rations for growing pigs.

Milk is a further source of both feed and drink for pigs. Liquid separated milk (about 9% dry matter containing 37% protein and 53% lactose) is very digestible and is well suited to liquid feeding systems. It is usually pickled with 0.1% formalin, which preserves it for about 7 days. Other milk-by products, such as whey – the watery residue from cheese making – can also be fed. Care should be taken to ensure an adequate water supply, because whey often has a high salt content.

Swill is a very variable product which is usually collected by the producer at source. Canteen wastes, for example, will contain meat and vegetable fragments, as well as confectionery residues and trimmings produced during food preparation. Swill must by law be boiled for at least 1 h. This is important for disease control purposes and may also improve the quality of the product. The plant which treats the swill must be licensed by the Ministry of Agriculture, and swill-fed pigs must be transported directly to a slaughterhouse. These stringent health and hygiene precautions,* coupled with the unpredictable composition of the feed which makes it difficult to feed with any accuracy, tend to make swill an unpopular feedstuff for most producers.

Growth-promoting substances

Various non-hormone growth-promoting substances are commonly included in rations for growing pigs; their use is controlled by EEC Directive 70/524. For example, some antibiotics, such as bambermycin, and virginiamycin, which are without any therapeutic value, are permitted for use as feed additives.† Therapeutic antibiotics, however, can be used only on veterinary prescription. The incorporation of copper sulphate in the diet will usually improve growth rate and food conversion ratio, but there are recent EEC Directives (83/615 and 85/520) which limit the inclusion rate to 175 p.p.m. for pigs up to 17 weeks of age, 100 p.p.m. for pigs from 17 weeks to 6 months, and 35 p.p.m. for pigs over 6 months and breeding stock. Nitrovin is another non-antibiotic growth promoter which may be used as a food additive (but not with antibiotics). All these substances

* Diseases of Animals (Waste Food) Order, 1973.

† Medicines (Exemptions from Restrictions on the Retail Sale or Supply of Drugs) Order, 1979.

must be used in such a way that unacceptable residues do not occur in the carcass, and any instructions for withdrawal must be acted on.

Some other substances are currently being considered as growth-promoters. For example, some naturally occurring intestinal bacteria (probiotics) such as lactobacilli, are being evaluated as an alternative to antibiotics, to prevent colonization of the intestine by pathogens and to enhance feed utilization. Porcine growth hormone (somatotropin) is also being examined as a growth-promoting substance which has no biological activity in humans.

Water

The importance of providing a fresh clean supply of drinking water for all animals cannot be overstressed. New-born pigs consume between 0.25 and 1 litre of liquid per day. Growing animals need at least 2–6 litres, and adult stock some 8–12 litres daily. Environmental conditions (such as poor heat control) may increase this requirement. It is important to ensure an adequate number of water points within a pen, if dry food is being offered, as animals will take in water several times during a meal, and an inadequate supply can lead to bullying and over-consumption of feed by dominant animals. If water is not available *ad libitum*, then it should be provided at a ratio of 2.5 units per unit of dry feed.

Design of diets

When designing diets, the system of production has to be taken into consideration, in conjunction with the following factors relating to the animal and to the feed.

Genetics. Different strains of pig may react in different ways to diet quality and pattern of feeding, but there is little evidence that any major variation exists among hybrid pigs from UK pig breeding companies.

Sex. This exerts an influence on growth performance and on amounts consumed. Given the same amount of feed of the same diet, boars grow faster and are leaner than castrated males; gilts have higher rates of lean tissue growth than castrates. Boars and gilts probably will respond, with extra lean growth, to higher levels of protein in the diet than will castrates. The superiority of the boar is only realized when the diet contains enough protein to support the high rate of gain of which it is capable. When offered feed *ad libitum*, castrated males will eat slightly more than will boars and gilts.

Behaviour. Feed intake and production performance can be affected by behavioural factors. High stocking density may lead to increased behavioural reaction between individuals and, at its extreme, aberrant behaviour such as tail-biting and cannibalism, with subsequent production loss.

Processing. Digestibility is improved when foods are processed (e.g. grinding, soaking, heating, drying, pelleting). However, very fine grinding increases the risk of gastric ulcerations. Also, overheating or overdrying reduces digestibility. Pelleting sometimes improves performance, possibly because the steaming process improves digestibility, and because there is less wastage.

Nutrient density. If the density of the diet is increased, then intake may be reduced. Bulking, using inert diluents, tends to increase intake, but reduces digestive efficiency.

Appetite. Appetite can be influenced by any of the above factors. In addition, an animal offered free choice of feeds may well choose those of which it has had previous experience. Reduced appetite is often an early sign of disease.

Feeding systems

In most commercial situations, pigs are housed and fed in groups. Individuals within groups show considerable differences in feed intake, and consequently in growth. The following are the most common methods of feeding growing pigs.

Controlled feeding. In this system the amount of feed given is under the daily control of the farmer. Amounts are restricted to less than the pig's maximum intake at each mealtime. Feeding is to a pre-determined feed scale which varies with either weight or age. It is the most common system of feeding.

Table 6.4
Guide to environmental requirements

Pig type	Live weight (kg)	Temperature (°C)	Space (m^2/pig)	Trough length (mm)
Suckling pigs	up to 5	23–30	0.07	
Weaners (flat deck)	5–18	21–28	0.07–0.25	100
Weaners (6 week +)	10–20	17–24	0.25–0.37	150
Early finishers	20–45	13–21	0.37–0.6	200
Late finishers	45–90	13–17	0.6–1.33	230–280
Sows in groups	150+	10–15	[a]	300[b]
Dry sows (boars) in stalls	150+	18–22	1.3	300[b]
Farrowing and lactating sows		15–21	1.6	300[b]

Average ventilation rate 1 m^3/h/kg of pig housed. Humidity 70–80% RH.
[a] No firm recommendations.
[b] 300 mm, but related to housing and method of feeding.

Ad libitum or semi-ad libitum. Pigs are offered continuous access to feed, or access for a fixed period of time. This is invariably less efficient in carcass terms than is controlled feeding. Up to pork weight the differences are slight, but above this, and particularly for bacon pigs, it is necessary to feed to a restricted scale in order to achieve good food conversion and leanness of carcass.

Self choice. This is common in the USA. Pigs have access to at least two types of feed and are expected to eat sufficient of each to satisfy their requirements. This system is usually inefficient when compared to any form of restricted feeding.

Alternating feeding. Access is provided to, say, barley meal at one feed and a protein supplement at the next. If both feeds are offered within a 24-h period, this has no effect on pig performance. With longer intervals between feeds, differences in the weights of tissues and organs are sometimes reported. There are obvious savings in the cost of mixing.

Frequency of feeding. There are no advantages in growth terms in feeding older pigs more than once per day, and to omit one feed in a 7-day period is generally acceptable. Behavioural problems may, however, occur in some pigs. Baby pigs and young animals require either frequent feeding or continuous access to feed.

Wet or dry feed. If given a choice, pigs prefer wet feed and are capable of eating it much more quickly than dry feed. In some piggeries, meal suspended in water is pumped to troughs. Dry feed is more commonly fed because it is easier to handle, the risk of bacterial infection is less, and capital costs are lower.

Trough or floor. If pellets are used, results are slightly better from trough feeding. If meal is used, results are definitely better with trough feeding, because there is less wastage.

Environmental requirements

Wild pigs will seek shelter from adverse weather conditions. Systems of outdoor pig production provide protection from the elements in the form of moveable field shelters, which the animals can take refuge in when they choose. Under most systems of production, however, pigs are kept wholly indoors, which improves the environment available to them, but does impose a social stress, because there is insufficient space for them to avoid their companions. The producer needs to be as generous as possible in allocating space, but is constrained by cost.

When animals are housed, it is necessary to provide living conditions which fit the pig's need at each stage in its life. The buildings themselves may cover a wide range of types, but they all seek to provide an insulated living space in which temperature, ventilation, bedding, etc., can all be wholly or partially controlled by the producer. The design should be such that both pig and operator find the accommodation easy to function in. Table 6.4 lists the environmental requirements of the various classes of stock.

The social and competitive problems which

may arise in intensive situations can be overcome by using individual feeding arrangements, or by housing in stalls, but, while such systems may eliminate problems such as bullying, this degree of restriction leads to considerable curtailment of the animal's freedom of movement, and can result in the development of abnormal behaviours which may in turn adversely affect production.

Temperature

The pig's thermoneutral zone is defined as the range of environmental temperatures within which heat production is independent of air temperature. At the lower end of this zone the pig must increase heat production to maintain equilibrium, at the upper end it runs the risk of overheating, with consequential problems. The size of pig is also important: small piglets with a high ratio of surface area to body mass run the risk of chilling; large pigs can tolerate cold but have difficulty in cooling when exposed to high temperatures.

Pigs need to feel comfortable. Body heat may be lost by convection and evaporation, the rate of loss being related to air temperature and air movement. The degree of floor insulation affects the rate of heat being transferred from pig to floor, and the stocking density in a pen will affect air temperature and the movement of heat between pigs. Heat-stressed growing pigs eat less and, in breeding stock, foetal growth and sperm quality may be reduced. Cold stress will be obvious to the stockman, because animals huddle more, erect their hair and, over a period, there will be an increase in hairiness. Rapid loss of body condition occurs if action is not taken to improve housing, and possibly feed supply, and so prevent feed being used for heat production instead of growth.

Ventilation

Ventilation is used to manipulate temperature, relative humidity and air composition at pig level. Over-ventilation should be avoided, because this can lead to a loss in production. Minimum recommended rates probably show the thresholds of heat and odour acceptable to the stockman, rather than showing any particular advantage for the pig. The problems of noxious gases, dust and microorganisms are often coupled with inadequate ventilation. Ammonia and hydrogen sulphide are very obvious to the stockman before they affect the pig, while carbon dioxide, being odourless, is not obvious, but under normal conditions does not occur in dangerous concentrations. Dust does arise from feed, whether fed as meal or pelleted, and as pigs and humans predisposed to pulmonary infection are at risk from this, all practical steps should be taken to keep dust levels as low as possible.

Ventilation may be achieved in a number of ways. Air may be displaced naturally through the effects of air movement and temperature, or it may also be forced in and out of buildings by fan systems. This air movement can take heat with it and also move dust and microorganisms around. Most attempts to control ventilation are based on the use of temperature-control mechanisms which switch fan systems on and off.

Relative humidity

Pigs can tolerate a wide range of humidity, and it is sometimes claimed that high humidity reduces the incidence of pneumonia. Certainly there appear to be fewer viable microorganisms in hot, humid environments. Low humidities have a tendency to dry out the pig's skin and also increase the dustiness of the atmosphere, which may in turn aggravate respiratory conditions for both pigs and stockmen.

Light

Experimental observations on the effects of light have given confusing results, probably because the pig can adapt to a wide range of lighting regimes with little effect on growth rate. Breeding females sometimes seem to be adversely affected by long periods of darkness, while semen production and quality in boars is adversely affected both by increasing day length and by total darkness. A 12–14 h day should be arranged for breeding animals whenever possible.

The *Codes of Recommendations* (MAFF, 1983) give the following advice:

> Pigs should not be kept permanently in darkness. Throughout the hours of daylight the level of indoor lighting, natural or artificial, should be such that all housed pigs can be seen clearly. Adequate lighting for satisfactory inspection should be available at any time.

Floors

Floors should provide maximum comfort for the pig to lie, stand and walk. The various forms of slatted floors all allow heat loss from the pig, and house insulation and ventilation are important in controlling this. Floor design in terms of slotted metal, slats or mesh must cater for the foot shape and size of the animal using it, but comfort and cleanliness needs have to be balanced against installation costs. Pigs between 4 and 8 weeks of age can use slats with gaps of up to 15 mm. Over this age the gap can be increased to 20 mm. Dry sows are usually accommodated in stalls, or tethered, with the front two-thirds of the body on concrete and the hind end on metal slats: sometimes slats form the whole of the standing/lying area. Farrowing crates may also have slatted floors, in which case there may be a risk of piglets trapping their feet. Slats for sows should be at least 100 mm wide, with 25-mm gaps: they should have rounded edges and there should be no sharp projections.

Methods of intensification, such as the use of slats and concrete for pigs to stand and lie on, often with little opportunity for movement, result in various arthritic and rheumatoid conditions in some older animals. In many cases, the movement of stock to a more generous form of housing such as a strawed yard will result in an improvement in condition. The importance of good floor design and maintenance is stressed in the *Codes of Recommendations*, and it is in the interests of all producers to provide flooring and accommodation of a high quality.

Pen shape

This must be considered in relation to the type of pig housed, feeding method and dunging provision. Rectangular pens (length 2–3 times the width) with solid insulated floors encourage animals to be clean and dry, probably because, in a pen of these proportions, they can more easily define and maintain a dunging area at the one end.

Feeding may be on the floor or from a trough. If a trough is used, then the length of this is important, and this controls the number of animals in a pen at a certain weight (Table 6.4).

Bedding

The latest *Codes of Recommendations* (MAFF, 1983) encourage the use of bedding for all classes of breeding stock, and strongly recommend the provision of straw or other material in the lying area of the pens. Bedding is necessary in the whole range of simple kennel-type housing, to keep the animals warm. A producer's decision to use bedding will depend largely on the availability of straw or wood shavings, the level of intensification practised on the unit, and the ease of disposal of either soiled bedding or slurry. Many units will use solid manure systems for some classes of stock and slurry systems for others. No reliable information is available on any changes in the attitude of producers to the use of bedding since the revised *Codes of Recommendations* were published.

Effluent disposal

The disposal of faeces, urine and washing water from any intensive unit can be a problem. Slurry-based systems ease the physical work in a unit, but increase the problems of storage and disposal. Absence of bedding may also lead to a greater incidence of traumatic injuries to limbs, particularly if the level of finish of slats and floors is poor (see above). Straw-based or litter-based systems ease the burden of stockmanship, but increase the work of littering up and cleaning out, but disposal is easier in that manure can be stacked and rotted cheaply and easily. Regardless of the system chosen, the final dispersal on to land is always a problem. Slurry can have an obnoxious smell, and a considerable acreage of land is needed if problems of pollution are to be avoided. Untreated pig slurry has a biological oxygen demand some 20 times greater than that of crude town sewage, and great care must be taken to ensure that it does not enter rivers and other water courses.

The problem of effluent disposal has still to be solved in a manner which is both acceptable to the public, and cost-efficient for the producer. It could become the greatest single limiting factor in the establishment of a pig unit in the future.

Management of boars

Breeding males are usually selected for introduction into a herd at 6–8 months of age. Selection will be based on an assessment of growth rate, feed conversion and carcass quality of siblings. Boars are usually bought into commercial holdings from breeders or breeding companies who specialize in their production and evaluation. The physical soundness of the animal is important, as well as a good level of libido. At this age, the boar is not yet fully sexually mature, and care must be taken not to over-use him. Four to six sows per young boar per 21-day breeding period is recommended. Inexperienced animals need to be supervised in use because their libido may be reduced through the bullying activities of females.

Both the fertility of a boar and the total quantity of semen produced builds up gradually over a long period. Maximal volumes of ejaculate are produced at about 30 months of age, while the densest sperm count might occur later than this. A 3-year old male is therefore more fertile than an animal of 8–12 months. On commercial units, one boar per 25 sows is usually considered adequate.

Young boars are frequently group-housed in straw yards. This appears to be of behavioural and management benefit, because such animals are more tractible to handle and less likely to show reduced libido than individually housed males. Older animals tend to be penned singly, but this is a reflection of the small number of boars which are kept on most units. These will be of varying ages, and strangers to each other. The risk of antagonistic behaviour is therefore avoided by individual housing.

Management of breeding females

Gilt selection

Gilts are selected for breeding on the basis of the growth rate, feed conversion ratio and reproductive history of their dams and sires, and on their own anatomical and production characteristics. They are fed in the same way as slaughter pigs up to about 100 kg live-weight (see p. 165) and are then brought into the breeding herd for some time before service so as to be exposed to the diseases present there. Bought-in gilts and boars need to be quarantined for a 21–28 day period, while at the same time being exposed to the manure produced by the indigenous stock.

Puberty in the gilt is affected by breed, genotype, nutrition and environment. Age at puberty varies between 135 and 250+ days. Traditionally, gilts have not been bred at their first oestrus, but increasing pressure on producers has led to a change in this view, and producers now attempt to mate gilts as early as possible. Contact with the male by sight, sound or smell is important in triggering the occurrence of first oestrus, and animals should be managed so as to achieve sudden exposure. Gilts which fail to come on heat by the age of about 8 months should be culled, as should any sows which produce poor litters, show unsocial behaviour to their young, such as savaging or cannibalism, or are unable to cope with housing conditions in an intensive unit or suffer arthritic limb or foot problems.

Oestrus

Oestrus in sows and gilts is characterized by restless behaviour, a variable appetite and a swollen, red vulva. The animal sniffs the genitalia of pen mates and allows herself to be mounted, or will ride others. There is a characteristic grunting, and she will actively seek the boar. If he is receptive, she will adopt a rigid, immobile stance. Pig producers can usually detect this phase, since the female will respond to any back pressure at this time. In the absence of males, synthetic boar odour aerosols containing a steroid similar to that produced by the boar are available to help trigger this behaviour.

Ovulation rates in sows are very variable. The number of ova shed per cycle increases over the first three parities, peaks for a further two or three parities, and then falls away. Ovulation rate may also be affected by energy intake, and short-term increases in the ration fed in the post-weaning period can increase the number of ova shed.

It is relatively unusual to synchronize oestrus in batches of sows and gilts by means of hormones or their derivatives. Gonadotropins can be used to stimulate ovulation, and progestagens may also be used for synchronization, but there is no ready information as to the level of use of these in the UK.

Mating

During mating, the boar approaches the sow, giving a characteristic series of grunts. He noses the vulva vigorously, champing his jaws and frothing at the mouth. Most boars mount once, and there is a rapid thrusting phase which ceases at ejaculation. The process lasts about 4–6 min.

Mating may be supervised (hand mating), by pen or batch mating, where a male has access to a number of females, or by artificial insemination. The optimum time for insemination is 6–12 h before ovulation. In hand mating, the sow is served twice, with the first service on the first day that standing oestrus is detected, and the second 12–24 h later. This technique allows spanning of the ovulation period and increases the chances of successful insemination. Pen mating, while encouraging the detection of behavioural oestrus, has the disadvantage that the male may continue to mount a particular female as long as she is receptive. Other females in the group may, in consequence, miss his attention, or be inseminated with an inadequate quantity of semen.

Artificial insemination

Artificial insemination (AI) of sows has been used commercially in the UK since about 1955. In 1965, a semen delivery service was introduced by the Pig Industry Development Authority and this service has been maintained and improved by the Meat and Livestock Commission. AI is availablc to any pig producer through this system and also through other schemes run by commercial organizations.

At the present time (1987), AI accounts for only about 5 per cent of all services in the UK, but its use is gradually increasing, though still far below that in some other European countries. For example, recent figures from the Netherlands (MLC, 1987) show 46 per cent of matings being carried out by AI. The reason for the comparatively low use of AI in the UK might lie in the wide distribution of pig holdings, with the associated difficulties in the delivery of semen. The increased rates of use of AI reflect improvements in the handling and storage of semen, and interest in the use of boar plus AI matings and "pooled" (multi-boar) AI matings.

The major reason for using AI is to improve performance and carcass quality in slaughter pigs. Only the top 5 per cent of tested boars enter stud, and offspring from these show all the characteristics of increased lean meat production and superior feed conversion. Other benefits of AI accrue from the fact that breeding boars do not need to be kept on individual holdings, thereby allowing either a slight increase in breeding sows, or a reduction in costs, and in reduced disease risks on units which do not need to buy in boars. Disadvantages of AI lie in the inconvenience to producers of having to order semen and carry out inseminations at the optimum time, and in the rather high rates for return to oestrus, 60–70 per cent conception rates being considered good. This may reflect lack of skill in oestrus detection in the absence of a boar, and in the insemination technique itself, although this is straightforward and easily learned.

The 50 per cent increase in pig AI reported by the MLC for the period 1982–1986 suggests that, after a long period of use by a small number of pig producers, many producers are now realizing the benefits to be derived. As pressure increases for efficient lean meat production it is to be expected that AI will be practised increasingly.

The greatest single factor responsible for poor conception rates is probably poor timing of insemination. Sows respond to touch, sound and smell. Ideally the sow should show the "standing reflex" when the pigman sits astride her. Not all animals do this and the presence of "teaser" boars, aerosols or synthetic pheromones and even the taped "love song" of a boar will all help in detecting service time. Stockmen should look for oestrus signs at least twice per day, and the use of two inseminations about 12 h apart should improve the chances of conception (Fig. 6.3).

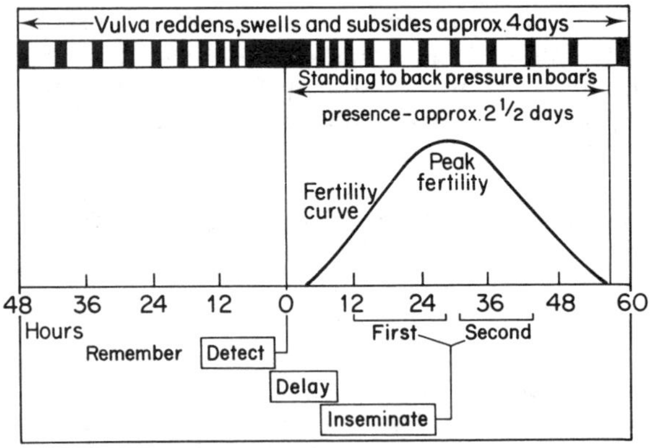

Fig. 6.3 Optimum time for insemination of the sow. *Source*: Meat and Livestock Commission recommendations for the artificial insemination of sows.

Insemination is carried out using a clean, spirally twisted catheter which is locked into the sow's cervix. The semen is allowed to run into the sow under gravity, the process taking some 5–15 min: it should not be hurried (see Hughes and Varley, 1980, for a full account of artificial insemination in pigs).

Pregnancy

Fertilization approaches 100 per cent of ova shed, but embyronic mortality is high and as many as 40 per cent may be lost. If the number of fertilized ova falls below five, then the pregnancy is terminated naturally, and the sow will return to oestrus some 25–30 days later. After 35 days dead foetuses tend to be retained until parturition, when mummified foetuses are delivered. All stresses in the first 35 days of gestation should be avoided, since this is the period when loss is most likely to occur.

Pregnancy diagnosis

Pregnancy diagnosis may be carried out using a variety of techniques: ultra-sound measurements through the uterus wall; oestrone sulphate estimations in blood or urine; enzyme-linked immune-serum assay (ELISA) using blood; rectal palpation. Close supervision of service, together with the precision of the sow's oestrus cycle, should mean that the efficient stockman has little need of these techniques other than in a relatively small number of cases of infertility. Of the diagnosis techniques available, the use of ultra-sound is the one in most common use at this time. A range of equipment exists, from the relatively inexpensive device which identifies the pulsations of the uterine artery, through to sophisticated scanners which produce an illustrated section scan through the uterus. It should be emphasized that efficiency of interpretation is poor before 30 days of gestation. No figures are available of the level of use of this equipment on pig units.

Feeding

The feeding of both male and female adult stock has to be judiciously assessed so as to maintain the animal in good condition for reproductive performance while not allowing it to become over-fat or too thin, since both conditions can result in reduced fertility. Sows may be condition scored visually, for example, at weaning (for details, see Brent, 1982). This technique needs to be carried out by an experienced stockman. Essentially one should not have animals which lack good cover over the pin bones each side of the tail, and which have hollow loins or prominent pelvic bones, prominent spine and ribs. Such an animal would be emaciated and Grade 0. Nor should the sow be so covered with fat that further cover seems impossible at any point (Grade 5). A moderate to good animal (Grade 2/3) has some cover at all points, with bone structure being felt only if firm pressure is applied.

A daily allowance of 2.0–2.5 kg of a medium-energy ration is generally thought adequate for pregnant females. This may be increased if sows are group-fed and competition occurs, or reduced if individual feeding is practised in an environmentally controlled house. The feed may be offered dry, wet or pelleted, the choice reflecting the type of housing and the producer's preference.

Boars are usually fed the same or a little more than sows, with feed levels being related to housing conditions. A boar housed in an individual pen outdoors will need extra food to compensate for winter heat loss, particularly if the housing is not insulated or little straw is used.

Housing

Housing ranges from simple kennel-type houses for either individuals or groups of animals, through yards, which may be strawed with or without an individual feeding system, to insulated force-ventilated buildings in which breeding stock are almost invariably individually penned. The important husbandry element is the provision of an adequate environment for the pig rather than the constructional variations which occur. Group-housing with individual feeding facilities is common between weaning and service, since this seems to encourage return to oestrus in the event of an unsuccessful insemination, and its behavioural detection. Once sows have been served, they may be either individually housed in stalls, or tethered. This allows precise allocation of food, permits close observation and eliminates bullying. Individual accommodation also minimizes space requirements and capital cost. The higher stocking density achieved in an insulated building can provide sufficient heat to maintain the temperature in the house with no further heat input. Temperature is then regulated solely by the use of fan ventilation. While the full stall and the stall and tether systems used for the housing of pregnant sows are efficient in production terms for the reasons mentioned above, the *Codes of Recommendations* (MAFF, 1983) do point out that their use deprives the sow of freedom of movement, denies her normal exercise and is the possible cause of abnormal behaviour such as bar gnawing, i.e. chewing the surrounds of her stall because it is the only physical outlet available to her.

In addition, the quality of the surroundings is important. Floors should not be so rough as to cause skin abrasions, but must be able to provide sufficient grip for the sow to stand easily. Metal work should have no rough edges or protrusions to injure the sow. The *Codes of Recommendations* encourage the use of bedding in sow housing, because this is thought to alleviate problems of lameness and skin abrasions. In addition, it provides another plaything for the sow during this enforced period of restriction. These recommendations seek to ease the lot of the restrained animal. The *Codes* do, however, go further, and strongly recommend that sows should be housed in conditions which allow behavioural and exercise needs to be met. A widespread return to yarded housing in its various forms is unlikely to occur voluntarily, because considerable amounts of space and finance are involved. A legislative ban on stalls and tethers throughout Europe seems the only way forward.

Parturition and lactation

The sow should be wormed about 14 days before farrowing, washed with a lice and mange solution and moved as a clean animal into clean farrowing accommodation some 4 days before farrowing. The boosting of the sow's immune status with products such as *E. coli* vaccines should also be complete by this time. Efficient service records, coupled with the relatively exact gestation length in sows (112–116 days), should make this preparation a routine on any well-run unit. As parturition approaches, the sow becomes restless and will attempt nest making even in the absence of bedding. Milk is usually present in many or all of the teats, and the udder is firm to the touch.

There may be some fluid from the vulva and many sows show tail twitching associated with abdominal contractions. Farrowing usually lasts for 2–3 h, with gilts taking longer than older sows. Piglets are born at about 15-min intervals, and expert help is seldom needed. Intervention is only necessary if the interval between piglets is more than 1 h. Such intervention should only be undertaken by an experienced person. The long farrowing period leads to a risk that early-born piglets may be trodden on by the restless sow and that later-born animals will be born dead due to anoxia. Provision of a warm, well lit creep area goes someway to reducing this risk for early born animals, but, on average, up to one piglet per litter may be born dead.

Little other supervision of farrowing is generally needed, apart from regular checking to ensure that the airways of the piglets are clear and that no animals are chilled.

Where precise records of service are kept, some degree of synchronization of farrowing may be practised. This may be to avoid the supervision of night or weekend farrowings and its consequential costs, or to practice balancing up of litters with piglets born at more or less the same time. Prostaglandin is the substance usually employed. If it is injected on the 111th day of gestation, farrowing occurs within 30 h with no deleterious effects on the piglets. Batch farrowing allows depopulation of the house, and the opportunity to clean and disinfect the farrowing area.

Farrowing crates

Under intensive housing conditions, it is usual to farrow sows in crates. These can reduce piglet mortality, particularly in the 48 h after the birth, and they allow a close watch to be kept on the sow, make routine handling of piglets simpler, and are an efficient use of space.

Crates provide a restraining structure for the farrowing sow so that she can stand or lie down, but without being able to turn round. They are batched in groups of about 10 in an easily cleaned temperature-controlled room. Room temperature needs to be in the range of 15–21°C for the benefit of the sows, and a further closed creep area (a warm protected box which allows piglets to huddle in a safe dry place) heated to about 21–27°C needs to be provided for the piglets, the higher temperature being preferable in the first few days of life. Farrowing crates come in a variety of designs, but are usually about 700 mm wide and 2400 mm long. The creep area may be at the front or side of the crate. There has been little alteration to basic crate design over the years, although attempts have been made to make modifications which will further reduce neonatal mortality. These include devices such as air blowers which are directed at the abdomen of the sow, providing a blast of air as she lies down, momentarily halting her descent and allowing piglets to escape from beneath her. If it were not for financial constraints, many producers would possibly abandon crates for a generous litter-filled space in which the sow could make her own farrowing nest, but the efficient use of such a system would need a high level of stockmanship and would be counterproductive in terms of the efficient use of expensive buildings.

Feeding during lactation

The quantity of milk the sow produces can be influenced by her nutrition during gestation as well as by the food she receives during lactation. The lactation curve of the sow has only a slight peak at 3–5 weeks. Milk production should not be stimulated either too early so that supply exceeds demand, nor so that supply is so generous in the latter part of lactation as to cause drying-off difficulties at weaning. Under-feeding also has to be avoided because, if the sow continues to use her own body reserves for milk production, consequential problems, such as a delay in returning to oestrus, can result. Feeding during lactation should therefore take account of previous nutritional status, housing conditions, number of piglets in the litter, and intended weaning age. Many recommendations exist for feed allowances at this time, but are usually within a range of 4–7 kg/day of a balanced concentrate ration.

The piglet

Suckling period

Piglets are born naked, with little hair, no fat,

little liver glycogen reserves and poor disease immunity. It is essential that they are born into a warm, dry, clean environment so that they escape chilling, and also that they rapidly find a colostrum supply. Colostrum is a source of immune globulins which are directly absorbed prior to gut closure, as well as being a rich energy source and gut stimulant. Piglets will find their way to a teat directly after being born. It is imperative that colostrum is then available, and the sow's teats should be checked to ensure this. Most piglets which die in the first 3 or 4 days of life have been shown to have starved or never sucked. Supervision which ensures suckling is therefore to be encouraged, and time spent in transferring weakly piglets to other sows, or feeding colostrum through a stomach tube, is probably well spent, since the weakly piglet will soon cope for itself and thrive if it survives through this period.

Piglets need further attention during the first 24 h of life. Eye teeth have to be clipped to avoid damage to the sow's udder and the faces of competing piglets, and the umbilicus may need trimming and should be disinfected. It is also common practice to ear-notch the piglets with a simple coding system for purposes of identification (see pp. 166–168 for further details of these procedures).

Sow's milk is low in iron, so piglets need supplementary iron to prevent anaemia. This is usually given as an injection, although there is interest in increasing the iron content of the sow's feed with a consequential increase in the iron content of her milk, which might eliminate the need for injections. Piglets may run the risk of becoming anaemic at any time before they are eating solid food.

It is usual to begin offering creep feed to piglets from about 4 days of age. Creep feed should be a high-density, highly palatable product and offered in small quantities at frequent intervals. This tempts the piglets to eat solid feed and aids gut enzyme development. A clean source of water should be available from about 7 days of age. The chosen system – nipple drinker, straw drinker or trough – should be placed at a suitable height for the piglets and should be the same type as that in use in the weaner house because, if the piglets learn to use a drinker in the farrowing pen, the risk at weaning of dehydration resulting from inadequate water consumption will be reduced.

Weaning

Under natural conditions, weaning does not occur until piglets are about 12 weeks old. By this time they are functioning much like an adult as regards digestion, and the natural lactating period of the sow is coming to an end. In terms of production, such a lengthy lactation is inefficient, because the lactating sow seldom comes on heat.

The earlier that weaning is practised, the greater the need for sophistication in housing, management, feed supply and disease control. Piglets can be weaned from 1 day of age, but such animals obviously need a very clean, warm environment and a manufactured food which is similar to sow's milk. Later-weaned, naturally-reared pigs are less costly to produce, because the producer is relying to a great extent on the sow to provide their nutritional requirements, and a less controlled environment is needed. The costs of very early weaning systems and the skills – both mechanical and husbandry – necessary to run them successfully, have reduced their popularity with producers.

Presently, most piglets are weaned between 3 and 5 weeks of age. This is a compromise position. Piglets of this age have a fair degree of acquired immunity, have adjusted to solid food and can be easily looked after. They will also be outgrowing the crate and creep area allotted to them in the farrowing house. The sow will need no special treatment for drying off, apart from a reduction in food, and access to water should be maintained. The removal of the suckling stimulus is adequate to begin gland reversion, secreted milk is reabsorbed and the process is complete within about 3 days.

Reconditioning of the reproductive system of the sow is complete within about 3 weeks after parturition and the sow is then ready to return to oestrus. While more pigs per sow per year may be reared the earlier that weaning occurs, sustained production is only achieved at a high management cost, and production time is lost in waiting for sows which were weaned earlier than 21 days to show oestrus. The *Codes of Recommendations* state that 3-week weaning should

Table 6.5
Some types of housing for growing pigs

Housing type	Controlled environment	Age of pig (weeks)	Group size	Litter/straw used	Comments
Weaners					
Multiplier pens	Yes	2–5	Whole litters	No	Used by a very small number of producers for early weaning. May have a future role for intensive care of small piglets
Flat-decks[a]	Yes	3–11	10–20	No	Popular system. Allows good observation of piglets, easily cleaned; slatted floors used
Weaner kennels[a]	Heating possible	3–11	10–20	No	Easily constructed, boxed units within an insulated building, with open dunging area discrete from lying area. Observation difficult
Weaner verandahs[a]	Heating possible	5–11	10–20	No	A kennelled undercover area with an outside slatted dunging area. Success depends on stocking levels and the frequent movement of the kennel lid to control temperature. Low capital cost, but high level of stockmanship needed for continued good production
Weaner pools[a]	No	5–11	Very variable	Yes	Any building can be used. Straw builds up for length of time pigs are penned. Fair level of observation possible. Disadvantages are increased workload (mucking out) and high building costs since much space is needed
Open-front monopitch	No	5–10	Whole litters	Maybe	Form of housing once popular for farrowing and follow on. Now pressed into use for weaners on some farms. Simple layout and good health control an advantage but high labour input and high level of stockmanship needed to maintain an optimum environment
Finisher pigs					
Enclosed and insulated[a]	Yes	10 weeks to slaughter	11–20	No	Usually with central feed passage. Pens slatted or part slatted along outside walls. Good observation possible.
Kennels with covered dunging area	No	10 weeks to slaughter	11–20	Yes	Popular in arable areas. High capital costs because of space requirement. Easy to manage
Open-fronted monopitch	No	10 weeks to slaughter	11–20	Maybe	Low capital cost. Stockman exposed to environment. High level of skill needed to maintain an optimum environment
Kennel and verandah	No	10 weeks to slaughter	11–20	No	Kennel for pigs with outside dunging over slats. Low capital cost but efficiency of use dependent on skill of stockman

[a] Illustrated in Fig. 6.4

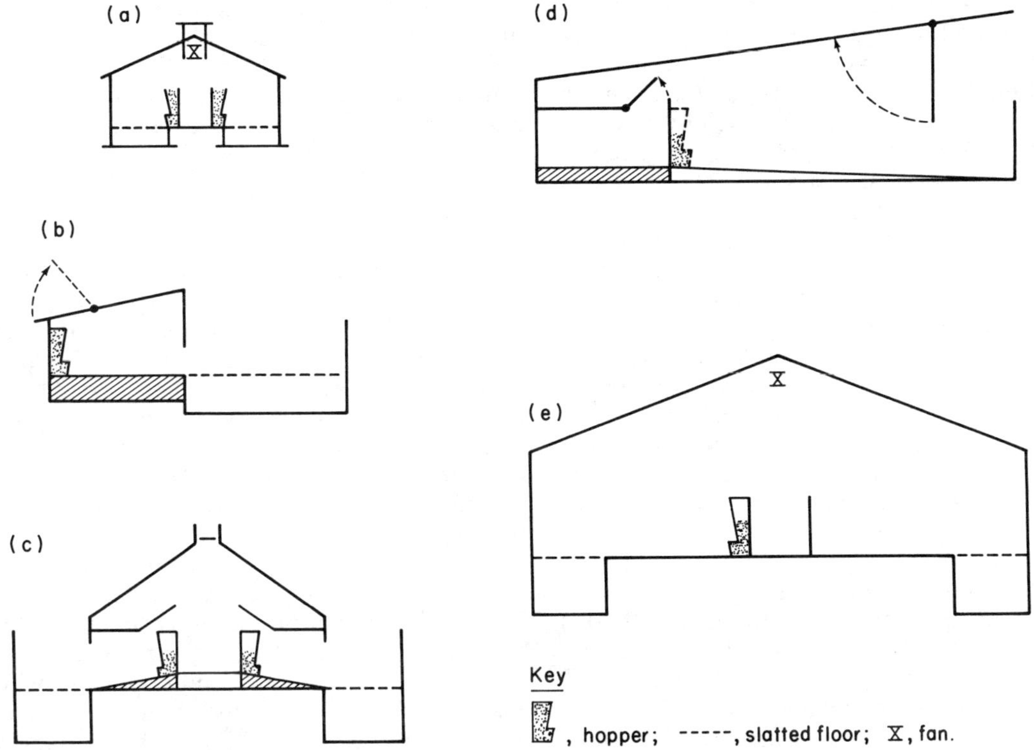

Fig. 6.4 Diagrammatic representations of housing types (not to scale). (a) *Flat deck rearing pens.* Insulated house. Piglets on slatted or perforated floors. Extractor fan in roof. Feed hoppers. (b) *Weaner kennel.* Insulated pen with hinged roof and insulated floor. Feed hopper. (c) *Kennel pens with verandah* within insulated building. Kennels with insulated floor. Feed hopper. Open ridge to roof. (d) *Weaner pool.* May be straw based. Insulated roof, floor and walls. Includes kennel with hinged, removable lid which pigs enter through a pop-hole. Usually hopper- or trough-fed. Of sufficient height to allow cleaning out from front with tractor loader. (e) *Finisher accommodation.* Insulated building. Complete environment control with fan and heater system. Pens with slatted or perforated floor dunging area. May be hopper-, trough- or floor-fed.

not be practised other than for piglets needing special treatment because of lactation failure or disease in the dam.

Meat production

All schemes for growing pigs aim to produce a lean carcass as efficiently as possible. The methods used depend on the resources available, but usually involve housing with some level of environmental control, together with the opportunity to manipulate feed intake.

Housing

Various forms of accommodation are used for growing pigs, all of which try to minimize housing costs while maintaining an adequate environment for efficient food use. Table 6.5 and Fig. 6.4 give details of some of the housing types which are available.

Feeding

Present-day pig types have been genetically

manipulated so that they reach a more advanced weight for age than earlier types before becoming fat. Pigs may be fed close to their *ad libitum* intake in the early phases of growth, up to about 60 kg live-weight, followed by a restricted scale as slaughter weight is approached. Ration scales are designed to cover frequency of feeding and size of individual meal. At the lowest level they must satisfy the maintenance need of the animal, at their highest level they represent either the appetite of the pig as regulated by physical gut capacity, or total energy limit. If rations of low nutrient density are used, then care must be taken to ensure adequate gut capacity for the bulk of feed needed. High-density diets often satisfy the pig without using the full physical capacity of the gut. This results in a reduced gut weight and a higher killing out percentage.

The ideal feeding system for the young pig would be *ad libitum* access, but the newly weaned pig may be unable to cope with this level of access to food and some level of restriction is usually practised. At 3-week weaning a pig will eat between 100 and 150 g of feed per day. Twice-daily feeding, associated with a gradual increase in the amount of food offered, should result in the pigs eating almost to appetite. The good stockman will be looking for complete consumption of a feed coupled with pigs keenly demanding feed at the next feeding time. A breakdown in this situation suggests overfeeding, ill-health or an environmental fault.

As pigs grow and increase in age, so the tendency to lay down fat increases and management begins to impose a more restricted level of feeding. Various forms of ration scale are encountered. Pigs may be allowed to eat to appetite up to a certain weight, then restricted either on a weight-based or a time-based scale, and finally given a flat rate of feeding until the slaughter weight is reached. Time-based scales are useful in that they have a smoothing effect on the fluctuations in growth which may occur. Also, the more frequently a scale is altered, then the more closely it matches the pigs' requirements. Examples of ration scales are shown in Tables 6.6 and 6.7. Individual pig producers manipulate the form of feed scale to take account of the genetic merit, sex and market outlet for their stock. Better strains of pig can produce good carcasses at high levels.

Entire males grow faster than gilts and castrated males, and many pig producers have abandoned castration of male pigs. This is recommended in the *Codes of Recommendations* and seems to be an increasing trend.

Handling, transport and slaughter

Most meat pigs are transported directly from the production unit to the abattoir. Handling, transport and slaughter should be carried out as humanely and quickly as possible to produce meat of a high quality. Stress and physical damage, such as bruising, must be avoided and well designed races and ramps with non-slip floors are essential. Male animals are liable to fight with strange pigs, and mixing should be avoided. Pigs should be held for the minimum length of time in lairage in order to keep weight loss to a minimum. Poor pig handling, particularly of very lean pigs, results in meat which may be either pale, soft and exudative (PSE) or dark, firm and dry (DFD). In both cases the meat is unpleasant in appearance and lacking in palatability. PSE is associated with some strains of pig, but more problems result from poor pre-slaughter treatment.

Outdoor production

Some 5 per cent of the national pig breeding herd is estimated to be kept outdoors. Successful outdoor production requires light, free-draining soils – chalk, gravel and sandy soils are all suitable. Areas of low annual rainfall and/or those where the winters are mild are preferred, and outdoor herds tend to be concentrated in the southern part of England, and East Anglia. However, outdoor herds are also found in north east Scotland. Large herd sizes are common with 1000 and more sows under one management. The success of the outdoor sow enterprise depends upon the sensible use of modern equipment and techniques such as electric fencing for paddocks and large plastic ear-tags for identification, and relatively small groups of animals (say 15–20), as well as skilled enthusiastic stock-

Table 6.6
Daily feed requirements of growing pigs based on live-weight

	Approx. daily gain 0.6 kg		Approx. daily gain 0.7 kg	
Live-weight (kg)	Medium-energy diet (kg)	High-energy diet (kg)	Medium-energy diet (kg)	High-energy diet (kg)
20	1.00	0.95	1.00	0.95
30	1.50	1.40	1.50	1.40
40	1.75	1.60	1.95	1.80
50	1.95	1.75	2.15	1.95
60	2.05	1.85	2.25	2.05
70	2.15	2.00	2.40	2.20
80	2.30	2.10	2.50	2.30
90	2.40	2.20	2.60	2.40

Source: MAFF (1980a).

Table 6.7
Daily feed requirements of growing pigs based on age (approximate daily gain of 0.6 kg)

		Feed requirements	
Age (weeks)	Approx. live-weight (kg)	Medium-energy (kg)	High-energy (kg)
9	20	1.00	0.95
11	27	1.35	1.25
13	34	1.60	1.50
15	42	1.80	1.65
17	51	1.95	1.75
19	59	2.05	1.85
21	68	2.15	1.95
23	76	2.25	2.05
25	85	2.35	2.15
27	93	2.45	2.25

Source: MAFF (1980a).

men. Such persons need to be committed to an outdoor system, if only to cope with the British weather, as well as being eager to record data and maintain stock and equipment under difficult conditions. Under good conditions of stockmanship, coupled with the forms of weaning practised under intensive conditions, sow performance is comparable with sows kept indoors.

Under this system piglets are usually weaned at 3–4 weeks and transferred to the types of weaner accommodation mentioned above (p. 162), where they can be reared under more controlled conditions. Piglets born under outdoor systems are frequently sold on for fattening at other units, in contrast to most indoor units which carry out all aspects of the production process.

Integration of an outdoor sow herd into an otherwise arable rotation can have beneficial effects on soil and crop performance. Normally, sows will occupy a pasture of, for example, a short-term grass ley for 12 months. They will then move on, to reduce the risk of enteric disease and internal and external parasites and before they have damaged the soil texture. After 1 year's grazing by sows, it is common to grow winter wheat. Stocking rates under these systems will vary with soil type and with land values, but densities of 10–15 sows/ha are common.

Feeding

It is essential that sows have access to a good water supply. As they are not able to consume sufficient bulky forage to satisfy nutrient needs for maintenance and foetal growth, a balanced concentrate ration must be provided, a medium-energy compound (12.5 MJ DE/kg) with a crude protein content of 16 per cent being considered adequate. This is fed in the form of a roll or cob, the relatively large size of which ensures minimum wastage. Generous feed scales are used so as to make allowance for the needs of gestation under doubtful weather conditions, and regular condition-scoring of sows is carried out to ensure that the animals are receiving adequate food. Under cold weather conditions, up to 4.5 kg of feed/day may be fed to a sow at service, rising to 8 kg during lactation. This quantity will include an element for litter size. A target feed figure for an outdoor sow is 1.25 tonne/year.

Arable products such as potatoes, carrots and other root crops may be fed to in-pig sows. The maximum allowances of these which are offered must be calculated in relation to the total appetite of the sow and the need for her to consume, in addition, sufficient concentrates to ensure a nutritionally balanced ration.

Housing

Simple half-round, timber-framed huts with solid backs and roofed to the ground with curved, corrugated iron are commonly used for both dry and farrowing sows. Large huts for up to five or six animals are used for pregnant sows, gilts or boars, while a smaller hut, incorporating a low boarded extension to retain the piglets, is often provided for the lactating sow.

General care and handling

Stockmanship

While breeding, feeding and housing may control the upper limit of production, the stockman will have the greatest influence on manipulating these factors. It is the stockman who ultimately carries out management decisions within the limitations of any unit and he must have the ability not only to carry out routine procedures such as ear-tagging, injections, teeth clipping, weighing and loading in an efficient humane manner, but also to recognize individual animals, to detect signs of unease or disease, and to take remedial action. He has to ensure that animals are as comfortable and contented as possible, while at the same time working to exploit them as efficiently as he can. The continuing process of intensification means that stockmen must increase their vigilance and their awareness of the signs of problems developing in animals. While catalogues may be made of the conditions that should be observed in animals (see p. 168), there is no substitute for training at a practical level – for example, the stock task courses run by the Agricultural Training Board and by local Education Authorities.

It is impossible to overstate the need for sound, efficient, humane stockmanship. It is only by paying regular attention to the well-being of his pigs that a producer obtains good returns and avoids the increasing opprobium of some parts of society not involved in production. Stockmanship has to evolve as production systems are modified and changed, although the basic skills required will not alter.

Routine procedures

Handling

Pigs are inquisitive animals, and quiet handling, using well-designed, well-constructed and well-maintained alleyways, passages and loading bays, will result in large numbers of animals being manipulated with little need for force, other than the occasional light tap from a stick, or encouragement with a handling board if a directional change is needed. Most units make regular use of fixed weighing, handling and loading facilities and, therefore, some time spent in planning the layout of a unit will save countless hours later on (Figs 6.5 and 6.6). At all times, handlers should remember that pigs are social animals and are more at ease in a fairly densely packed space than when separated from other pigs.

The traditional ways of handling individual pigs are very much related to their size. Piglets may be cradled in the hands or arms, while small to medium-sized pigs can be restrained by one person standing astride them with the animal perched on its hams and the front legs held. Large pigs need to be controlled by a snout noose placed around the upper jaw behind the canine tooth. Pigs are always noisy when restrained: stockmen need to be aware of this and should try to restrain animals quickly and quietly.

The loading area should be at the periphery

of the unit, with sufficient pens to avoid mixing groups, and with the entrance separate from the exit to the loading ramp (Fig. 6.6).

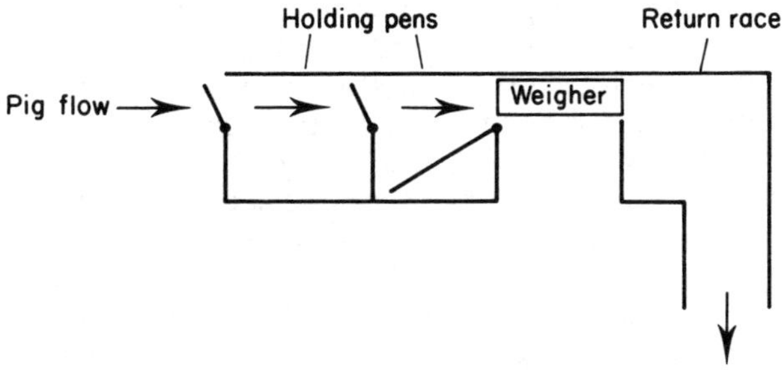

Fig. 6.5 Weighing and handling facilities.

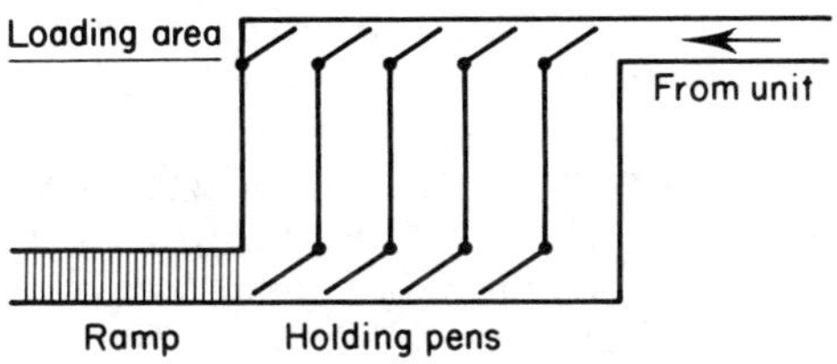

Fig. 6.6 Loading area.

Teeth clipping

Most breeders clip the incisor teeth of new-born piglets, since these might lacerate the sow's udder and also abrade the faces of other piglets during competition for teats. Such lacerations can easily become infected with bacteria. Tooth clipping is mentioned in the *Codes of Recommendations* which advise that it should be performed by a veterinary surgeon or a competent trained operator. No comment is made on the age at which this procedure should be carried out, but it should perhaps not be a routine task but confined to those litters where damage is seen to occur.

Docking

Tail-docking of piglets, leaving only a short stump, is routine on some units, usually taking place before the piglets are 7 days old. The procedure is used where there is a history of tail-biting in growing pigs. The *Codes of Recommendations* discourage the technique and stress the need for it to be carried out by a trained operator. Legislation* requires that after 7 days of age docking must be carried out by a veterinary surgeon, using an anaesthetic. All management and environmental factors should be considerd if tail-biting occurs on a unit. At various times, many factors have been implicated in its incidence, ranging from breeding factors to stocking densities, sudden temperature fluctuations and iron deficiency. Despite many investigations, the situation is still inadequately explained, and the removal of a piglet's tail does not necessarily cure the problem, because docked pigs will often show increased levels of neck and shoulder biting. The producer should attempt to correct his husbandry defects rather than mutilate the piglets.

Castration

This is an increasingly uncommon practice on pig units. Male piglets were frequently castrated at 1 day or so of age when they were small enough to be easily handled by one operator. The testicles are usually removed through low-placed scrotal skin incisions, each testicle being slowly pulled until the arteriovenous plexus breaks. Legislation forbids the castration of pigs with elastrators unless applied within the first week of life, and also requires that after the age of 2 months, the operation must be carried out by a veterinary surgeon, using anaesthesia.†

Iron injections

Piglets usually receive an intramuscular iron injection within 48 h of birth, although there is some interest in eliminating this handling operation and possibly reducing production costs by increasing the iron content of the sow's diet and so increasing the level of iron available in her milk. For any injection, it is essential to either

* Docking of Pigs (Use of Anaesthetics) Order 1974; Welfare of Livestock (Prohibited Operations) Regulations 1982.

† Protection of Animals (Anaesthetics) Acts 1954 and 1964; Veterinary Surgeons Act 1966.

restrain the individual animal or pack the group of animals tightly together. A poor injection technique can lead to abscesses, carcass staining and broken needles becoming embedded in the animal, all of which are causes of carcass rejection and down-grading.

Identification

Individual identification of pigs is important if accurate records of performance are to be kept. New-born piglets can be ear-notched or tattooed and/or tagged with a range of metal or plastic tags to aid identification in later life. Animals destined for slaughter should be slap-marked over the shoulders. This tattoo makes carcass identification possible so that complete records of an individual pig's performance can be achieved.

With all the above-mentioned procedures, stockman competence, together with a high level of equipment maintenance and general cleanliness, are essential for the efficiency and smoothness of the operation. All of these procedures cause some stress to the animals and it is usual to carry out as many as possible at the same time. Hence, ear-notching, teeth clipping, iron injection, tail clipping and castration may be done together within 48 h of birth. This blocking-up of procedures reduces handling and seems to have the least unsettling effect on the piglets.

Welfare considerations

The essential factors which influence the welfare of pigs have been summarized by Sainsbury (1984) (Table 6.8).

Health and disease

General health and hygiene precautions

In an ideal situation, a pig unit would be totally closed to the entry of animals from other sources. Feed, transport and personnel would be required to go through a full hygienic preparation before entry, and the pigs themselves would live in a germ-free environment. This ideal is difficult to sustain, even under experimental conditions, and the commercial pig producer can only aim for a clinically normal animal maintained in a state of balance with itself and its environment. To achieve this, he prevents infectious organisms from entering the unit as far as possible, by paying rigorous attention to washing out and sterilizing buildings and by taking such action as may be necessary in the use of routine vaccination and medication programmes. In practical terms, the unit needs to be isolated from all but essential inputs. Siting and layout should be designed to limit the risk of disease transmission within the unit.

Prevailing weather conditions must be taken into consideration, to limit airborne infection, and the reduction, if not elimination, of small mammals and birds must be taken in hand. The accidental introduction of a major infectious disease into a modern intensive unit can result in great economic loss. The effects of many subclinical infections interacting with an inadequate environment can be equally loss-causing, since affected animals fail to perform at their optimum biological levels.

Early and rapid detection of disease is essential if production losses are to be avoided. This will depend on efficient observation of all pigs, and the competent stockman will inspect his animals several times a day. Healthy pigs are characterized by a keen interest in feed: they should be bright, lively and alert, and look clean and sleek. Variations from this should be noted quickly and an attempt made to find an explanation. This explanation should be sought not only in a more detailed examination of the pig but also in a consideration of management actions and records, since the cause might be traced to some past event.

Unfortunately, as herds become larger, so the effects of infection increase, because animals often share the same air space, or come from the same susceptible population. Large units tend to have a continuous through-put of pigs and a low labour input. This too can lead to a time-lag in detecting infection, with consequential production loss. Paradoxically, a concern with separating pigs from their excretion products has probably not helped the development of adequate immune status in younger animals.

Table 6.8
Essential factors influencing the good welfare of pigs

1. Consider limiting the overall size of breeding units to 250–500 sows	9. There must be space for the essential behavioural needs of the pigs: freedom to stand up, lie down, stretch their limbs and groom themselves
2. Ensure all housing of young or growing pigs can be depopulated, cleaned and disinfected after each batch	10. Where there is no bedding there is a real danger of discomfort and injury for the pigs unless great care is taken with the design and construction of the flooring
3. Limit number of pigs in one common air space – about 16 sows and piglets, 200 weaners and 500 growing pigs, but smaller numbers are always advantageous	11. The building must be well lit to ensure good management, hygiene and disease control
4. Maintain temperature, humidity and ventilation rates at recognized ranges for age	12. Competition for feeding, watering, resting, exercising and dunging areas should be removed
5. Reduce the opportunity of a "build-up" or of "cross-infections" by separating pigs from their respiratory or intestinal excretions as quickly and completely as possible	13. There must be full provision to cope with mechanical breakdowns of essential equipment and emergencies such as fire
6. Ensure stocking rates are as generous as possible and the number of pigs in each pen are minimal to reduce competition and assist in the uniform distribution of pigs	14. No housing or management system should be provided which makes the pigs dependent on continuous drug administration
7. Whenever possible provide some bedding, preferably good quality straw	15. Nutritional essentials for pigs include enough fibre and bulk to ensure the comfort and well-being of the intestinal functions, and the housing design must provide uniform access to their feed for all the pigs
8. Separation of various age groups is a great advantage in helping growth and reducing disease incidence	

Reproduced from Sainsbury (1984)

The commercial producer aims to reduce disease levels so that they are not a drag on production, by using preventive hygiene and medicine. Preventive hygiene relies on the thorough cleaning of buildings between batches of pigs. High-power pressure washers are usually used, together with Ministry-approved disinfectants,* sometimes followed by fumigation. A stringent programme of washing and disinfection is dependent on the regulation of breeding programmes and the consequent supply of growing pigs so as to allow periodic depopulation of buildings between successive batches of animals. It is comparatively simple to apply such a programme to farrowing accommodation and housing for weaned pigs and finishers. The problem becomes more difficult with adult stock because housing for these is seldom completely emptied of stock. Here the use of preventive medicine, in the form of routine vaccinations and other treatments, as listed in Table 6.9, is important in keeping infection at a low level. Such treatments would include vaccination against swine erysipelas, atrophic rhinitis, and *E. coli* and *Pasteurella* infections, and inclusion in the feed of products for the control of swine dysentery and intestinal parasites. Some of these products can be routinely included only when covered by a veterinary prescription. Again, if such products are used, it is

* A regularly updated list of products suitable for disinfection and sterilizing procedures is available from the Animal Health sections of MAFF.

Table 6.9
Common disease conditions: their prevention and treatment

Class of pig	Housing	Common conditions	Routine control/treatment
Gilts	Yards/stalls	Endemic virus infections	Acclimatization + vaccination before service. Always quarantine *bought-in* gilts for 28 days before acclimatization
		Helminths	Anthelmintics at regular intervals
		Lice and mange	Regular lice and mange washes
		Coliforms	Vaccination
Pregnant sows	Yards/stalls	As for gilts	As for gilts, with rational use of booster vaccines
Lactating sows	Farrowing house	Lice and mange	Suggests inadequate late pregnancy treatment. Repeat wash if necessary
		Mastitis/metritis	Rapid treatment with antibiotics *plus* feeding attention for piglets
		Agalactia	Improve farrowing house conditions. Pay attention to feeding
		Hip and foot problems and lameness	Suggests poor pregnant sow management. Check records and accommodation. Cull severe cases
Piglets	Farrowing house	Chilling/anaemia	Correct the housing conditions. Iron injections
		Coliform scours	Routine use of vaccines. Treat with antibiotics
		Joint ill	Correct flooring defects if necessary. Query cleanliness of stockman's equipment
		Other enteric problems	Antibiotic treatment
Weaners	Follower housing	Coliform scours	Vaccination should have been carried out earlier. Treat with antibiotics
		Salmonella *Pasteurella* Atrophic rhinitis Pneumonic conditions	Improve environment. Use vaccines and antibiotics if necessary
		Swine dysentery	Medicate feed
		Erysipelas	Vaccinate at weaning
		Helminths	Anthelmintics usually administered in feed
Fatteners	Various house types	Problems as for weaners	Prior preventive measures, and the use of clean buildings should be adequate for this stage
Boars			Treat as for sows. Routine vaccines and regular use of lice and mange washes and anthelmintics

N.B. Always try to isolate infected animals. If curative treatment fails, or response is poor, then slaughter.

important that any instructions regarding withholding time prior to slaughter should be adhered to.

Health schemes

In addition to the health maintenance which individual producers may operate on their own units, in liaison with their veterinary surgeons, other nationally based schemes exist which strive to raise the health status of pigs countrywide. Schemes which merit especial mention are those run by the Ministry of Agriculture, and by the Pig Health Control Association, which in addition seeks to register and make known herds which are free of specific infections. Herds within schemes are regularly monitored by independent inspection and sampling of blood, etc.

Common disease conditions

These, together with methods of prevention and treatment, are listed in Table 6.9.

Notifiable diseases

Notifiable diseases which may occur in pigs include anthrax, Aujesky's disease, foot-and-mouth disease, swine fever, swine vesicular disease, Teschen disease and rabies (see Table 6.10 for details of those which are most likely to be seen in the UK). The presence of any of these, even if only suspected, must be reported immediately to the police or to the Government Department or local authority concerned. They can usually be controlled only by the most stringent measures (including slaughter, possibly of whole herds) and are liable to put whole populations at risk.

Zoonoses

Anthrax and rabies have already been mentioned under notifiable diseases. Probably the main danger to humans from pigs in the UK is *Salmonella* infection. Helminths such as *Trichinella spiralis* and *Trichuris suis* affect humans as well as pigs, but are not a major health problem in the UK.

On-farm slaughter

On all units there is frequent occasion to kill animals. These are most often small, runt or deformed piglets and, less frequently, older animals which are poor doers, or which have some traumatic injury such as a broken limb. Often these animals are of no commercial value, and arrangements must be made for disposal of the carcasses.

It may also be necessary to dispose of older pigs from time to time. For example, sows kept in close confinement and/or on slats may become so lame and in obvious pain that there may be no other recourse but to send them to a slaughterhouse immediately, assuming that they are still able to walk. If the condition of the animal is such that it has to be slaughtered on the farm, a veterinary certificate will have to be obtained before the carcass can be accepted at an abattoir. The good stockman should be aware of the need to dispose of animals for humane reasons before emergency slaughter becomes necessary.

While slaughter should be carried out as humanely as possible – the pig should not feel menaced or mishandled – the physical methods available to producers are limited and sometimes may seem brutal. Piglets are swiftly dispatched by a clean sharp blow to the back of the skull with a hammer. Growing pigs and adults are best stunned with a captive-bolt and pithed. The stockman carrying out these tasks needs training as much for this as for his more usual activities. A twelve-bore shotgun, a heavy-bore rifle or a revolver might also be used for killing pigs in an emergency. In this case the gun is aimed behind the ear of the animal ensuring that the shot passes forward through the brain, in contrast to the captive bolt which is placed about a finger's width above eye level, midway across the forehead and aimed well up into the head.

Disposal of carcasses

The disposal of dead pigs and foetal membranes

Table 6.10
Main symptoms of the most common notifiable pig diseases

Disease	Causal organism	Species affected	Symptoms
Anthrax	*Bacillus anthracis*	All animals including man	Listlessness, throat swelling, deaths from choking in 8–16 h
Aujeszky's disease	Virus	Mainly pigs, occasionally cats, rats, dogs, cattle or sheep	Sneezing, coughing, high temperature, trembling and/or convulsions
Foot-and-mouth disease	Virus	All cloven-hooved animals	Blisters on mouth and feet, lameness
Swine fever	Virus	Pigs	*Acute*: high temperature, arched back, greenish scours, reddish tongue *Chronic*: ill-defined symptoms, hind-quarter weakness, high temperature
Swine vesicular disease	Virus	Pigs	Similar to foot-and-mouth disease

is an ongoing problem on pig units. There are two methods which are used on larger units:

1. Incineration (usually fired by propane gas) which is suitable for smaller pigs, dismemberment being necessary for larger animals. Running costs are high and there is some odour.
2. A pit in which microbial digestion of carcasses can occur. It can be constructed from a large-diameter concrete pipe, with an earth floor and an airtight lid (see MAFF, 1980b). Planning consent from the Local and Water Authorities is usually needed.

The way ahead

The pig farmer, in common with other livestock producers, has responded to consumer demand for cheap foodstuffs by intensifying his production methods. As consumers have become more detached from animal agriculture, and as the quantity of money available to them has increased, so their needs have changed: they have become more critical of the methods used to raise animals and have begun to re-evaluate the product. The pig industry needs to respond to the consumer in a number of ways: it should be more open about how pigs are produced; it must explain that traditional methods of production are not automatically the best; it must be responsible in its use of drugs and, wherever possible, avoid the use of other feed additives. Farmers need to produce products the consumer wants; this implies that production methods must change, but not necessarily regress to more traditional technology, and that pig production must be seen to lose its apparently inhumane aspects.

References and further reading

Agricultural Research Council (1981). *The Nutrient Requirements of Pigs*. Farnham Royal: CAB.

Baxter, S. (1984). *Intensive Pig Production: Environmental Management and Design*. London: Granada.

Brent, G. (1982). *The Pigman's Handbook*. Ipswich: Farming Press.

Brent, G. (1986). *Housing the Pig*. Ipswich: Farming Press.

Bourne, F. J. (1986). Pigs. In Cole, D. J. A. and Brander, G. C. (eds), *Bioindustrial Ecosystems*. Amsterdam: Elsevier.

English, P. R. and MacDonald, D. C. (1986). *Animal behaviour and welfare*. In Cole, D. J. A. and Brander, G. C. (eds), *Bioindustrial Ecosystems*. Amsterdam: Elsevier.

Fox, M. W. (1984). *Farm Animals: Husbandry, Behaviour and Veterinary Practice*. Baltimore: University Park Press.

Hughes, P. and Varley, M. (1980). *Reproduction in the Pig*. London: Butterworth.

Lawrie, R. A. (1985). *Meat Science*, 4th edition. Oxford: Pergamon.
Meat and Livestock Commission (1987). *Pig Yearbook*. Bletchley: MLC.
Ministry of Agriculture, Fisheries and Food (1980a). *Nutrient Allowances for Pigs*. B 2089. London: HMSO.
Ministry of Agriculture, Fisheries and Food (1980b). *A Simple Disposal Pit for Foetal Material and Small Carcasses*. L 648. London: HMSO.
Ministry of Agriculture, Fisheries and Food (1983). *Codes of Recommendations for the Welfare of Livestock: Pigs*. L 702. London: HMSO.
Ministry of Agriculture, Fisheries and Food (1984). *Pig Production and Welfare*. B 2483. London: HMSO.
Muirhead, M. R. (1983). Pig housing and environment. *Veterinary Record* **113**: 587–593.
National Office of Animal Health (1987). *Animal Medicine Record Book*. London: NOAH.
Pond, W. G. and Houpt, K. A. (1978). *The Biology of the Pig*. Ithaca: Cornell University Press.
Ridgeon, R. F. (1984). *Pig Management Scheme Results for 1984 Agricultural Enterprise Studies in England and Wales*. Economic Report No. 92. University of Cambridge.
Sainsbury, D. W. B. (1984). Pig housing and welfare. *Pig News and Information* **5**(4): 377.
Sainsbury, D. W. B. (1986). *Farm Animal Welfare: Cattle, Pigs and Poultry*. London: Collins.
Taylor, D. J. (1983). *Pig Diseases*. Published by D. J. Taylor. Distributed by the Farming Press, Ipswich.
Walton, J. R. (1987). *A Handbook of Pig Diseases*, 2nd edition. Liverpool: University of Liverpool Press.
Whittemore, C. T. (1980). *Pig Production: The Scientific and Practical Principles*. London: Longman.
Whittemore, C. T. (1987). *Elements of Pig Science*. London: Longman.

7 Rabbits

J. O. L. King

The industry in the UK

Structure

Rabbit farms vary markedly in size. There are some specialist units with over 200 breeding does, and there are also enterprises with as few as 30 does, providing part-time occupation for the owner. On nearly all these farms the rabbits are bred and fattened and sold live to specialist packers who slaughter them and market the carcasses. Rabbit production and consumption in the UK is below that of most European countries and there are export outlets for quality carcasses, which could be developed.

During the Second World War, rabbit breeding in hutches reached a peak because families bred and fattened a few rabbits to provide meat for home consumption, but after the war ended this back garden industry declined rapidly. At the end of the 1950s, the importation of strains which had been developed for rapid growth and economic meat production led to the establishment of rabbit farms. In the 1970s, crossing of the imported strains was widely practised and, in the 1980s, hybrid lines were developed.

Rabbit farming is recognized as an agricultural enterprise by the Ministry of Agriculture, Fisheries and Food. The Commercial Rabbit Association, which was founded in 1960, is the body which looks after the interests of all those engaged in the rabbit industry.

Productivity

Forty years ago, the diet of rabbits consisted mainly of green foods in the summer, and hay, roots and cereals during the winter. Rabbits are now fed on pelleted complete foods which provide all the nutrients required. Wooden hutches with solid floors have been replaced by wire cages with mesh floors through which the faeces and urine fall. Does of the breeds formerly used produced about 20 young a year, but does of the modern strains can rear over 60 per annum. Some pathological conditions, such as mucoid enteritis and respiratory disease, have become more prevalent with the new methods, but others, for example coccidiosis, are less frequently encountered.

Breeds

Two pure breeds have been developed for meat production, the Commercial White, derived from the New Zealand White, and the Californian.

The Commercial White. This albino breed is the more popular. The bucks weigh about 4.5 kg and the does about 5 kg. The young grow rapidly and are efficient converters of food into meat.

The Californian. These rabbits have white bodies with black markings on the nose, ears, feet and tail, and pink eyes. Adult animals weight from 3.5 to 4.5 kg and are blocky in type. The young have a good dressing-out percentage.

Improvement schemes

Quality hybrid rabbits have been produced by

crossing individuals from three or more unrelated, inbred strains and then crossing the best progeny from these lines to produce a generation of superior animals of a uniformly high standard.

Artificial insemination has been performed successfully experimentally, but its use in commercial units is not justified economically because of the high labour requirements.

Outside pressures

Codes of Recommendations for the Welfare of Livestock: Rabbits have recently been published (MAFF, 1987). Those concerned with animal welfare are critical, in particular, of rough handling during transportation from the rearing farms to the packing stations.

There are relatively few environmental problems on well run farms. There are no noise problems, as rabbits are normally mute animals and, if the faeces and urine collected under the cages are removed regularly, unpleasant odours are rarely a cause of complaint.

Under the Agricultural Improvement Regulations, 1985, certain EEC-assisted capital grants for rabbit farm buildings and equipment are made available to farmers who have an approved improvement plan. Details are complicated and full particulars should be obtained from the Ministry of Agriculture, Fisheries and Food.

Natural history and normal behaviour patterns

The wild rabbit, *Oryctolagus cuniculus*, is the ancestor of all domestic rabbits. Wild rabbits live in burrows tunnelled in slopes. Although rabbits can be found on most types of soil they prefer dry areas and, to maintain good health, captive rabbits must be housed in well ventilated buildings and kept in cages providing dry floors. Wild rabbits live in colonies, but there is a social hierarchy with a small number of dominant bucks fighting for the best grazing areas. Thus, to prevent fighting, commercial stud bucks are always caged individually. Wild does give birth in short tunnels some distance away from the main site and farmed breeding does are kept in separate cages. In the wild, does make nests consisting of fur plucked from their breasts, leaves and grass, and in captivity they are given bedding with which to prepare similar nests in nest boxes.

The alimentary canal of the rabbit is particularly well adapted for the digestion of the large quantity of forage typical of an herbivorous diet. For this reason some, though not all, rabbit farmers supplement the pellet diet with good meadow hay. On welfare grounds its consumption may be beneficial by providing some activity for the rabbits.

Adult wild rabbits weigh about 1.5 kg, while domesticated meat breeds average from 3.5 to 5 kg. These heavier animals must be fed a more nutritious diet than the grass and weeds available to wild individuals. Wild does usually mate at about 4.5 months and bucks at 5.5 months of age, but the larger commercial animals are generally first bred when slightly older. Although rabbits can live for up to 9 years few wild animals reach this age, and most commercial breeding stock are killed when about 3 years old because of reduced prolificity.

Reproductive and breeding data

The mating of does should not be delayed for too long after they reach the age of 5 months (Table 7.1) because, if unbred, they may lay down internal fat, with a resultant failure to conceive.

The signs of oestrus given in Table 7.1 are not always shown, and some does will mate successfully when the vulva appears normal. Ovulation is induced by the mounting of the buck at mating. Does kept under good environmental conditions will remain in oestrus throughout the year, but anoestrus is likely to occur in does living in cold houses which are not provided with artificial lighting to extend the day length. A mating between the third and seventh days after parturition generally leads to conception, but this practice as a routine is not acceptable on welfare grounds.

Weaning young rabbits at 3 weeks of age can be successful, but is likely to check their growth, and is opposed because it causes unnecessary hardship.

Table 7.1
Reproductive and breeding data

	Buck	Doe
Age of onset of puberty	5 months	4 months
Age at which first bred	6 months	5 months
Duration of oestrus	—	Long periods. No definite cycle.
Signs of oestrus	—	Vulva purplish in colour. Slightly enlarged
Gestation period	—	31 days $\pm$ 2 days
Age of natural weaning	—	6 weeks
Age of commercial weaning	—	Usually 4 weeks
Normal breeding life	3 years	2 years
Average number of young	—	8–10
Seasonality of breeding	—	All year
Ratio of males to females for natural mating	1 : 10–15	

Although the average number born to does of a commercial strain varies from 8 to 10, litter sizes may range from 2 to 16.

The ratio of males to females depends on the breeding policy adopted on the farm; care must be taken not to overwork a buck because poor fertility can result. The maximum number of matings which should be allowed per buck is 4 per day and 12 per week. If it is the practice to mate a number of does on the same day so that several litters are born at about the same time, a ratio of 1 buck to 10 does is a common figure; otherwise 1 to 15 is probably adequate.

Nutritional requirements

The rabbit is, whenever possible, a frequent feeder, and its digestive system works most efficiently when there is a steady progression of food material through the alimentary canal. Compared with other mammals, the digestive transit is remarkably rapid. The stomach is thin-walled and is never completely empty in a healthy animal. To facilitate this, pelleted food in hoppers is continually available for caged rabbits to eat. The practice of coprophagy also helps to ensure that the digestive system always has partially digested contents passing through. The rabbit forms hard pellets during the day, which are voided, and soft pellets during the night, which are eaten directly from the anus. The soft faeces are swallowed in larger numbers after a period of fasting, in a caged animal during the night, and they are then moved through the alimentary canal as food is again eaten. The rabbit has a large caecum in which there is microbial digestion of starch and cellulose. Another advantage of coprophagy is that previously undigested components of the diet which have been broken down by the micro-organisms in the caecum are passed through the alimentary canal again and so utilized.

Commercial rabbits are almost always fed all-purpose pelleted foods as the sole diet, although a small quantity of hay may be given in addition. All-purpose rations generally contain from 60 to 70 per cent of cereals or cereal by-products – barley meal, ground oats, maize meal, weatings and bran – 10 to 20 per cent of grass meal, and about 20 per cent of high-protein foods such as fish meal and soya bean meal, together with a mineral and vitamin supplement. The minerals which are nearly always included are calcium, phosphorus and sodium chloride, together with vitamins A and D. Growth promoters and antibiotics are rarely used, but a coccidiostat is almost always added to reduce the incidence of coccidiosis in growing animals.

The pellets are fed *ad lib.* to most of the stock, but the quantities given to stud bucks and barren does are adjusted according to their requirements. A constant supply of clean water is always supplied, because the amounts drunk will vary with the individual, its size and age, and the temperature and humidity of the building.

Environmental requirements

It is possible to farm rabbits in outdoor hutches with solid walls and floors, but this is a much less efficient system than indoor housing and is rarely practised. Food consumption is increased to supply the additional energy required to maintain body temperature in cold weather and bedding is needed to provide warmth. Many does will not come into oestrus during the winter months and, if litters are born, the losses of baby rabbits from chilling are likely to be high. If outdoor farming is attempted, shelter from rain, wind and bright sunlight must be provided, this being best effected by extending the roof of the top hutch in a tier forwards over the gangway. Protection from vermin and predators must also be ensured.

On most rabbit farms the animals are housed in substantial buildings. The provision of several small units, rather than one large undivided space, is favoured, so that periodic cleaning and disinfection can be more easily carried out. Optimum environmental standards have not yet been clearly defined, but the figures given in Table 7.2 have proved satisfactory in practice. On most farms, all classes of stock are housed in similar buildings, although weaned young grow faster and utilize their food more efficiently at temperatures slightly higher than those advised for breeding stock. The higher temperature needed by baby rabbits is attained by providing nest boxes in which the does can make warm nests.

The buildings must be well insulated to retain the body heat coming from the rabbits in winter, because the provision of artificial heat is uneconomic. A regular supply of fresh air is essential, but draughts must be avoided. A natural ventilation system is adequate and economic, with outlets in or near the roof which allow the lighter, hot, exhaled air to escape as it rises and so facilitate the admission of heavier fresh air through inlets sited low down in the external walls. The building doors may need to be left open in very hot weather, but the entrance of vermin through the openings must be prevented. Poor ventilation leads to the accumulation of water vapour, carbon dioxide and ammonia, and this could render the rabbits more susceptible to disease.

Table 7.2
Environmental data

Air temperatures	
breeding stock	10–15°C
weaned young	16–18°C
young in nest	30°C
variation over 24 h	Not more than 5°C
Relative humidity	Under 80 per cent
Light intensity	About 10 lux
Lighted period in 24 h	14 h

The building should be well lit. Natural lighting coming through the windows should be fully utilized, but artificial light must be provided so that the rabbits can be seen clearly at all times. Light plays an important part in regulating the activity of the pituitary gland. Under natural lighting conditions sexual activity falls during the autumn and winter when the hours of daylight lessen and rises in the spring as the days lengthen. The use of supplementary lighting to provide a uniform light period of 14 h/day throughout the year is a great economic advantage because this encourages the regular production of litters and avoids the reduction in reproductive activity and the resultant shortage of young rabbits which occurs during the winter when reliance is placed on daylight alone.

Planning and general management

Site

The most important consideration for those producers who do not kill their rabbits and sell the carcasses is that the farm must be near a packing station prepared to purchase throughout the year all the fattened rabbits as soon as they have reached the required weight. A provisional agreement should be entered into with the selected packer before starting to farm.

Farmers designing a new building should first ascertain whether planning permission for such structures is necessary in their particular area, and should obtain building regulations approval before starting to erect the building. The site must be approached by roads wide enough for the lorries which will deliver food and collect rabbits. An electricity supply should be near to keep the cost of installation low and an ample quantity of wholesome water, preferably mains water, must be available.

Type of housing

Substantial houses are required. Disused farm buildings are utilized by stock farmers seeking an alternative source of income because of the reduced profitability of their present farming system. In adapting such buildings attention must be paid to the ventilation, because in many, particularly those standing in exposed positions, the existing method is inadequate for the needs of the number of rabbits to be kept. The installation of an electrically powered ventilation system may prove to be economic in buildings of this type.

When new premises are being designed, it is essential that the walls and roof are well insulated. Because of the high cost of building, many prospective farmers buy second-hand units, often surplus wooden poultry houses, but money spent in providing good environmental conditions should yield a financial return through the attainment of high production levels. On free draining soils, rabbit houses may have earth floors, but these are not satisfactory on a long-term basis, and the installation of concrete floors, which can be regularly cleaned and disinfected, is strongly advised. A separate food store for the storage of pellets and hay in dry conditions is necessary. Although expensive, the erection of a strong fence round the premises to keep out dogs and foxes is advised.

Effluent disposal

If a sewer system is not available for the removal of urine and the water used for washing down floors and cages, properly designed septic tanks must be installed so that these liquids do not contaminate the water in ditches and streams. Rabbit faeces are an excellent fertilizer for spreading on farm land and, if the rabbit farm does not cover a sufficient area on which these can be spread, an arrangement should be made with a local farmer to remove and utilize the faeces produced. Although rabbit manure can be heaped, it is advisable to have it removed regularly to minimize fly problems.

Some rabbit farmers have started breeding earthworms of a large, productive strain in the manure pits under the cages, claiming that this reduces offensive odours. The worms are sold to fishermen, fish hatcheries, aquarium keepers and zoos, and their sale can be quite profitable. A market for regular supplies should be obtained before such a project is initiated.

Cages

On commercial farms the rabbits are kept in wire mesh cages. These can be arranged in a single tier (Fig. 7.1) but, to make better use of the space in a building, they may be stacked in two, or even three tiers. The cages in the upper tiers are set back so that no cage is directly above another, to allow the urine and faeces to fall into the pit below. The selected arrangement must permit the staff to handle and feed all the rabbits easily. The cages are made of galvanized wire mesh and the joints must be smooth. An even floor surface is essential to avoid causing abrasions on the under surfaces of the hind feet. A square mesh not exceeding 19×19 mm is favoured as this allows faecal pellets to pass through and does not cause foot injuries.

The floor area must allow all the rabbits in the cage to lie comfortably and move around without disturbing the others, and eat and drink without difficulty. For rabbits over 12 weeks of age, the height should be not less than 45 cm. A cage of this height, with a floor space of $0.56\ m^2$ can accommodate one breeding buck or doe, a doe and litter up to 4 weeks of age, or eight weaners until marketed at 10 weeks old, and many farmers now use cages of this standard size only. The *Codes of Recommendations* (MAFF, 1987) give further details of the space allowances which are in use commercially for all types of stock in both cages and hutches, emphasizing that these should be regarded as absolute minima.

Each cage has a pellet hopper hooked to the front and, if the farmer feeds hay, a hay rack is also fitted outside the cage. An automatic watering system with a nipple drinker in each

Fig. 7.1 Welded wire mesh cages with food hoppers and automatic watering.

cage is an essential provision. The pipes supplying the nipples are connected to a header tank fixed inside the building so that the water is warmed by the heat in the building before it reaches the rabbits.

Management of breeding stock

Stud bucks and breeding does are always housed individually and must be carefully rationed to prevent over-fatness. Most bucks, and does which are not heavily pregnant nor lactating, require about 110 g of pellets/day, but any which fail to maintain good bodily condition must have this allowance increased. Pregnant does are fed *ad lib.* and during the last week of pregnancy their consumption of pellets will rise to 200–225 g/day. Water must always be available, and an adult will drink about 0.28 litres and a pregnant doe about 0.57 litres daily. The exact amounts will depend on the temperature and humidity of their surroundings.

Mating

For mating, the doe must always be placed in the buck's cage, because serious fighting can occur if the buck is taken to the doe. The doe should be placed in the cage with her back to the buck because some keen bucks mate at once. A doe which is anxious to be mated will raise her hind quarters ready for the buck to mount, but some only adopt this positon after the buck has started mounting. It may be necessary to hold a young doe during her first mating. One hand should be placed over the ears and shoulders and the other used to support the body, raising the hind quarters so that the doe is in a suitable mating position.

After mating, the buck falls sideways, often emitting a slight scream. Many breeders allow a buck to mate the doe twice before removing him from the cage. If there is no mating within a few minutes the doe should be taken away and re-introduced about 6 h later. If a doe which is clearly in oestrus fails to accept the buck she may be placed in a cage previously occupied by a buck and left there for 1–2 days. This may stimulate sexual desire.

If artificial insemination is to be attempted, expert instruction must be sought, because the technique must be carried out efficiently if it is to be successful. Semen is collected twice a week in an artificial vagina as the buck mounts an immature or a dummy doe. The semen can be diluted 10 times with normal saline, but will only remain viable for up to 24 h. Between 2 h before and 3 h after inseminating a doe, a dose of luteinizing hormone must be injected into an ear vein to induce the ovulation which normally follows the mounting of a buck. A pipette containing diluted semen is inserted into the doe's vagina and the semen solution is expressed. Although the semen from one ejaculation will suffice for up to 25 inseminations, the conception rate is only about 60 per cent of normal.

Pregnancy diagnosis

Pregnancy diagnosis by abdominal palpation should be carried out at between 14 and 16 days after mating. A doe is placed gently on a non-slippery surface so that she is relaxed, because if she is frightened the abdominal muscles will be tensed, making palpation difficult. By placing a hand under the body slightly in front of the hind legs, the embryos can be felt as marble-sized ovals slipping between the thumb and fingers on either side of the mid-line of the abdomen. If a doe is not pregnant she can be re-mated, preferably on day 18 or 19 after the unsuccessful mating. These days are selected because a barren mating may result in a pseudo-pregnancy which lasts for 16 or 17 days, at the end of which there is usually a period of greater fertility.

Parturition and suckling

About 1 week before a doe is due to give birth, she is provided with a nest box and nesting material (usually hay) which she will supplement with fur plucked from her breast. Most nest boxes are made of wood and are disinfected before being used again, but some breeders favour stout cardboard boxes which are only used once. The great disadvantage of these is that some does eat away the walls and so reduce their efficiency. A nest box (Fig. 7.2) should be about 40 × 25 cm. It must be large enough for the doe to feed the litter easily and small enough to keep the young together for warmth. Three walls should be 25 cm high and the front 15 cm. The sill in front enables the doe to enter without difficulty, but prevents the young from being drawn out on a teat while suckling if the doe is suddenly disturbed. The nest boxes are placed on the floor of the cage and most have open tops to allow the young to be inspected easily. On the day after parturition the litter should be examined and any small or dead young removed. Occasionally a doe will urinate in the nest and, if this occurs, dry new hay should be substituted for the damp material; all dry, uncontaminated fur should be put back in the nest. As a doe only feeds her young once in 24 h, breeders with poorly insulated buildings may, during particularly cold weather, remove nest boxes containing very young rabbits after the doe has suckled the litter, keep them in a warm room, and return them each day for a sufficient time for the doe to feed them. The nest boxes are taken away as soon as the young leave the nest, to allow more floor space for the doe and litter.

Fostering

If a number of does have given birth at about the same time, when the litters are examined on the day after parturition, some youngsters from the larger litters can be fostered on to does with small litters. Ideally, each doe should have about 10 young to suckle. Before introducing new young, the doe is removed from the cage and the nest is carefully opened from the top so that the sides are not disturbed, and the fur lining is left in its original position. Some breeders rub the young to be fostered with faeces from their new mother before placing them in the nest, but with placid does this is not necessary. Attempts to foster animals more than 4 days old are almost always unsuccessful.

Fig. 7.2 Californian doe with young in nest box. (Courtesy of J. C. Sandford.)

Weaning

The usual age for weaning the young is 4 weeks, and the doe is re-mated when the young are 3 weeks old, at which time her milk yield is rapidly decreasing. This allows 3 weeks in which the doe's body tissues can be built up before the birth of the next litter and enables her to produce five, or possibly six, litters per year. An alternative, practised on some farms, is to wean the young at 3 weeks, having re-mated the doe between the third and seventh days after parturition. This cannot be advised, because of the stress caused by full lactation and early pregnancy occurring at the same time. This leads to a shortened productive life for the does and a lack of condition in the young. A compromise, which is being employed increasingly, is to remate the doe at 2 weeks after parturition and still wean the litter at 4 weeks of age. This enables the doe to give birth to eight litters per annum. If the stock is carefully supervised and very well fed, satisfactory results can be obtained, but on welfare grounds this practice can only be condoned if great attention is paid to the condition of each doe throughout the year.

After a litter has been weaned, the doe's teats should be examined to see that the milk supply has dried up and the mammary glands are regressing normally without any heat or swelling. If there is any congestion of the glands, their activity may be lowered by reducing the concentrate ration and feeding hay to compensate. Re-mating the doe at once also helps. In severe cases, fomenting the glands with hot cloths will relieve the discomfort.

Food and water

The pellet consumption of the doe rises markedly during lactation and she will consume 450 g/day or even slightly more. It is essential to ensure that pellets are always available. A lactating doe with a large litter close to weaning may drink as much as 4.5 litres water/day. If she is unable to obtain all the water she requires, her milk yield will be reduced and the young will be underfed. On economic, as well as welfare grounds, great care must be taken to see that the nipple drinkers in the breeding cages are working efficiently.

Meat production

On nearly all rabbit farms, the animals sold have been bred and reared on the premises. Attempts have been made to establish fattening units for rabbits collected at weaning from their breeders, but the outbreaks of disease resulting from the mixing of rabbits from several sources have rendered this system uneconomic. Because of the suffering resulting from this high incidence of disease, the practice must be condemned for humanitarian reasons.

Before planning to sell any rabbits to a packing station, it is essential to ascertain the weight of rabbit required, because there is no standard grading system, and different packers have slightly varied market needs to satisfy. However, the favoured range is from 2.2 to 2.5 kg, weights which should be reached at 8–10 weeks of age. As rabbits may lose as much as 0.1 kg during transit from the farm to the packing station, the desired weight should be slightly exceeded before dispatch.

In conformation, a blocky type of rabbit is desired, with well muscled hindquarters and a short chest area. The killing out percentage at slaughter is a good guide to body type in life. Reports from a packing station are valuable indications of conformation quality and should be obtained and studied carefully. The flesh must be free from bruising, necessitating careful handling when crating rabbits for transport. As the liver is a valuable by-product, the herd must be kept free from hepatic coccidiosis, a disease which causes white spots on this organ, leading to down-grading and rejection.

The rabbit breeder must try to provide a regular supply of rabbits throughout the year. As indicated earlier, it is more difficult to breed during the autumn and winter when the demand for carcasses is highest. To some extent this discrepancy can be evened out by freezing carcasses during the summer and selling them in winter, but this adds to the cost of marketing, and frozen meat is not liked by some consumers.

Farmers with large units sometimes slaughter their own animals and sell the meat to large food firms either as whole carcasses or as prepacked joints. The demand for attractively packaged joints ready for cooking is increasing, and the higher price obtained more than covers the cost of preparation. Before adopting this marketing method, the local Environmental Health Officer should be contacted to ensure that the slaughtering conditions comply with the Food Hygiene Regulations.

The usual method of slaughtering meat rabbits is to administer a heavy blow to the back of the head, and then to decapitate the animal. For the small specialist market which requires a carcass complete with head, either dislocation of the neck, or electrical stunning followed by cutting of the throat, will ensure that the skull and brain remain intact.

Most rabbits are fattened in groups of about eight, in cages of the standard type, but it is possible to run larger groups in colonies on wire mesh floors. The additional exercise taken by the occupants of the more spacious areas increases food consumption and there is a greater risk of bullying. It is also not possible to arrange colony type pens in tiers.

The amount of food consumed by a growing rabbit in the period from weaning to sale varies between 100 and 115 g/day. In view of the high cost of pelleted foods, the food conversion rate is an important economic measure and should be about 3:1, although in some herds a narrower ratio is achieved.

Production of pelts

The young age at which rabbits are slaughtered for meat reduces the value of their pelts, because these are not sufficiently mature for manufacturing into garments, but the hairs are used in the production of felt. Pelts from older rabbits, after tanning, can be made into coats, coat collars or gloves. White furs are preferred to coloured because they can be dyed. The low value of rabbit skins is the reason why no rabbit farms have been developed in the UK with pelt production as the primary objective.

Wool production

Angora rabbits have been selectively bred to produce long, dense wool which is spun, knitted and made up into a variety of fashion garments. These are soft, silky, warm and washable. Angora wool is only produced on a small scale in the UK, where many breeders spin and knit the wool from their own stock, but there are larger industries in France, Russia and China.

Most Angora rabbits are white in colour and their wool is usually dyed, but they are now being bred in a number of shades including golden, chocolate, smoke and blue. Their management is labour-intensive, as they must be caged individually. Their coats require periodic grooming, but there are strains which carry a type of coat that is comparatively free from matting. The wool is clipped or plucked every 3 or 4 months and economical yields should be produced between the ages of 9 months and 4 years, after which the rate of wool growth slows considerably.

Rearing of replacement stock

Breeders of pure Commercial White or Californian rabbits usually breed their own replacement stock, occasionally buying in a new buck, to avoid excessive inbreeding. Those who favour cross-breeding keep animals from the lines used for crossing and breed replacements from these. Crossbreds should not be used for breeding as their progeny will not be uniform in type. Large producers with hybrid stock will also retain animals from their carefully selected, inbred parent lines.

The choice of the rabbits to be used for future breeding is a very important part of a rabbit farmer's work because the future success of the herd depends, to a large extent, on the quality of its breeding stock. The aim must be to select animals likely to improve the productivity of the herd.

The first step is to make a careful study of the herd records. The rabbits to be considered should be the progeny of mature does which have bred several litters. These does should have been successfully re-mated at the required times, particularly during the winter months. Litter numbers should have been uniformly high with few deaths between birth and weaning. At weaning, all the young should have been uniform in size and should have then grown steadily to attain the required killing weight at the earliest possible age for the strain, having efficiently utilized their food as shown by the food conversion ratio. The opinion of the packer should have been sought to ascertain that the carcass quality of rabbits from earlier litters was good. Prospective replacements should have been sired by bucks which had mated keenly and produced good sized litters with a high carcass quality.

Individuals from a selected litter should be examined carefully at killing weight and any which show any weakness in conformation must be discarded. A blocky appearance is most important in a buck, and a doe must have at least eight teats. Those chosen should then be caged individually until they reach breeding age – those which do not grow evenly should be rejected. The handling which regular inspections entail is advantageous, because the rabbits become accustomed to human contact. During this period, the quantity of pellets consumed per day will increase to about 140 g, but this figure should not, normally, be exceeded because, when they reach breeding age, stock animals should be fit, not fat.

General care and handling

Stockmanship

The success of a rabbit farm depends on the efficient management of highly productive animals. A skilled stockman can rear rabbits with few losses in moderate surroundings, while one who is not properly trained or is inefficient will not obtain good results even in excellent buildings. It is essential that all the rabbits are carefully inspected at least once a day to make certain that they are in good bodily condition and that they have ample supplies of food and water. The suffering caused by faulty husbandry and disease must never be underestimated. Anyone considering commercial rabbit production should attend a short course of instruction, for example one of those organized annually by the Commercial Rabbit Association, and then, if possible, work as a pupil on a successful farm. Many rabbit farms fail because the owner has had little experience and does not appreciate the difficulties likely to be encountered. A good stockman must know what action to take in an emergency – for example a fire, the freezing of the water in an automatic drinking system or a failure in the delivery of pelleted food.

Handling

Rabbits should be lifted by placing one hand over the ears and firmly grasping the skin over the shoulders, at the same time slipping the other hand under the body to bear its weight. They may be carried in a similar manner. Occasionally young animals are lifted by grasping them gently round the loins with one hand.

Routine procedures

Toe nail trimming

The toe nails of adult rabbits must be shortened periodically to avoid overgrown nails catching on the cage floors. The rabbit is placed on a flat surface and each foot is held firmly while the claw end is cut with strong nail clippers. In white-footed breeds the red quick shows clearly and must not be cut. In black nails the quick is not obvious, so, to avoid causing bleeding, only the sharp tip should be removed.

Sexing

It is possible to sex rabbits soon after birth, but a more common age is when they are weaned. The rabbit is held with one hand grasping the ears and underlying skin while the other is used to apply gentle pressure to the sides of the reproductive orifice. In the male this will express an immature penis, while in the female a V-shaped slit appears.

Record keeping

Accurate records of stock performances must be kept to enable poor producers to be eliminated and good ones used for replacement breeding (see p. 184).

Identification

Records which are of any value can only be kept if each breeding rabbit can be easily identified. The most satisfactory method is to tattoo an individual number in the ear. This is performed with a set of tattooing forceps, numbers and ink. Aluminium ear-tags can be used but these may be torn off by the rabbits with resultant tissue damage. The closed leg-rings used by fancy rabbit breeders should not be employed. They are expensive and, as the leg sizes of commercial rabbits are not standardized, some rings will be too small and will cause pain and swelling through pressure.

Weighing

The regular weighing of stock at different stages of their development is desirable, and accurate scales with a deep pan into which the rabbits can be placed are an essential piece of equipment.

Administration of medicines

The best, and most widely used, methods are to include the drug in the food, in the case of disease prevention (this can be done by the compounder) and, for curative measures, in the drinking water. It is possible to administer liquid medicines to individual rabbits, but this is rarely practised. Medication of individuals by injection is preferable and there are three methods: subcutaneously behind the shoulder, intramuscularly into the muscles at the top of the hind leg, and intravenously into one of the ear veins.

Transport

Adult rabbits must be moved in individual containers, to avoid fighting. Young animals are normally transported in wooden or metal crates, preferably in batches of about 10. These crates must be soundly constructed and adequately ventilated and the rabbits must not be overcrowded. The doors must be sufficiently large to allow the easy passage of the rabbits and, because the crates are usually stacked on the transporting vehicle, the floors must be solid to prevent the soiling of the animals in the lower units by faeces and urine from those carried above. As it is not practicable for food and water to be supplied in the crates, rabbits should not be confined in them for more than 8 h. If the time between leaving the farm of origin and slaughter exceeds this period, meat rabbits should be moved into metal cages, again in batches of about 10, and supplied with food and water.

Health and disease

Signs of health

A healthy rabbit is alert with bright eyes, wide open. It is active and has a good appetite. The body should be well fleshed and evenly covered with fur and the faecal pellets hard. Disease is any adverse deviation from this normal state. Thus, listlessness, eyes half closed, a huddled, tucked-up posture, a grinding of the teeth and a loss of appetite are indications of ill-health. Sneezing is a sign of respiratory disease, and head shaking may denote ear canker. These variations from normality can be observed while the animal is in its cage, and should be noticed by the attendant on his or her daily inspection. For a more detailed examination it is necessary to handle the rabbit. The nose should be dry, the inside of the ears clean, the fur sleek without bare patches and the undersides of the hocks free from sores. The fur round the anus must not be matted with faeces and the mammary glands should not feel hot. Common diseases are listed in Table 7.3.

Hygiene and disease prevention

When establishing a new herd, very great care must be taken to acquire disease-free rabbits, and the purchase of animals from a breeder who has been accredited by the Commercial Rabbit Association is advised. Steps must then be taken to prevent the introduction of bacteria, viruses or parasites. It is also important to avoid poor nutrition and bad environmental conditions which may reduce a rabbit's resistance to disease. On large farms, a closed herd can be maintained into which new rabbits are not introduced, but most farmers need to buy rabbits occasionally in order to maintain stamina and increase performance levels. These purchases must come from a reliable source.

Infectious agents may also be introduced by visitors, food, water, wild rodents and birds, and flies. Visitors who have been in contact with rabbits should be made to wear disposable overalls and shoe covers; food, particularly hay, should not have come in contact with wild rabbits; water should be obtained from a pure source; and rats, mice and wild birds should be prevented from entering the buildings. Flies can carry disease organisms and fly sprays should be used to control their numbers. Rabbits suffering from an infectious disease should be killed promptly because the value of individual animals is so low that treatment is rarely an economic proposition. Should a number of deaths occur, veterinary advice must be sought.

Disinfection

As a regular hygienic measure, and certainly after a disease outbreak, the buildings should be disinfected. All the surfaces with which the rabbits have come in contact must first be thoroughly cleaned and then the inside of the building scrubbed with a chemical disinfectant approved for general farm use under the Diseases of Animals (Approved Disinfectants) (Amendment) Order, 1975.

Isolation

Newly purchased rabbits, even if coming from a

Table 7.3
Common rabbit diseases

Disease	Clinical signs	Prevention and treatment
Coccidiosis		
intestinal	Diarrhoea in young. Sudden death	Wire cage floors. Coccidiostat in food. Sulphamezathine (POM) in water
hepatic	Often no signs in life but may be loss of condition. At post-mortem white spots on liver	
Mucoid enteritis	Mucoid material instead of faeces in young	Increase fibre in diet
Typhlitis	Diarrhoea with mucus. Oedema of caecum on post-mortem	Oxytetracycline (POM) in water
Ear canker	Yellowish-brown exudate in ears	Benzene hexachloride (GSL) in ears
Chronic rhinitis (snuffles)	Sneezing, bilateral nasal discharge	Antibiotics (POM), possibly vaccination
Mastitis	Mammary glands hot and swollen	Improve hygiene, antibiotics (POM)
Sore hocks	Necrosis on underside of hind feet	Avoid rough cage floors. Antiseptic dressing

Legal classification of veterinary medicines: GSL, general sale list; POM, requires a veterinary surgeon's prescription.

reliable source, should be isolated for 2 or even 3 weeks, because the stress following the move may allow the development of disease conditions which lay dormant on the farm of origin. Should the treatment of a rabbit suffering from an infectious disease be attempted, it must be strictly isolated to prevent the spread of the condition to other animals during the time required to effect a cure. Rabbits in isolation must be attended by a worker wearing a waterproof overall and gum boots kept for use in this area only.

Emergency slaughter

Young rabbits may be killed by manual dislocation of the neck (Fig. 7.3). With the left hand, the rabbit is held firmly by the hind legs just behind the hock, while the right holds the neck behind the ears and gives a sharp pull with a downward and backward twist. Adult rabbits are best killed by a blow with a heavy stick at the back of the neck. The animal may be held by the hind legs as before or placed on a flat surface and lifted by the ears until its front feet are just clear of the stand. A check must be made to ensure that the rabbits are killed by the blow and not merely stunned. The carcasses should be incinerated.

The way ahead

As the success of a rabbit farm depends to a great extent on the skill and dedication of the staff, the facilities for training stockmen need to be improved. The possibility of holding courses of specialized training at Colleges of Agriculture should be investigated.

The universal use of wire cages has led persons concerned for the welfare of commercial rabbits to be critical of the design of some cage floors, and an alternative should be sought. Perforated sheet-iron has been tried and discarded as inefficient and costly. Attempts to return to solid floors, or the use of bedding on wire floors, have failed, largely because of losses through coccidiosis. Improved drugs to control this disease might be developed.

An investigation into the optimum environmental conditions for rabbits of different ages might lead to better housing systems, and could be linked with studies to increase the efficiency

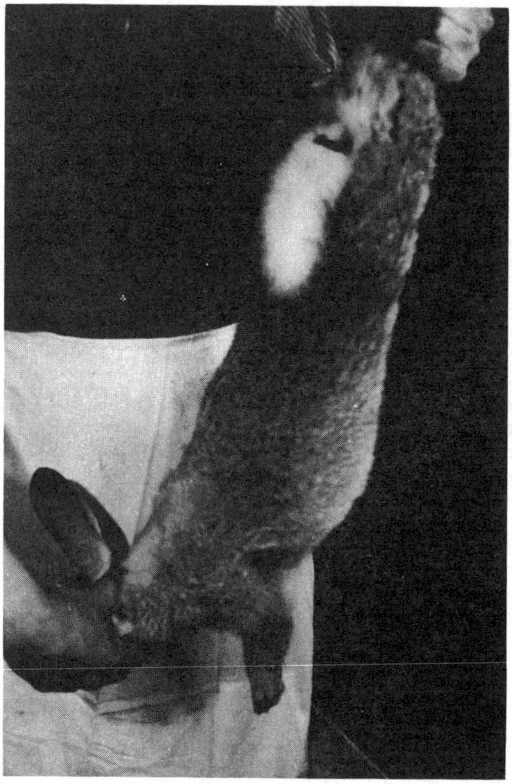

Fig. 7.3 Killing by manual dislocation of the neck. (Courtesy of J. C. Sandford.)

of winter breeding and the rate of growth of fattening rabbits. Attempts are sure to be made to raise the number of young weaned per doe per annum. This should only be acceptable if it can be achieved without adversely affecting the health of the mother.

On the marketing side in the UK, a publicity campaign to advertise the particular merits of rabbit meat could increase sales. The meat is white, fine-grained, nutritious and easily digested, and has a high protein, low fat content. Abroad, the enlargement of the UK market within the European Economic Community depends on having a stable and regular number of high quality carcasses for sale and the employment of experienced salesmen specializing in the selling of rabbit meat.

Further reading

Prospective rabbit farmers are advised to join the Commercial Rabbit Association, 172 Reading Road South, Church Crookham, Nr Aldershot, Hants GU13 0AE. They will receive a copy of the Commercial Meat Rabbit Producer's Handbook. The association also publishes occasional bulletins, a Code of Practice and a Journal.

References

Adams, C. E. (1987). The laboratory rabbit. In Poole, T. (ed.), *The UFAW Handbook on the Care and Management of Laboratory Animals*. Harlow: Longman.

Aitken, F. C. and King Wilson, W. (1962). *Rabbit Feeding for Meat and Fur*. Farnham Royal: Commonwealth Agricultural Bureaux.

Ministry of Agriculture, Fisheries and Food (1976). *Feeding of Meat Rabbits*. Leaflet No. 562. London: MAFF.

Ministry of Agricultre, Fisheries and Food (1977). *Breeds of Rabbits*. Leaflet No. 565. London: MAFF.

Ministry of Agriculture, Fisheries and Food (1978). *Breeding Principles and Systems*. Leaflet No. 556. London: MAFF.

Ministry of Agriculture, Fisheries and Food (1979). *General Management and Housing*. Leaflet No. 544. London: MAFF.

Ministry of Agriculture, Fisheries and Food (1985). *Commercial Rabbit Production*. Bulletin No. 50. London: HMSO.

Ministry of Agriculture, Fisheries and Food (1987). *Codes of Recommendations for the Welfare of Livestock: Rabbits*. 938. London: HMSO.

National Research Council (1977). *Nutrient Requirements of Rabbits*, 2nd edition. Washington, D.C.: National Academy of Sciences.

Netherway, M. E. P. (1974). *A Manual of Rabbit Farming*. Idle, Bradford: Watmoughs.

Portsmouth, J. I. (1962). *Commercial Rabbit Meat Production*. London: Iliffe.

Sandford, J. C. (1971). *Reproduction and Breeding of Rabbits*. Idle, Bradford: Watmoughs.

Sandford, J. C. (1986). *The Domestic Rabbit*, 4th edition. London: Collins.

Wiseman, J. (ed.) (1986). Nutrition of rabbits. In *Feeding of Non-ruminant Livestock*. London: Butterworths.

8 Structure of the UK poultry industry

C. M. Haan

Historical background

The structure of the UK poultry industry in the second half of the 1980s is the result of a number of closely interacting forces, the impact of which has been more dramatic than in any other major sector of British agriculture. The trend has been in the same direction for egg and poultry meat production as for primary breeding, stock multiplication and the growing of replacement stock. Before outlining the structure as it is, a summary of the forces that have operated to create the present pattern will help to improve perspective.

As wages in real terms increased during the 1950s and 1960s particularly, there was an inevitable and progressive replacement of labour by capital investment in housing and equipment. The changes in systems which were made possible by technological improvements arising from research and development, as well as by simple "trial and error", enabled labour productivity to improve dramatically. For example, one man was able to look after about 1000 layers on the free range system as practised in the 1950s. Today, one man can easily manage a flock in excess of 30 000 in a properly mechanized caged layer house.

The adoption of intensive systems, and the progressively greater use of mechanization, favoured the emergence of large units, and also the grouping and integration of those units. It should be noted that systems in which the birds were housed all the time were often preferred by staff because they were less physically demanding and there was protection from the weather, and because problems, when they arose, tended to be more easily managed.

At its peak in the 1950s there were around 800 poultry breeders in the UK actively selecting for improved stock performance – now there are less than a dozen of serious commercial significance by present-day standards. The principal reasons for this are, first, the changes in the market which has become dominated by the requirements of large production enterprises. These need large consignments of uniform stock which are impossible to provide unless the supplier is also operating on a large scale. Secondly, a few enterprises have been more successful than others in combining the application of sound genetic principles (where there are particular scale advantages), effective organization of stock multiplication, careful marketing and successful promotion. Several have been so successful that they are important in worldwide terms as well as nationally. In addition to the undoubted genetic gains through selection, the high standards of performance now being achieved commercially are the outcome of several other factors, notably increased knowledge of nutritional and environmental requirements and measures to control disease. Table 8.1 serves to show the scale of improvement that has been achieved over the past 30 years.

The market and the consumer

During the last 40 years, consumers in the UK

Table 8.1
Typical performance levels in 1987 compared with 1957

Type of production	Performance factor	1957	1987
Commercial eggs	Eggs/bird housed in 52 weeks	180	265
	Ratio of feed to egg mass (excluding rearing feed)	4.2	2.7
Broiler growing	Days to reach 1.8 kg live wt.	68	38
	Ratio of feed to live-weight (at these ages)	2.7	1.8
Turkey growing	Live-weight at 16 weeks		
	male birds (kg)	13.5	18.5
	female birds (kg)	9.9	13.9
	Ratio of feed to live-weight		
	male birds (16 weeks)	3.0	2.5
	female birds (16 weeks)	3.6	3.0

have moved from the stage of having to endure chronic shortage to one in which there is an element of continuous oversupply. In terms of eggs this led to a progressive increase in consumption up to the mid-1960s with a subsequent decline from the mid-1970s, due to changing eating habits and the availability of an increasing range of competing convenience foods (Fig. 8.1).

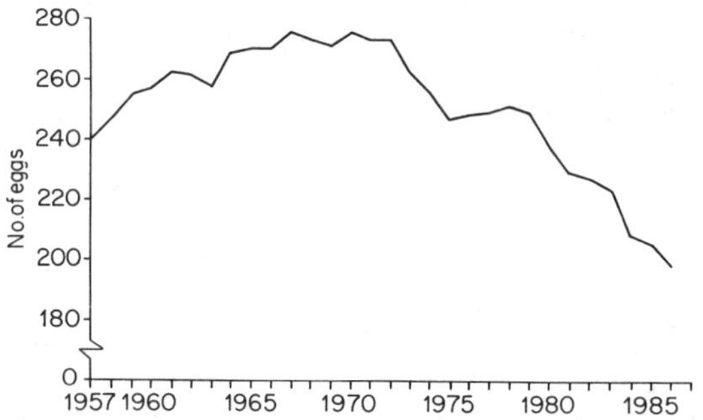

Fig. 8.1 Estimated annual egg consumption per head in the UK, including eggs consumed in products, 1957–1986. (From MAFF and Egg Authority data.)

Poultry meat was rather different. Typically regarded as a luxury in pre-war years, it was not until the latter half of the 1950s that poultry meat started to become available in quantities and at prices that attracted real consumer interest. From then on the growth of the broiler industry and, a few years behind, turkey production followed a remarkable course, with per capita consumption today some 10 times greater than in 1955 (Fig. 8.2).

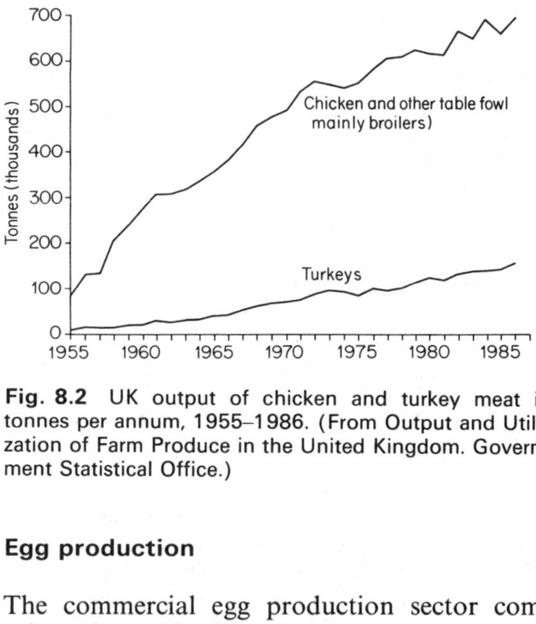

Fig. 8.2 UK output of chicken and turkey meat in tonnes per annum, 1955–1986. (From Output and Utilization of Farm Produce in the United Kingdom. Government Statistical Office.)

Egg production

The commercial egg production sector comprises about 41 million layers at any one time, with a further 2–3 million birds kept in domestic flocks not covered by the annual MAFF Agricultural Census. Although a year-to-year fluctuation of a million or so birds occurs, the population trend is towards a smaller national flock to accommodate rising egg output per bird and the downward trend in egg consumption over the past 10 years.

On the basis of the MAFF Agricultural Census Returns, it is possible to get a reasonable idea of the sizes of egg production enterprises found in the UK (see Table 8.2). In terms of ownership or management control, many of the

Table 8.2
Distribution of holdings and laying fowls by holding size groups in the UK, June 1986

	Birds per holding						
	1–49	50–499	500–4999	5000–19 999	20 000–49 999	50 000 and above	Total
Holdings	36 290	5533	1572	830	232	142	44 599
Holdings (%)	81.4	12.4	3.5	1.9	0.5	0.3	100
Laying fowls	581 400	660 819	3 190 571	8 275 564	7 068 163	18 050 715	37 827 232
Laying fowls (%)	1.5	1.8	8.4	21.9	18.7	47.7	100

Source: MAFF, DAFS and DANI statistics, June Agricultural Census, 1986.

large holdings are grouped within large companies. It has been estimated that one company accounts for 2–3 million layers and another owns or controls about 2 million. There are perhaps 20 companies, each with between 100 000 and 1 million birds.

In large enterprises there are typically a number of production sites, in some cases supplying separate packing facilities; and usually several sites at which replacement laying stock is brooded and reared to about 16 weeks. It is at this stage that the pullets are usually transferred to the laying accommodation. In smaller-scale enterprises, replacement pullets may be reared by the producers themselves, or purchased from specialized growers.

Mention needs to be made of egg packing and grading because of the inevitable close commercial arrangement that has to exist between production and subsequent stages in the marketing chain. All the large-scale producers either own or have a close relationship with an egg packing station. Here the eggs are graded according to the EEC Egg Marketing Regulations and then packed, typically in boxes of 6 or 12. In the UK there are also two large Co-operative Egg Packing Organizations which are owned by the producers who supply them.

Most commercial egg production is from birds kept in cages in windowless houses fitted with power ventilation and electric lighting. The number of birds per house typically ranges between 10 000 and 40 000. The use of battery cages reached a peak of popularity in the early 1980s, when about 96 per cent of the commercial laying flock was kept under this system. Since then, there has been a marked increase in demand for eggs produced from birds in non-cage production systems, principally free range, but including deep litter, percheries and aviaries (see Chapter 9). From the late 1950s to the early 1980s, free range egg production, and (latterly in that period) non-cage systems, were virtually the preserve of small-scale producer marketeers, typically selling in localized areas. In the last few years, the prospect of substantial price premiums for free range eggs in particular has resulted in:

1. An expansion of production by a number of those producers who were already producing such eggs.
2. Many newcomers from within and from outside agriculture.
3. A number of existing and often large-scale cage production enterprises diversifying into free range and sometimes other systems of production.

Some of the units established have involved many thousands of birds. Often these are kept, somewhat optimistically, in large flocks so as to facilitate capital and labour efficiency through the use of mechanized feeding and egg collection. It is estimated (1986–1987) that, in total, the number of commercial laying birds housed in non-cage systems of production, principally free range, is of the order of two million.

Producer marketeers

Because eggs are produced by the birds ready packaged in a form that is directly saleable without a processing stage, there has always been a substantial sector of the egg market – now probably of the order of 30 per cent – that has been supplied by eggs sold more or less directly by the producer to the consumer (e.g. from farm shops, in markets or from small retail outlets supplied directly by the producer).

Poultry meat

In some respects the structure of the poultry meat industry is more simple than the egg sector. Virtually all meat fowl, turkeys, ducks and geese are sold "oven ready" whether the plucking and evisceration are undertaken in large processing plants – the great majority – or on a smaller scale, on farm. These days, the proportion of birds sold to retail butchers in feather or plucked but uneviscerated (leaving the butcher to carry out these operations for his customers) is small and amounts to less than 1 per cent of total sales. It is helpful to deal with the species separately.

Broilers and other table fowl

About 465 million broilers are produced each year, although strictly a proportion of these – perhaps amounting to 2 or 3 per cent – are grown to greater weights outside the broiler live-weight range, which is normally taken to be between 1.0 and 2.5 kg. Virtually all fowls grown for meat are raised on the litter system in large flocks, typically 20 000–30 000 birds in a house.

It is not now possible to grow true capons legally. However, because modern meat type fowls grow rapidly, large tender roasting birds up to about 4 kg live-weight can be produced simply by feeding male broilers, or broiler type birds, beyond the normal killing age. There seems to be a growing market for them with a current production probably of the order of between 20 and 30 million per annum.

As broiler growers are able to place six, and sometimes even seven, crops per year in each broiler shed, Agricultural Census Returns cannot give a direct picture of annual output, but they do provide a sample at one point in time that shows the pattern of holding size (see Table 8.3). As in the case of the egg sector, the data serve to illustrate the dominance of large-scale holdings.

It is estimated that between 60 and 70 per cent of broiler production is under the complete control of the processing companies and that most of the remaining 30–40 per cent is in the hands of individual growers producing under contract to these same processors. In both cases a high degree of planning and co-ordination is essential to ensure that resources – especially the high capital investment and large labour force involved at the processing stage – are used efficiently. This applies to the supply of 1-day-old chicks to the growing sites according to a carefully established schedule, to the supply of the correct type and quantity of feed at the right times at each site and to the arrangements for catching and transporting the birds in an orderly way so that there is a continuous supply at the processing plant. If delays occur in transit, there is a risk of financial loss and a risk that the welfare of the birds may be adversely affected.

Increasing attention has been paid in recent years to the need to improve the whole process of broiler catching and transportation and this has resulted in the development of a number of new systems designed both to improve efficiency and to reduce the risk of injury and stress to the birds. As in the case of house cleaning and preparation that takes place on production sites between each crop, the catching and transportation of broilers when they reach marketing age may be organized through independent contractors or by teams directly employed by the company.

Current market trends are for a continuing growth in consumer demand for fresh and chilled – as distinct from deep frozen – poultry and poultry meat, and also for a wide variety of further processed products ranging from raw poultry portions to complete pre-cooked meals in which chicken meat features as a main item or a component.

Turkeys

An important characteristic of turkey production is the large peak in demand that occurs at Christmas time – a period during which

Table 8.3
Distribution of holdings and broilers[a] by holding size groups in the UK, June 1986

	Broilers per holding						
	1–999	1000–9999	10 000–19 999	20 000–99 999	100 000–499 999	500 000 and above	Total
Holdings	821	257	216	569	136	14	2013
Holdings (%)	40.8	12.8	10.7	28.2	6.8	0.7	100
Broilers	75 431	1 333 273	3 037 395	24 812 185	22 551 666	11 485 847	63 295 797
Broilers (%)	0.1	2.1	4.8	39.2	35.6	18.2	100

[a] It should be noted that the Agricultural Census collects data relating to the number of stock on holdings at a particular time. Because it would be normal to raise between five and six crops of broilers per year, the data are not a reliable basis on which to calculate national annual production of broilers.
Source: MAFF, DAFS and DANI Statistics, June Agricultural Census, 1986.

approaching half the annual output of turkeys and turkey meat is purchased and consumed (much smaller peaks occur at Easter and Whitsun). Because the demand at Christmas includes a strong requirement for fresh, whole birds, met most easily by small- to medium-scale producers geared to supply it, the sector has become strongly polarized. There is hardly more than a handful of large or very large producers who grow and sell on a year-round basis and who have become increasingly involved in the development and marketing of cut-up portions and prepared turkey meat products. This group accounts for between 70 and 80 per cent of annual turkey production. The remaining 20–30 per cent of annual output is in the hands of 1 000 or so small- to medium-sized growers, many of whom only produce for the Christmas market. There are several factors that have helped such producers to maintain their market share, particularly:

1. Their ability, usually by employing temporary labour, to cope with the hand plucking that is closely associated with traditional farm fresh turkey production.
2. The use of housing that at other times of the year is used for other purposes. This, together with the use of family labour or farm labour that is not as fully occupied as at other times of the year, helps to reduce the overhead costs.
3. The ability to promote the "farm-produced" image that large-scale producers selling through multiples and supermarkets find more difficult to convey.
4. The facility with which a limited number of customers can be provided with a high quality and personal service that the large-scale enterprise cannot expect to match.

For the large turkey production company there is considerable similarity with the broiler sector in the need for careful planning and effective co-ordination of the inputs and output if the enterprise is to be efficient.

Ducks

The duck sector has been one of the most stable in terms of output and production for many years, ranging between about 7 and 9 million birds per year. Until the 1960s, the Brecklands of Norfolk and Suffolk were the home to most of the ducks produced in the UK. Now over 70 per cent are produced by one large enterprise in North Lincolnshire, while most of the remainder are still produced in Norfolk.

Traditional duck rearing involved brooding on wire floors or litter and then a transfer to grass pens, often in flocks of many hundreds, at 3 weeks or so where they remained until marketing age at 7–8 weeks. This pattern of production was characteristic of the sandy Brecklands and is still used there to some extent. In recent years, the main development has been in the use of housed systems with either solid (littered) and/or wire floors. Improved stock through selection, and better understanding of nutritional requirements, has resulted in average liveweights of around 3 kg at 7 weeks of age.

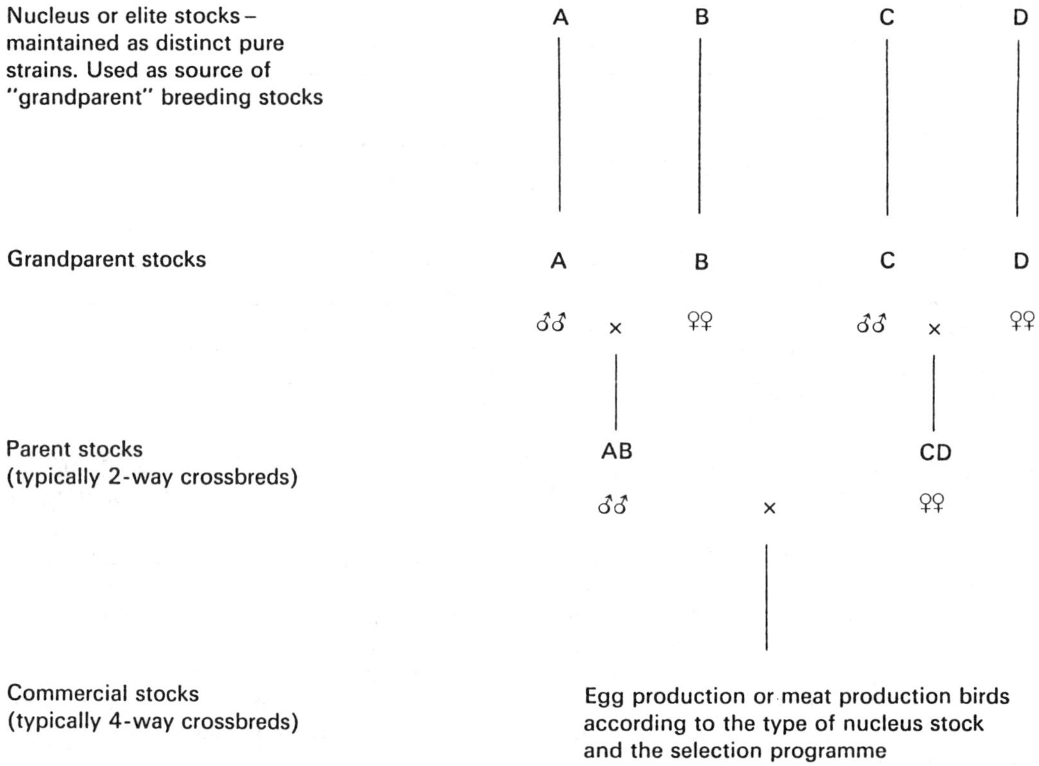

Fig. 8.3 Structure of poultry breeding programmes.

Primary breeding and stock multiplication

Whether producing replacement layer chicks, broiler chicks, turkey poults or ducklings, the general pattern – in large measure dictated by the characteristics of the stock – is very similar between one breeding organization and another. Primary breeding, involving the maintenance and improvement of elite stocks, is conducted under conditions of tight security to minimize the risk of disease outbreaks and to prevent competitors getting hold of a valuable resource. It is normal practice to maintain a number of strains, applying different levels of selection intensity to a range of traits. In the production of the commercial egg or meat type bird, there are a number of intervening stages between these elite stocks and the end-product.

Two objectives dominate, the first of which is to achieve a desirable balance of performance characteristics by the careful combination of one base stock with another. For example, in egg production there needs to be a balance between the number of eggs produced and the size of those eggs. Thus, by crossing a strain with high yield potential but moderate egg size with another which may be less productive but has better egg size potential, the progeny would be expected to show a balance between these two production factors. In practice the breeder has to pay attention to various other important performance characteristics as well. Typically, the end-product bird results from four basic strains (Fig. 8.3). The other major objective is to multiply the stock efficiently from the small source populations. This is achieved either by an expansion of the pure strain base population, or by introducing a multiplication stage after the first cross between the pure bred strains. Whatever the procedure used, in order to maintain security, it is strain-cross birds and not pure strains which are placed on multiplication farms to supply the hatching eggs which produce the end-product stocks.

High standards of hygiene are essential on the supply farms that maintain breeding stock. The hatchery normally co-ordinates the placement of breeding birds and organizes two or three collections of hatching eggs each week. The eggs, which in some cases are fumigated using formaldehyde gas on the farm, are transferred to the hatchery, where they are graded and frequently subjected to further fumigation before being set for incubation. At hatching, chicks are removed from the incubators, graded and, particularly in the case of egg production stocks, the male and female chicks indentified and separated. The use of specially designed boxes with compartments which are able to hold 25 or 50 chicks enable them to be dispatched safely on journeys that may take up to 24 h and sometimes longer for consignments sent to countries overseas.

Summary

The dramatic changes that have taken place in the UK poultry industry have been the result of a response by producers to the demand for poultry meat and eggs. In the past, this has applied particularly to the quantities of these products produced. Increasingly, product quality and product diversity will shape the course of the industry's future development as the consumer is offered an apparently ever widening choice.

On the farm, the roles of the manager and stockman have enhanced importance because of scale, the considerable amount of vertical and horizontal integration and the high levels of technology employed. However, the need for sound management and good stockmanship is vital in both the large, intensive and often highly mechanized enterprises and in the demanding circumstances of free range and other non-cage systems of production. Although there is a lack of consensus between producers and those concerned primarily with animal welfare issues on the components making up acceptable environments for stock, there are fortunately elements of agreement on an association between good stock performance and basic stock well-being.

9 Laying hens

A. H. Sykes

The biology of the fowl

The life history of the fowl starts with the laying of the fertile egg. Fertilization would have occurred about 26 h previously and, during the time in the oviduct, at a body temperature of 41 °C, development would have commenced, but it is suspended by the ambient temperatures experienced after laying; this enables a group or clutch of eggs to be laid before the start of incubation. In the wild, when a clutch has been laid, further ovulation ceases and the hen goes into a state of broodiness. A nest site would have been chosen earlier and the eggs would now be incubated for 21 days by the hen. However, selection for egg production has virtually eliminated broodiness and the use of the broody hen has been replaced almost entirely by accurately controlled, warm air incubators. On hatching, the chicks must be kept warm either by the hen, or by the provision of warm brooding conditions by man, for the 3–4 weeks it takes for the chick's temperature-regulating system to mature. The egg yolk remaining in the chick at hatching provides nutrients for about 1 day, but food and water are essential after that. Chicks grow rapidly for several weeks and then more slowly for a total rearing period of about 18 weeks. Reproductive maturity is now imminent, and the female chicks (pullets) are transferred to laying quarters at this time. The age at which the first egg is laid depends primarily upon the daylength to which the birds have been exposed. In summary, long or increasing daylengths advance maturity, short or decreasing daylengths retard it. The control so exerted is only partial; maturity may be as early as 17 weeks or as late as 24 weeks, but within this span the exact age at which laying begins can have a considerable effect on the economics of production and thus control of daylength is an important aspect of poultry management (see p. 206).

The onset of maturity involves profound physiological change. Under the stimulus of hormones from the anterior pituitary gland, the Graafian follicles of the single (left) ovary develop into the future yolks of the egg. The yolk is a single ovum vastly distended by phospholipids synthesized in the liver, transported in the blood and deposited through the greatly enlarged vascular system of the ovary. The follicles form a graded series of sizes, the largest being the first to be shed from the ovary (ovulation). There is also concurrent growth and differentiation of the oviduct, in which will be laid down the albumen (egg white), the two shell membranes, additional water, and the calcareous shell which contains colouring pigment in certain strains (Fig. 9.1).

The vent, or cloaca, becomes wider and more supple and the pelvis wider to allow easier passage for the egg. There are considerable metabolic changes associated with the synthesis and transport of phospholipids and the absorption, storage and transport of calcium for the shell; these are brought about by the sex hormones (oestrogen, androgen and progestogen) secreted by the ovary. Food intake increases to meet the demands of egg production and a strong specific appetite for calcium becomes evident.

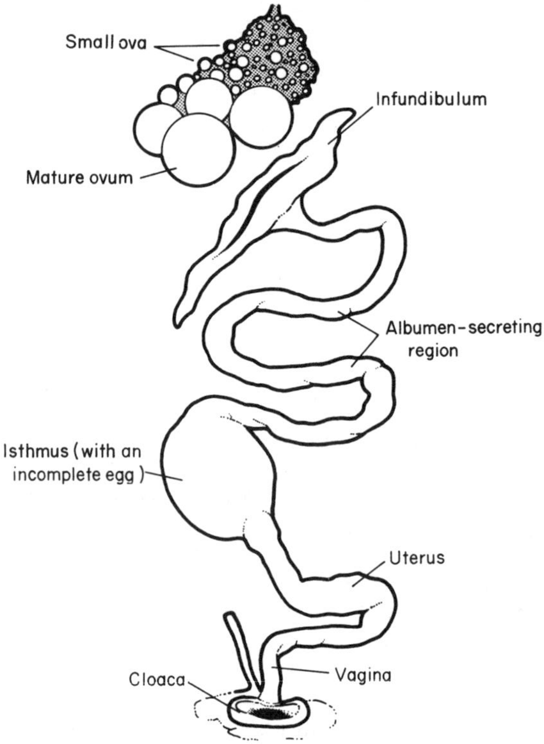

Fig. 9.1 The reproductive system of a mature hen.

The formed eggs are typically laid during daylight and at a progressively later hour on each succeeding day until darkness supervenes, when this pattern of laying is broken for a day (or more) until a new period of egg laying, called a sequence, starts once again in the early morning. In high producing hens the sequences are very long and so the period of egg formation is more nearly 24 h. Except for the first egg of the sequence, ovulation occurs within 15 min after laying (oviposition) but the two events are not related. The first egg of a clutch is ovulated after midnight some 28 h before it is laid. Ovulation is brought about by the pulsed release of ovulating (luteinizing) hormone from the pituitary gland governed by a timing mechanism that is influenced by daylength and the ovarian hormones.

As the hen ages, the rate of ovulation, and therefore egg production, becomes slower and the oviduct functions less well, resulting in a more watery albumen and thinner shells. Eventually, the ovary becomes less active, ovulation ceases and the consequent withdrawal of sex hormones then acts as a stimulus to feather growth which hitherto has been held in abeyance. The new feathers force out the older, worn ones and the whole process of feather loss and regrowth is called moulting. Reproductive activity is resumed when moulting is complete and the period of rest has restored, to a marked extent, both rate of laying and egg quality. A second laying period is followed by a second moult and this cycle can go on for several years.

However, towards the end of the first year the cost of egg production exceeds the income some time before moulting is due to occur; at this point, between 30 and 50 per cent production, according to prevailing prices, the producer either disposes of the flock or initiates the moulting procedure himself. To bring about such a forced moult, it is necessary to inhibit ovulation by some form of stress. Deprivation of food and water accompanied by greatly decreased daylength was the usual practice. However, this is considered inimical to sound welfare, and water restriction is particularly hazardous in hot weather; it was previously recommended that deprivation of food and water should not exceed 24 h but the latest *Codes of Recommendations* (MAFF, 1987b) advise against this method of force moulting completely. Withholding calcium from the diet or reducing the daylength are less stressful methods of terminating egg production.

After between 50 and 60 weeks of laying, the flocks are disposed of either for human consumption (boiling fowls or in prepared forms) or for pet foods.

The life of the fowl is, however, more than the biochemical conversion of nutrients into body tissues and eggs. In the wild and in the barn yard the fowl seeks out food, protects itself and co-exists with others of its own species. It possesses a well-developed nervous system, acute sight and hearing; it is sensitive to touch, temperature and pain; it perceives some tastes and smells. It communicates vocally and by posture and reacts to predators and fright by fight or flight. Its behaviour in response to the needs of self-preservation and reproduction is not random but, on the contrary, consists of specific, reproducible patterns which are inherent even if not always relevant to the man-made environment in which the fowl now lives. Thus chicks will cower in the presence of a model hawk; preening and dust bathing keep the plumage in good order, although the protective function of feathers may no longer be necessary with modern poultry housing; the characteristic pacing movements immediately prior to laying and the act of seeking a nest are still displayed and may have had some survival value in the wild; the ritual combats by which position in the social order is established (the peck order) may have ensured that only the fittest survive to reproduce, whereas today weak as well as strong must be allowed to fulfill their genetical potential for egg production; perching or roosting at night kept fowls away from predators which, except perhaps in the farmyard or on free range, are no longer a threat. All these inherent behaviour patterns can still be expressed by the modern, high producing fowl; only maternal behaviour (broodiness and the care of young chicks) appears to have been almost completely lost. It is held by some that the inability to express the patterns of behaviour described above causes unacceptable distress; this inability arises from overcrowding and from a bare environment (a laying cage) which does not provide dust bath, perch or nest. Science may eventually reveal more about the nature of animal suffering and the means by which it can be measured. Meanwhile it must be accepted that sound poultry husbandry should include due consideration of behaviour as laid down in the *Codes of Recommendations for the Welfare of Livestock: Domestic Fowl*, (MAFF, 1987b).

Breeding and genetics

The advent of the thermostatically controlled incubator replaced the broody hen and led to breeding as a specialist activity. As a result there are now only six companies in the UK which supply chicks for the egg producer (see also Chapter 8). There remain a number of breeders of the traditional breeds of fowl now seen mainly at agricultural shows but this activity, although giving much pleasure and interest, has little relevance to the industry.

Breeding biology

The male matures at about 18 weeks, but is not used for mating until the pullets, hatched at the same time, are in full production. Mating takes place after courtship display and the semen is deposited into the everted vagina of the hen. The spermatozoa can live for at least 3 weeks in the oviduct but mating is much more frequent than this. In breeding flocks, 7–10 males are allowed for every 100 hens. In order to retain the advantages of laying cages and also to exploit more fully the selected males, artificial insemination (AI) may be used (MAFF, 1978).

At the start of laying the eggs are too small to produce vigorous chicks, but eggs over 48 g are collected and stored at 15°C and 75% RH for, ideally, not longer than 1 week. Eggs can also be too large; they will not then readily fit into the incubator trays and, if they contain double yolks, will not hatch viable chicks; an upper limit of 65 g is often set. In order to prevent the growth of microorganisms, the eggs are disinfected (by washing or dry cleaning and fumigation) prior to setting, and sometimes they are fumigated during incubation.

Incubation is for 21 days at an initial temperature of 37.8°C. The incubator must be well ventilated since not only is gaseous exchange important but also because the embryos produce much heat which needs to be circulated and integrated with the thermostatic controlling system. After a few days it becomes possible to determine which eggs are fertile by inspecting them with a strong light placed underneath (candling). In small-scale incubation the infertile eggs may be removed but it is too time-consuming with commercial incubators holding up to 100 000 eggs and so all the eggs which fail to hatch, whether fertile or not, are removed at the end of the set incubation period. During incubation the position of the eggs must be changed several times daily, otherwise the development of the embryo does not proceed normally and many do not hatch. The broody hen turns her eggs when sitting; in domestic incubators each egg is moved by hand; in commercial incubators the trays of eggs are automatically rotated through an angle of 90° every few hours. After 18 days the eggs are transferred to a hatching compartment, or to a different incubator, set at 36.9°C and with a higher humidity (60%). On day 21 the trays are brought out and the newly hatched chicks removed. The eggs that have not hatched are treated as waste although they may include some chicks that were late in hatching or unable to complete the process. It is recommended in the *Codes of Recommendations* that all such waste should be speedily macerated to kill any living embryos.

The newly hatched chicks are now sexed either by manual inspection of the vent, an exacting task which requires specially trained operators, or by the visual recognition of the down or of the wing feathers of strains which incorporate these sex-linked characters. Since only females are required in laying flocks the male chicks are a waste product and should be disposed of humanely (see p. 218).

Female chicks are then vaccinated by injection against Mareks disease, which necessitates careful handling, and against Newcastle disease and infectious bronchitis by aerosol spray (see p. 217). The beak may also be trimmed at this time and sometimes the comb removed with sharp scissors. This is supposed to reduce the risk in later life of injury from pecking which can lead to outbreaks of cannibalism. The beak, sometimes only the upper mandible, is cut back about one-third of its length, often by an automatic guillotine or a heated cautery blade. When done carefully, there is little reaction from the chicks and their subsequent growth and performance is excellent. The beak has a nerve supply and although it may be trimmed, like fingernails, without harm, if cut too far back pain will be caused as well as bleeding and sometimes depressed food intake.* Removing the comb appears to be less destructive than beak trimming but it must be performed at 1 day old; later the comb develops an extensive blood and nerve supply and becomes extremely sensitive to touch and pain.† These procedures are time-consuming and are no longer performed routinely by breeders. They may eventually cease altogether once it has become generally appreciated that they confer little if any benefit to the producer.

The chicks are now placed in rigid-sided, ventilated boxes in groups of up to 100 and transported in air-conditioned lorries to the producer. They can withstand travel well for 24 h provided they are not exposed to climatic extremes. The breeding operation is now complete and the producer takes over the brooding and rearing of the stock to point-of-lay. There are specialist rearing farms which then sell the stock to egg producers, or the latter may have their own rearing operation to produce their replacement stock.

Genetics

Sex determination and sex linkage

The male is the homogametic sex and therefore all spermatozoa carry the same sex chromosome (Z). The female is heterogametic producing ova in approximately equal numbers with and without the Z chromosome. At fertilization the zygote becomes either ZZ (male) or Z- (female). Certain other characters are linked to the Z chromosome and if such a character is carried only by the dam, not the sire, then only the male progeny will exhibit that character. The female progeny (Z-) inherit the Z chromosome from the sire which is not a carrier. Thus the barred pattern of the Plymouth Rock breed – a white spot on the down of the head – is displayed only by the males when a Plymouth Rock female is crossed with a Rhode Island Red male. Similarly, the dominant gene for slow feathering is passed only to male chicks when slow feathering females are crossed with rapid feathering males. The difference in length of the wing feathers is obvious at hatching and allows easy sexing. Genetic sexing should always be better than vent sexing but unfortunately sex-linked genes cannot necessarily be incorporated into a breeding programme without some loss of production as an associated response, which is why it has not been universally adopted since its discovery at the turn of the century.

* The *Codes of Recommendations* advise that "beak trimming should be carried out only as a last resort, that is, when it is clear that more suffering would be caused in the flock if it were not done . . . . If beak trimming is necessary, it should be done by a skilled operator and as prescribed in the Veterinary Surgeons (Exemptions) Order, 1962."

† Dubbing of birds older than 72 h must be carried out by a veterinary surgeon (Veterinary Surgeon Act, 1966 (Schedule 3 Amendment) Order, 1982).

Genetic improvement

The fowl is a single species and, therefore, despite the great diversity of breeds, following centuries of selection by man, fertile mating between them is still possible. The older breeds were selected for their plumage, body weight and comb appearance and to perpetuate a number of single gene effects such as the abnormal feathers of a frizzle fowl or the short legs of a Scots Dumpy. Some were also selected for performance, for example, the White Leghorn for egg production or the Plymouth Rock for body weight, but they all retained strong breed characteristics (appearances). The recognition of the Mendelian laws of inheritance, although important for better understanding, did not bring about dramatic improvements in production because the traits of commercial importance, such as egg number, egg size, growth, food conversion efficiency, are not single genes inherited in an all-or-none way. They are quantitative characteristics which respond to selection but not in a predictable Mendelian manner and a new science of quantitative or population genetics has evolved around them. Breeding for production requires the productive capacity of the stock to be known or accurately assessed before selection can take place. This could mean keeping layers for a year on trial and then breeding from the selected ones in the second year. Moreover, selection is exercised on families or lines rather than individuals and this adds to the cost and complexity facing the breeder. Selection based on production results for only part of the year is now used, which speeds up the breeding programme.

The final commercial chick is unlikely to be the offspring of a single cross between two inbred lines; its parents themselves may be hybrids thus giving three-way or four-way crosses. The parent and grandparent lines which gave rise to the commercial chicks are retained by the breeding companies.*

Nutrition

Food accounts for 70 per cent of the costs of egg production and an adequate diet is essential for economic production and for sound welfare; this demonstrates the importance of nutrition in the management of laying flocks.

The digestive system of the fowl (Fig. 9.2) consists of the mouth and pharynx; the oesophagus with its thin-walled storage compartment, the crop; the glandular proventriculus which secretes hydrochloric acid and digestive enzymes; the thick-walled, muscular gizzard containing grit or stones to assit with its grinding function; a small intestine consisting of duodenum, into which opens paired bile ducts and triple pancreatic ducts, and upper and lower loops from which absorption takes place; and paired, blind-ended caeca which arise at the junction of the small and large intestine which opens into the cloaca. The ureters also open into the cloaca and thus faeces and urine are mixed and voided together as droppings. The alimentary canal is relatively short and food can pass through within a few hours.

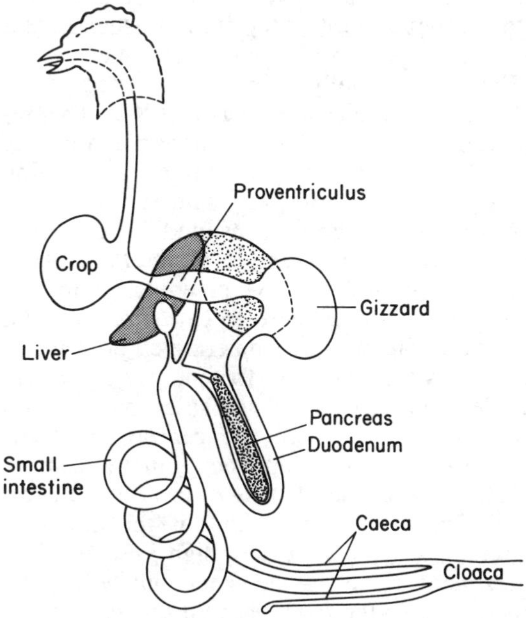

Fig. 9.2 The digestive system of the fowl.

The fowl is omnivorous except for an inability to digest cellulose and similar fibre. The caeca contain microorganisms that can produce small amounts of volatile fatty acids, the products of the digestion of cellulose, but these are of no nutritional significance.

* See also Chapter 8, p. 194 for details of primary breeding and stock multiplication.

Nutrient requirements

Water

This is the most obvious, but the most underrated nutrient. Diets contain about 10 per cent of free water and some is synthesized during metabolism, but an external source is essential and must be available at all times. There is little by way of bodily reserves and, since the fowl uses evaporative cooling (panting), water is particularly important in hot weather. Food intake and then egg production decline rapidly if water is unavailable for more than about 1 day.

Energy

This is derived mainly from carbohydrates (cereal grains and their by-products), though protein can supply energy but at higher cost. Fats are also an excellent source, but because of difficulties with rancidity and with mixing them into compound feeds, they are not usually present as more than 5 per cent of the diet.

Energy is usually expressed as megajoules (MJ) of metabolizable energy (ME). The requirements for growth, maintenance and egg production have been determined by feeding trials, and tables of ME content have been compiled which can be consulted when formulating diets (see, for example, Scott *et al.*, 1982).

Poultry generally (heavy breeding strains are an exception) are allowed to eat *ad libitum* and the major factor which determines how much they eat is their demand for energy. To satisfy this demand the bird will adjust its consumption according to the energy concentration of the diet, eating less of a high-energy diet and more of a low-energy diet. It takes a few days for this regulation to take place and if the diets are extremely low or extremely high in energy the adjustment might not be exact, but with most practical diets the fowl maintains a remarkably constant energy intake (Fig. 9.3). This is one of the basic laws of poultry nutrition and thus other ingredients have to be included in the formulation in quantities which will meet requirements given the expected daily food intake.

Nutrient requirements should be quantified in terms of daily intake, but, because of the primary importance of energy, other nutrients are often expressed in relation to it, e.g. g protein/MJ ME. In practical feed compounding, percentage inclusion rates, or g/kg or kg/tonne are used.

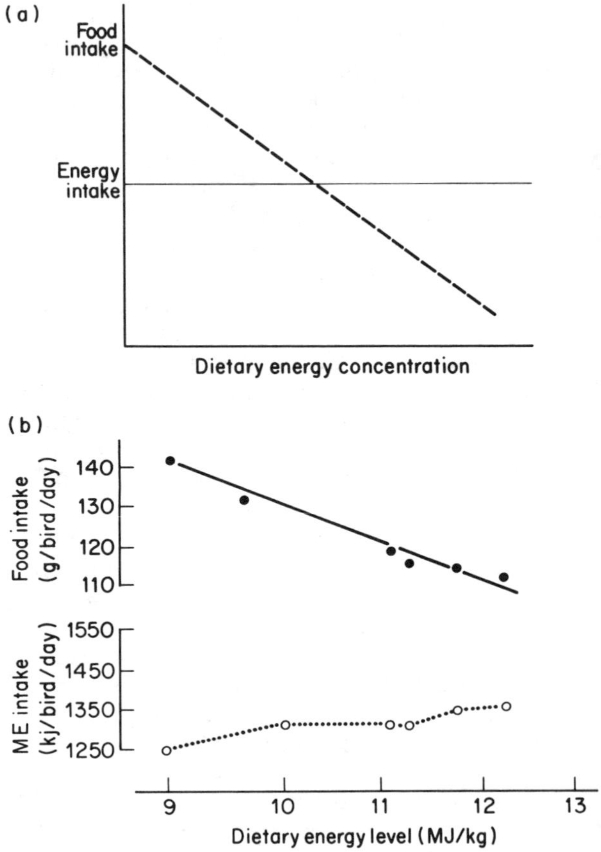

Fig. 9.3 Relationship between dietary energy concentration and food intake. (a) Theoretical relationship necessary if energy intake is to remain constant. (b) Food and energy intakes of laying hens fed diets of varying energy concentration.

Protein

Of the total of 19 amino acids, 12 are essential for the fowl, in that they cannot be synthesized from others but must be provided from the diet. From their distribution in foodstuffs they can be ranked in order of priority; thus lysine is the first limiting amino acid followed in turn by methionine, cystine and tryptophane. If the diet supplies enough of these then, from the chemical composition of most proteins, all the remaining essential amino acids will be provided. Thus diets are formulated to contain a given percentage of protein, and usually only the percentages of lysine and methionine are quoted. Cystine may also be specified – it is a source of sulphur-containing amino acids, but methionine alone can provide the cystine if present in large enough quantities. Some animals can synthesize non-essential amino acids from simpler sources of nitrogen such as urea; the fowl has virtually no capacity to do this.

Fat

The fowl can utilize fat as a concentrated source of energy, but in large quantities its digestibility is reduced and problems with rancidity may arise. Most of the constituent fatty acids can be synthesized but there is a dietary requirement for linoleic acid for growth and for maximum egg weight in layers; most practical diets will supply enough.

Vitamins and minerals

There are 13 vitamins which the fowl must obtain from its diet. These are the fat soluble vitamins A, D, E and K and the water soluble vitamins thiamin, riboflavin, nicotinic acid, folic acid, biotin, pantothenic acid, pyridoxine, choline and cyanocobalamin (B12). A deficiency of any of these vitamins can lead to poor growth, loss of egg production, lower hatchability and generally poor health but, with modern compound diets, which include a vitamin supplement, this should never occur.

The main inorganic elements which are required in the diet are calcium, phosphorus, sodium, potassium, magnesium and chlorine. Calcium is particularly important for laying hens since the production of one egg entails the loss of 2 g of calcium, and up to 3 per cent of the diet must be made up of calcium to provide for this. Phosphorus must also be provided, often as dicalcium phosphate, and sodium chloride is usually added separately as well. In addition, very small quantities of a number of other minerals are required for sound health and production; these so-called trace elements include copper, iron, zinc, manganese, selenium and iodine.

The main ingredients of a diet contain some vitamins and trace elements but these are not taken into account; instead, the requirement for them is met in a single supplement which has been accurately dispensed by a specialist supplier.

Additives

These comprise substances which are not strictly nutrients but which are nevertheless beneficial, if not essential. Coccidiostatic drugs might be included here. Certain antibiotics (virginiamycin, zinc bacitracin, olandeomycin), used as growth promotors for broilers, may also improve egg production marginally.* Layers are often provided with additional yolk pigments, from natural or synthetic sources, if they do not get enough from their diet to provide the depth of colour sought by the consumer.

Practical feeding

Knowing the requirements of the stock and the composition of the ingredients available, a dietary formula is then calculated which will supply the ingredients at least cost but take into account practical constraints such as those mentioned below. A simple example is given in Table 9.1

Cereals may be ground before or after mixing; fats are liquefied by heating before being added to the diet. The compound may be put through a pelleting press although this is less common for laying stock. The final product may be weighed into sacks or stored for delivery to bulk bins. The provision of a single feed, containing all the nutrient requirements, except water, suits a highly mechanized industry. A dry meal can be easily transported, stored and distributed by chains running inside the feeding troughs. Automatic censors on the bins enable consumption to be recorded. Flint grit is occasionally added to the feeding troughs for

* The inclusion of such antibiotics, at specified levels, is permitted under the Medicines (Exemptions from Restrictions and the Retail Sale or Supply of Veterinary Drugs) Order, 1979. Other antibiotics, which are used therapeutically, can only be administered with a veterinary prescription.

Table 9.1
The calculation of a diet for laying hens

Material	Composition		Inclusion rate (g)	Contribution	
	Crude protein (g/kg)	Met. energy (MJ/kg)		Crude protein (g/kg)	Met. energy (MJ/kg)
Wheat	104	12.2	433	45	5.3
Maize	82	13.2	300	25	4.0
Fishmeal	655	11.2	25	16	0.3
Meat and bone meal	515	11.0	40	21	0.4
Soya bean meal	442	9.3	120	53	1.1
Limestone	—	—	78		
Salt	—	—	2		
Supplement	—	—	2		
Total			1000	160	11.1

From more detailed tables of feed composition the contents of lysine (0.72%), methionine (0.3%), methionine + cystine (0.59%), calcium (3.6%), phosphorus (0.57%) and salt (0.4%) are determined. The calculated composition of the diet should be confirmed, as far as is practicable, by chemical analysis.

the birds to use in their gizzards as an aid to digestion. This is especially useful if whole grain is given.

Many potentially valuable feed ingredients contain deleterious substances which limit their usefulness. Soyabeans contain an anti-trypsin factor which reduces its digestibility unless the factor is destroyed by heat treatment; cassava can contain hydrocyanic acid unless this has been removed by washing, and it may also contribute to dustiness in the feed; oilseed meals that have been overheated during extraction can have the protein denatured; cottonseed meal contains gossypol which can cause discoloration of the eggs, so that this otherwise useful ingredient is avoided in laying diets; groundnut and other proteins may become infected during storage with moulds which produce very toxic compounds; meat meals may become contaminated with *Salmonella* bacteria if not properly heat-treated at source.

The compounded, complete diet is not the only method of feeding. Poultry possess what might be termed nutritional wisdom. Given the opportunity they can select from a range of foodstuffs those which will supply their nutrient requirements, although not necessarily on a least cost basis! They will consume grains to satisfy their energy needs but not to the exclusion of protein if sources are present. The farm yard or range will provide some nutrients although not enough to sustain maximum production. The provision of extra grain may be the only supplement that is provided. The principle of self-selection has been used in intensive husbandry; two troughs are used to provide either whole grains or a high protein supplement, from each of which the hen selects according to its appetite. This is not as economical as a single diet but nutritionally it is sound.

If calcium (limestone, oyster shell) is provided separately it will be consumed selectively whether or not the diet already contains calcium; there does not have to be a deficiency in order to demonstrate this strong, specific appetite.

Food should be palatable as well as nutritious: a coarse grist is preferred to a fine powder; fat increases palatability; and strong tastes and odours are aversive. Staleness gives rise to rancidity of the fats and the growth of moulds. Freshness is obviously appreciated; newly milled cereals are very eagerly consumed. It is good practice to allow the food in troughs to be completely consumed from time to time; fine textured ingredients may collect at the bottom and be missed. Over-filling of troughs leads to wastage. Above all else, the primary nutrient, water, must never be forgotten. Troughs, cups and nipples must be kept clean and in good working order at all times.

The feed compounding industry is bound to

comply with certain Statutory Declarations of the contents of their feeds. The list of ingredients so covered includes protein, oil, fibre, ash, vitamins A, D and E, selenium, copper, colouring agents, antioxidants, growth promoters and medicinal compounds such as coccidiostats. Metabolizable energy, as calculated from sugar and starch, is to be added to this list.

Environmental requirements

Poultry can live under a wide range of environmental conditions but experience has shown that they thrive best, in terms of welfare and production, when certain conditions are met.

Temperature

For convenience this is taken to consist solely of the air temperature (°C) but the total thermal environment is affected by wind speed, humidity and the presence of local radiant sources or sinks.

Chicks

Brooding temperature at 1 day old should be 35°C; this is reduced to 32°C by 7 days, to 27°C by 14 days and to 24°C by 21 days. The chicks may be kept in cages, often several tiers high, either with independent heating elements and thermostats on each tier, or the whole cage or groups of cages may be in a thermostatically controlled room. Alternatively, the chicks may be kept on litter with either focal sources of heat (gas or electric) or with general space heating, often supplied by ducted air from a central appliance. Although external heat is essential, chicks possess an ability to regulate their body temperature more precisely by behavioural means; they will huddle together near the source of the heat if chilled or disperse some way from it if they become too warm (see also Chapter 10, Fig. 10.1). It is important that they are given enough space to move and that this behaviour is observed by the poultryman so that if one or other type of behaviour persists he may make the appropriate adjustment to the brooding temperature. It should not be forgotten that the broody hen, although no longer in common use, can still provide satisfactory brooding conditions.

Growing and laying stock

The aim here should be to keep the temperature within the thermoneutral zone where metabolism is lowest; strictly this is more a simple point in the temperature scale than an extended plateau but the range of 15–24°C will suffice. Lower temperatures than this are not harmful, unless very severe, but metabolic rate and thus food intake increase. The temperature should not be allowed to fall below 5°C, otherwise there is a risk of the water supply freezing. High temperatures are more of a threat to survival; the physiological defences include the adoption of a typical posture, with outstretched wings, and evaporative cooling by panting. The normal body temperature of 41°C may rise to a higher but stable level; if it rises to 44°C external cooling is almost certainly required; at 45°C there is an imminent danger of heat stroke and death unless measures are taken. Panting is not in itself distressful but it is a valuable indicator for the poultryman to heed; ventilators and fans may need to be adjusted and the water supply checked. In the UK, and temperate countries generally, poultry should be housed in insulated buildings whether on free range, deep litter or in cages. Roof and wall insulation helps to maintain the minimum temperature in winter and to insulate against external heating in summer. Space heating is usually considered to be uneconomic and the minimum temperature is achieved by metabolic heat.

Ventilation

Ventilation is necessary for part of the year to prevent overheating. The number and size of the fans should ensure high enough ventilation rates to maintain the normal house temperature. This is usually achieved automatically by the use of thermostats to control ventilation (Table 9.2). It is prudent to plan for high fan capacities rather than to risk losses by overheating. Inlet ventilators are also required. There are several ways of arranging the ventilation system. Fans are usually extractor in operation but the reverse pressurized method is also used. Fans may be mounted on the roof, on an end wall or on one or both side walls. Inlet ventilators may be on side walls or on the roof ridge.

Table 9.2
Ventilation requirements for laying hens

Body weight (kg)	Minimum requirement (m^3/h/bird)	Maximum requirement (m^3/h/bird)
1.2	0.8–1.3	10
2.0	1.3	12
2.5	1.5	14
3.0	1.7	14
3.5	2.0	15

Source: MAFF (1976). Crown Copyright.

Ventilator openings can be under mechanical control but generally they are set manually according to season. A useful emergency system based upon electromagnetic locks allows ventilators to fall fully open if a power failure stops the fans. Powered ventilation is not solely for temperature control; ammonia is formed from the excreta and high levels can be noxious to fowl and to man; poor ventilation is considered unhealthy and conducive to disease; more specifically, ventilation is necessary, even in winter, to remove moisture. In deep litter houses, moisture causes the litter to become rank and offensive and encourages the survival of coccidial spores; in laying cage houses the excreta will not dry out and can become a breeding ground for flies. There is less of a problem in houses for poultry maintained in pens or on range, when natural ventilation from floor inlets to roof outlets is usually adequate.

Humidity

There are no realistic requirements for humidity and it is not a factor which can be easily controlled. High humidity, when combined with high temperature, reduces the birds' capacity to lose heat by evaporation and so adds to the thermal stress. High humidity also creates wet litter and the problems which arise from it. If water sprinklers (producing a fine mist) are used for cooling (unusual in the UK) the disadvantages of high humidity are usually outweighed by the benefit of a lower air temperature.

Wind speed

Within a given housing system local draughts or their opposite, pockets of still air, should be avoided. Ventilation inlets should be provided with baffles to distribute the air flow evenly. A handheld electronic anemometer is useful for measuring local conditions.

Light

The term daylength is taken to mean the total illuminated period within any repetitive cycle of light and dark, usually but not always the solar, 24-h day. Daylength may be natural, artificial or a combination of the two. There is a minimum requirement for light but in all practical situations this is more economic than biological. Under conditions of natural light, chicks will grow well and, as hens, lay well, but the number and size of the eggs can be affected by daylength as explained on p. 197.

In practice, the best economic return is found when maturity is delayed until pullets are 22 weeks old, by manipulating the amount of light available per day. The standard pattern used consists of:

0–1 week old	23 h light
2–18 weeks old	8 h light
19–22 weeks old	increase by 45 min/day
23–49 weeks old	increase by 20 min/day

The wrong lighting pattern, such as when daylength is reduced near the onset of maturity, can cause a considerable reduction in egg production. Often, the fault is in inadequate light-proofing in either the laying or the rearing house.

Stock on range or in pens, and therefore subject to natural daylength, can have their maturity delayed if the total daylength is first increased by additional, artificial light which is reduced later. This will shorten the daylength even in spring when natural daylength is increasing. The *Codes of Recommendations* (MAFF, 1987b) specify a minimum period of

8 h light per day for birds which do not have access to daylight.

Ahemeral light cycles

These are light/dark cycles with a periodicity different from the normal 24-h day. Periods shorter than 24 h have no economic advantage but longer periods, typically of 28 h, can result in larger, but fewer, eggs, which may be more profitable. An example is 11 h light and 17 h dark. The timing of ovulation and oviposition soon adjusts to the new pattern of light and dark. It is important to realize that the total light period is reduced and the total darkness is increased under this system. Conversely, on moving back to a 24-h day, daylength is increased and darkness is reduced. In practice, the laying house needs to be fully light-proof since the 28-h cycle will, from time to time, require darkness during the natural day. Servicing the house would have to follow the 28-h cycle but the problems that this would cause are avoided by having bright (40 lux) and dim (0.04 lux) periods instead of the light and dark; the hen reacts to dim lights almost as it would to complete darkness. Sometimes a sequence of bright, dim and dark periods is used to allow for servicing during the stockman's normal working day. It is usual to wait until laying is well established, say by 28 weeks old, before applying ahemeral lighting.

Intermittent light

Another lighting programme for layers, which has economic avantages, is for the normal illuminated period of a 24-h day to be broken into hours consisting of 15 min of light followed by 45 min of darkness. To facilitate servicing, the first 2 h of daylength are entirely light. Intermittent lighting can reduce food intake without lowering egg number or quality. A different version is to have two periods of light within 24 h: thus 8 h light, 8 h dark, 2 h light, 4 h dark. The total daylength is only 10 h but the long dark period is broken into two shorter periods which are less inhibitory.

Both ahemeral and intemittent programmes bring a saving of electricity as well as improvements in production but neither intermittent nor ahemeral lighting has yet been widely adopted (Michie, 1985).

Light intensity

There is a biological requirement for light to be of a minimum intensity, but in practice this is set more by the needs of the staff for inspection than by the needs of the fowl for perception. High intensities are associated with outbreaks of cannibalism and the use of low intensities may prevent this and certainly saves electricity. No absolute intensity is specified in the *Codes of Recommendations*, only that it should be adequate for inspection (to be able to read newsprint is a guide); a level of 10–15 lux is adequate throughout rearing and laying, although higher intensities are often used. There should always be a period of darkness in each 24 h; this accustoms the birds to sudden darkness, when a switch is thrown, and thus a power failure is less likely to cause widespread panic and damage.

Floors

The nature of the surface will depend upon the type of housing that is chosen (see below). High production can be obtained with all types of floor. The gauge of the wire used for battery cage floors has not been found to be important; wire netting offers as good a support as heavier gauge mesh. It is very important that the floor material should always be dry. Perches are desirable on welfare grounds.

Housing and management

Except for the barn yard fowl, some form of housing is necessary, and the type will depend upon the economic scale of the enterprise. For a domestic supply of eggs, a hut for night-time protection and a nesting site is adequate, but for economic poultry farming there are three main systems: extensive, intensive non-cage systems, and intensive in cages. This classification applies to both rearing and laying phases and more than one may be employed during the life of the fowl.

Table 9.3
Summary of the assessment of egg production systems carried out by the Farm Animal Welfare Council

Welfare		Commercial	
Battery cages			
Pros:	*Cons*:	*Pros*:	*Cons*:
1. Easy to control environment	1. No separate nesting accommodation	1. Ease of mechanization, management control contribute to feed and labour efficiency	None given
2. No exposure to risk of predation	2. Ability to stretch wings and the range of other exercise is limited	2. Overall cost of production lower than for other systems	
3. No exposure to extreme climatic conditions	3. No opportunity to dust bathe	3. Clean eggs produced	
4. Small colony size	4. Cannot escape other birds	4. Low disease risk	
5. Birds separated from droppings – disease control	5. Cage structure may give rise to feather damage		
	6. Confinement on a sloping floor		
	7. Cannot influence overgrowth of claws unless an abrasive strip is present		
	8. Environment is physically barren		
Wire or slatted floors with or without litter			
Pros:	*Cons*:	*Pros*:	*Cons*:
1. Ease of environmental control avoids exposure to climatic extremes	1. Large colony size increases risk of social disharmony	1. By using vertical space to achieve higher stocking densities the capital cost per bird begins to approach comparability with the cage system	1. Husbandry and management are more demanding than in the cage system
2. No exposure to predators	2. If bullying, feather pecking or cannibalism occur, control may be more difficult	2. At higher stocking densities environmental control can maintain optimum temperature and hence contribute to feed efficiency	2. The structures provided for the birds make routine tasks and cleaning more difficult
3. Freedom to move within the house area	3. Structures provided for the birds impede observation and inspection	3. Eggs can command a premium	3. Production costs tend to be higher than for cages
4. Opportunity to stretch wings to full extent and to exercise in a variety of other ways	4. Unless manure belts are provided between tiers, birds will get soiled from the droppings of birds above		

Pros:	Cons:	Pros:	Cons:
Deep litter pens			
1. Varied physical environment including litter making dust	1. Stocking density insufficient to achieve optimum environmental control throughout the year	1. Eggs can command a premium	1. Requires greater capital investment per bird than in cages because of the housing space requirement
2. No exposure to predators	2. Large colony size increases risk of social disharmony		2. Food consumption tends to be greater than in cages
3. Freedom to move within the house area	3. If bullying, feather pecking or cannibalism occur, control may be more difficult		
4. Opportunity to stretch wings to full extent and to exercise in a variety of different ways	4. Access to droppings in the litter increases risk of enteric diseases		
5. Opportunity to use nest boxes and achieve privacy when laying	5. Ammonia gas concentrations may be troublesome		
6. Birds easier to inspect than in tiered systems			
Semi-intensive pens and free range			
1. Freedom to move within the house and on the range area	1. Mortality may be higher due in part to predation	1. Premium prices for eggs	1. Food consumption, labour requirements and management demands are considerably higher than for other systems leading to greater production costs
2. Varied physical environment with ample opportunity to exercise and dust bathe	2. No control over environmental conditions on the range. Access to water may be a problem in winter		2. Generous land requirement
3. Opportunity to use nest box and therefore to get away from other birds when laying and achieve privacy	3. Access to droppings increases the risk of enteric disease as does contact with wild birds		3. Annual egg production likely to be lower than for other systems
4. Opportunity to graze on vegetation and to augment and vary diet in other ways during the warmer months of the year			4. Dirty eggs tend to be a problem

Source: Farm Animal Welfare Council (1986). Crown Copyright.

Table 9.4
Classification of egg production systems as contained in the EEC Egg Marketing Regulations (1943/85)

For labelling purposes only, Commission Regulation (EEC) No. 1943/85 (O) (NO. L181 of 13 July 1985) requires that poultry enterprises must comply with the following minimum criteria in order to mark their small egg packs with the terms shown:

1. **Free-range eggs**
   Hens to have continuous daytime access to open-air runs
   The ground to which hens have access is mainly covered with vegetation
   The maximum stocking density is not greater than 100 hens/ha of ground available to the hens or 1 hen/10 m^2[a]
   The interior of the building must satisfy the conditions at (3) or (4)

2. **Semi-intensive eggs**
   Hens have continuous daytime access to open-air runs
   The ground to which hens have access is mainly covered with vegetation
   The maximum stocking density is not greater than 4000 hens/ha of ground available to the hens or 1 hen/2.5 m^2
   The interior of the building must satisfy the conditions at (3) or (4)

3. **Deep-litter eggs**
   The maximum stocking density is not greater than 7 hens/m^2 of floor space available to the hens
   At least a third of this floor area is covered with a litter material such as straw, wood shavings, sand or turf
   A sufficiently large part of the floor area available to the hens is used for collection of bird droppings

4. **Perchery eggs (barn eggs)**[b]
   The maximum stocking density is not greater than 25 hens/m^2 of floor space available to the hens
   The interior of the building is fitted with perches of a length sufficient to ensure at least 15 cm of perch space for each hen

[a] According to the *Codes of Recommendations for the Welfare of Livestock: Domestic Fowl* (MAFF, 1987b), stocking rate in the UK should generally be lower than this (see also p. 211).
[b] Eggs from the so-called *aviary* system will probably be classified as perchery eggs.

Table 9.5
Egg production costs in different systems

System	Space[a]	Cost[b]
Laying cage	450 cm^2/bird	100
Laying cage	560 cm^2/bird	105
Laying cage	750 cm^2/bird	115
Laying cage	450 cm^2/bird + perch	100
Laying cage	450 cm^2/bird + perch + nest	102
Shallow laying cage	450 cm^2/bird	102
Get-away cage, 2-tier aviary	10–12 birds/m^2	115
Aviary and perchery and multi-tier housing	20 birds/m^2	105–108
Deep litter	7–10 birds/m^2	118
Straw yard	3 birds/m^2	130
Semi-intensive	1000 birds/ha	135 (140)[c]
Free range	400 birds/ha	150 (170)[c]

[a] Space refers in cages to cage floor area, in houses to house floor area and in extensive systems to land area.
[b] Taking battery cages as the 100% base cost.
[c] Includes land rental.
Source: Elson (1986).

The main systems, as used for laying hens, are shown diagrammatically in Fig. 1.1 (Chapter 1, p. 8), and their advantages and disadvantages, both commercially and from the welfare aspect, are listed in Table 9.3. The EEC Egg Marketing Regulations (No. 1943/85) lay down minimum criteria for the various non-battery systems, to ensure correct labelling of eggs sold in small packs (Table 9.4). Production costs for the different systems are given in Table 9.5.

Brooding and rearing

The requirement for external heat during brooding and the types of housing that are available have already been mentioned. Rearing from 4 to 18 weeks may be on the range, with moveable arks or more permanent houses for night-time accommodation, indoors on litter, or indoors in tiered cages. In all systems water must be readily available from open troughs or from cups or nipples supplied from low-pressure water cisterns. With cages, cups and nipples must be so arranged that each cage has access to at least two. Food is provided *ad libitum* in hoppers, in open troughs or from moving chain feeders.

Outdoor systems of egg production

Free-range

The hens are provided with a house for night-time accommodation and for egg laying. *The Codes of Recommendations* (MAFF, 1987b) give advice on management as follows:

> Land on which range birds are kept for prolonged periods may become 'fowl sick', i.e. contaminated with organisms which cause or carry disease to an extent which could seriously prejudice the health of poultry on the land. The time taken for land to become fowl sick depends on the type of land and density of stocking. A European Community Regulation (Table 9.4) on the labelling of eggs requires that eggs labelled 'free-range' must originate from a range system with a stocking rate not exceeding 1000 birds per hectare. In addition, the Regulation also requires the ground to which such birds have access to be mainly covered with vegetation. The stocking rate to be used in this country should generally be lower. Factors such as soil type, drainage and size of colony and frequency of flock rotation are very important in deciding the number of birds that a particular area can carry. Heavy, poorly drained soil can carry fewer birds than land which is light and well drained. In general land can be stocked more heavily by birds in small flocks of 100 or so when accommodated in well spaced and regularly moved houses than when kept in larger flocks in static houses. Flocks and portable houses should be moved regularly to avoid fowl sick or continuously muddy conditions leading to ill-health or discomfort of the birds.
>
> It is important to ensure that the land to which the birds have access is adequately covered with suitable, properly managed vegetation.
>
> Precautions should be taken to protect the birds against foxes, other predators, dogs and cats.
>
> Shelter from rain and sun should always be available. Windbreaks should be provided on exposed sites.
>
> Housing used by range birds should be of a sufficient standard to ensure that the birds are not subject to distress caused by extremes of temperature.
>
> When birds are transferred to range houses, precautions should be taken to avoid crowding and suffocation, particularly during the first few nights. Cannibalism is a danger under this system, and birds should not be confined for too long during hours of daylight or subjected to direct sunlight during confinement.
>
> All birds must have ready access to range and there should be sufficient openings so spaced and of sufficient size to allow a reasonable proportion of the birds to enter or leave at any one time.
>
> Unless the house is moved frequently it is good practice to protect the ground immediately adjacent to it, e.g. by providing slatted or wire mesh platforms, covered verandahs or areas of gravel.
>
> Feed and water should never be allowed to remain in a stale or contaminated condition. In freezing conditions, particular attention should be given to the provision of water.

Semi-intensive

Housing is usually static, and outside access is restricted to wired enclosures which are used in rotation.

Straw yards

These are protected by a roof, but are otherwise open to the weather, and often make use of existing farm buildings. A liberal provision of litter, usually in the form of chopped straw, is a necessary condition. Nest boxes and perches are provided. Eggs from straw yards can be classified as deep-litter within the EEC Egg Marketing Regulations (Table 9.4), provided the criteria laid down for deep litter systems are met.

Indoor systems of egg production

Deep litter

The flock is housed entirely indoors with nest boxes and perches provided. The floor area is usually covered with a deep layer of wood shavings or sometimes peat. Part of the floor may be partitioned off with raised slats over a pit; food and water supplies are available here and the hens roost on the slats which means that a substantial proportion of the droppings are deposited in the pit, thus keeping the litter drier. Nest boxes may be accessible from a central or side corridor for easy collection or they may be provided with sloping wire mesh floors which allow the eggs to roll on to a conveyor belt for mechanical collection. Deep litter housing is used particularly for breeding flocks so that males can mingle with the flock and fertility is achieved by natural mating. A disadvantage of the system is that eggs are often laid on the floor instead of in the nests; these require special collection and are often soiled and must be cleaned before being sent to the packing station or hatchery. Compared with cages, birds have more space and more freedom of movement but, because they can mingle freely together, social groupings (peck orders) are established, and this can result in low-ranking birds being deprived of food or even physically injured (see p. 217).

Laying cages

About 90 per cent of commercial laying hens are now kept in battery cages.* The system offers the highest stocking density and degree of automation combined with high levels of production, factors which make for a better economic return than with the other systems. Cage sizes vary but usually they hold up to six hens; each cage is provided with two or more nipple drinkers or cups or, less often, open water troughs running along the front or back. Continuous food troughs, usually supplied by chain feeder, run along the front. The floor slopes to allow eggs to roll to the front for easy collection by hand or by conveyor belt. Cages are usually in rows, back-to-back, and in tiers of up to six. Underneath each tier is a base of toughened glass on which the excreta collects and which is cleaned periodically by a moving scraper. Alternatively, a moving plastic belt may be installed. A simpler system, now widely used, is to arrange the cages step-wise so that there is a clear space under the wire floor, allowing droppings to accumulate in a pit below; this avoids daily reliance on machinery, and cleaning can be carried out less often. Another variation,

* *Editor's note*: The Welfare of Battery Hens Regulations, 1987 stipulate (para. 2) that from 1 January 1988 for new or reconstructed cages, and from 1 January 1995 for all others, the following requirements must be complied with:

(a) the minimum cage area, measured in a horizontal plane, for each laying hen shall be–
  (i) 1000 cm^2 where one hen is kept in the cage,
  (ii) 750 cm^2 where two hens are kept in the cage,
  (iii) 550 cm^2 where three hens are kept in the cage, and
  (iv) 450 cm^2 where four or more hens are kept in the cage;
(b) the minimum cage area for each laying hen shall be capable of being used without restriction and may include the area where the non-waste deflection plate (otherwise known as the egg guard) is placed so long as that area is capable of being used;
(c) a feed trough of a minimum length of 10 cm multiplied by the number of hens in the cage and capable of being used without restriction shall be provided;
(d) except where nipple drinkers and drinking cups are provided, the cage shall have a continuous drinking channel which shall be–
  (i) of the same length as the feed trough mentioned in paragraph (c) above, and
  (ii) capable of being used without restriction;
(e) where drinking points are plumbed in, there shall be a minimum of two nipple drinkers or two drinking cups within reach of the cage;
(f) the height of the cage, measured vertically from its floor to the nearest point in its roof, shall be not less than 40 cm over 65% of the minimum cage area and not less than 35 cm at any point over that area;
(g) the floor of the cage shall be constructed so as to support adequately each of the forward facing claws of each foot;
(h) the slope of the floor shall not exceed 14% or 8 degrees, when made of rectangular wire mesh, and 21.3% or 12 degrees for other types of floor.

seen more often in the USA, is to have the whole floor area of the house covered with a single tier of cages over a droppings pit. The passages are replaced by cat-walks on top of a flat-deck arrangement.

Modified cage and deep-litter systems

The battery cage is the system most criticized on welfare grounds because of the considerable restriction it imposes on the birds' movement, together with the occurrence of breast blisters, lameness and widespread loss of feathers. In order to meet these criticisms and yet retain some of the advantages of cages (automation, low disease risk) other types of intensive systems have been devised. *Shallow cages* provide the same floor area as a conventional cage, but the depth is reduced and the length increased; this gives a longer food trough and therefore more trough space per bird, which reduces competition. *Get away cages* are larger cages which provide nest boxes and perches. The hens are kept in larger groups which allows peck-orders to develop.

The aviary or perchery has been developed in order to make better use of the vertical space in a deep litter house. This has several slatted floor areas at different levels interconnected by ladders. Sometimes a distinct first floor with belt cleaning underneath is created. In both types, nest boxes are provided. Another development is to have an A frame carrying perches at different heights over a central droppings pit and deep litter passages, with nest boxes on either side. Food and water are accessible from the perches.

Because these recent developments are organized in three dimensions, many more birds can be kept in a set volume of accommodation. This allows the birds to maintain a warmer environment, and they show a better feed conversion efficiency than those kept in the floor only, two-dimensional system.

Stocking rates

Naturally the poultry farmer wishes to obtain a good economic return from his enterprise and will favour high stocking rates, but overcrowding can be detrimental and certainly causes an unacceptable loss of welfare. The *Codes of Recommendations* provide a guide to the maximum stocking rates for the different systems (Table 9.6).

Routine duties

The day-to-day management of the stock consists of a number of routine but important duties. These are:

1. To ensure that food and water are provided.
2. To check the environmental controls. This includes brooder and house temperatures, ventilator apertures, fan speeds, and lighting controls. Thermometers or thermographs can add precision to one's judgement of the environment.
3. To collect eggs by hand, or from the delivery point in mechanical systems. Soiled eggs can be cleaned but very dirty or cracked eggs must be removed. Eggs should be removed from the laying house and stored in a room kept at 10°C with a humidity of 85 per cent.
4. To inspect stock at least once a day and remove dead birds. Sign of disease should be noted (coughing, respiratory distress, listlessness, unusual posture) and advice sought if widespread. Ailing or moribund birds should be humanely destroyed by dislocation of the neck (if veterinary advice is not sought).
5. To keep records of food consumption, egg production, mortality and any events of note. These, together with details of age, strain, and immunization record, may be important in the diagnosis of disease as well as for economic assessment.
6. To review emergency procedures at regular intervals. This includes stand-by generators, self-opening ventilators and fire and intruder alarms. In winter, the risk of freezing water pipes and non-delivery of food should be considered.
7. To these routine procedures must be added an interest in the well-being of the stock. Keen observation based upon sight, sound and smell is an essential quality of stockmanship.

Table 9.6
Maximum stocking rates

Systems	Density (live-weight in relation to floor area)	Qualifications
Cages		
Birds being reared for laying	250 cm²/kg	For birds between 1 and 2 kg live-weight.
Adult laying birds	These are subject to legal minimum requirements with effect from 1 January 1988[a]	
Deep litter		
Birds being reared for laying	17 kg/m²	
Adult laying birds	17 kg/m²	No more than 7 birds/m²
Table chickens	34 kg/m²	
Straw yards		
Birds being reared for laying	10 kg/m²	
Adult laying birds	8 kg/m²	No more than 3 birds/m²
Housing for free-range birds	As for deep litter systems	

[a] See footnote on p. 212.
Source: MAFF (1987b). Crown Copyright.

Handling and transport

At the end of laying, adult fowls have to be removed from cages or from a deep litter house and placed in crates for transport to a processing station. To clear 30 000 hens by hand is a considerable task and not only distress, but bruising and fractures are easily caused, especially in multi-tiered battery cages, which often have small doors with access impeded by food troughs. A great deal of suffering could be avoided if designers of cages were to give some thought to facilitating the removal of the birds from them at the end of their laying life.

Ideally, when removing a hen from a cage, it should be picked up with both hands clasping the sides of the body and the thumbs placed over the wings to hold them close to the body and prevent flapping. The bird should be held firmly, but not squeezed. To pick a bird up from the floor, one hand should be slid under the front of the body, palm uppermost, so that the fingers trap the legs, while the other hand is placed over the back to prevent flapping. The bird is lifted and cradled in the arm, the legs being held in the hand. Chicks can be picked up in one hand, with the palm over the back and the fingers around the body. They must not be squeezed.

Transport by lorry can result in exposure to rain, wind or strong sunshine, and food and water are not available during the journey.* If, on arrival, the birds cannot be slaughtered immediately, they must be kept in good climatic conditions and, if the delay is long, food and water must be provided.†

Health and disease

The health of the national poultry flock is remarkably good. High levels of production are

* Conditions to be observed during the transport of live poultry are laid down in the *Conveyance of Live Poultry Order, 1919*.

† The Slaughter of Poultry (Humane Conditions) Regulations, 1984.

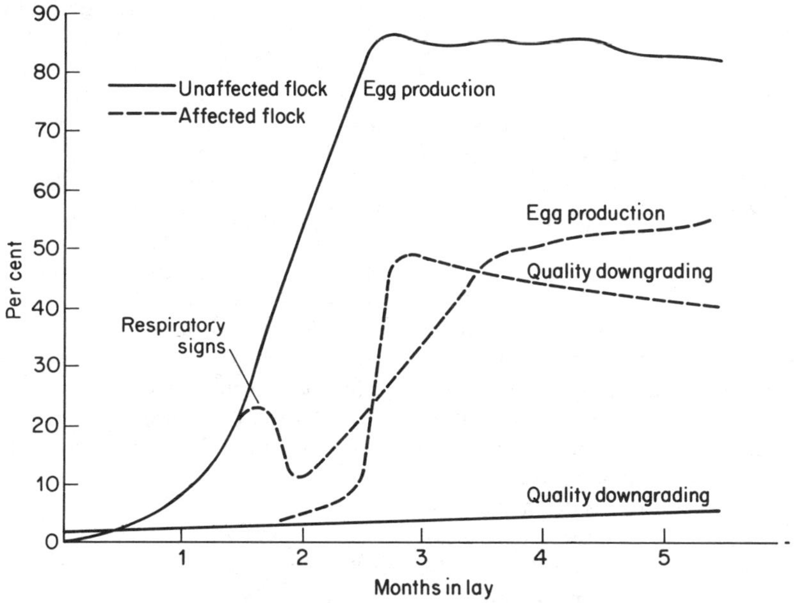

Fig. 9.4 Egg production and egg quality in a flock infected with infectious bronchitis. *Source*: Gordon and Jordan (1982).

found with mortality of only 1 per cent up to 18 weeks old and only 5 per cent for the ensuing laying year. However, this has not happened by chance and this section outlines the principles of poultry preventive medicine and lists the major diseases that still threaten the industry. The example of economic loss which can be caused by a typical respiratory disease, shown in Fig. 9.4, emphasizes the importance of preventive medicine in the poultry industry.

Good health is a consequence of the continuous application of the preventive measures of *hygiene*, *prophylaxis* and *eradication*. Hygiene is entirely in the hands of the producer; prophylaxis is available for him to use as he wishes but for veterinary scientists to provide in the first place; eradication is a matter of policy (national or commercial) based upon what is biologically and economically possible. To this list must be added *inspection*, for it is only by constant observation of the flock, at all ages, that a departure from good health can be detected. The physical appearance of the flock and its environment should be noted and the record of weight, food intake and egg production carefully scrutinized daily. If all is not well expert *diagnosis* should be sought. Because of the small value of individual birds and because there are no specific cures for the more important diseases, *treatment* is far less important than *prevention*. In view of its overriding importance in the poultry industry the principal preventive measures will be considered in greater detail.

Hygiene and management*

Measures include cleaning, disinfection, fumigation, isolation, periodic destocking, limiting access to the site and ensuring that adequate physical conditions (ventilation, temperature, dryness) are maintained. A minimum amount of ventilation is required, even in winter, and this is particularly important in deep litter houses in order to keep the moisture content of the litter acceptably low.

* For a more detailed treatment of this subject see Chapter 10, pp. 227–228.

Prophylaxis

This means the administration of products which specifically inhibit the development of pathogenic microorganisms before they can establish an infection. These products are either immunizing agents (usually vaccines) or synthetic drugs (usually coccidiostats).

Vaccines

Some poultry diseases may be controlled by live vaccines and others by inactivated ones, whereas in some cases there may be a choice between the two types. An important difference between the two is that most live vaccines can gain entrance to the body through the mucous membranes of the eye, the mouth or the respiratory tract, whereas killed vaccines have to be injected into muscle tissue in order to gain access to the blood stream. In practice the mass administration of live vaccine by aerosol spray or in the drinking water makes this a very attractive proposition compared with the injection of individual birds with inactivated vaccine, and the handling this entails. However, the overall level of protection is what matters in the end. Marek's vaccine, although live, is never the less administered by injection since it does not readily gain entrance through the eye, the respiratory system or the mouth.

The *Codes of Recommendations* advise that vaccination, by injection or otherwise, should be undertaken by trained operators and that care should be taken to avoid injury and unnecessary disturbance of the birds.

It is possible for maternal antibodies to pass from the hen to the chick via the egg, giving a certain degree of passive immunity. This can happen, for example, with Newcastle disease.

Antibiotics

The inclusion in the feed, at low levels, of specified antibiotics, is permitted (see p. 203). These may have an inhibitory effect on some disease organisms.

Coccidiostats

For poultry reared on range or litter, and therefore having access to droppings, protection against coccidiosis is usually considered essential. This is achieved by incorporating in the food one of the many coccidiostatic drugs that are available. These are very effective at therapeutic levels but the flock does not acquire any immunity to the parasite and it is more economical for laying flocks to have long-term immunological protection rather than to bear the costs of continuous medication throughout the laying period. Therefore, coccidiostats are included at a level which allows some coccidia to survive and thus stimulate the immune reaction. This immunity is species-specific and the flock remains susceptible to other species if these are introduced subsequently.

Eradication

This would appear to be the best long-term goal but it is not one which can be achieved quickly. It is expensive, and is pointless if new infections can arise from imported stock or wild birds. Pullorum disease has successfully been eradicated, but Newcastle disease, despite a lengthy slaughter policy has not. Many diseases are now far less common but this reflects mainly changing husbandry and good hygiene. There is a genetic factor in disease resistance which may be selected unconsciously when breeding for high production. Deliberate selection for disease resistance has not so far had any notable success but it remains a possibility. With mycoplasmosis, carriers can be identified and it has been possible to breed from mycoplasma-free hens thus preventing egg transmission of the pathogen to replacement chicks.

The main infectious diseases of layers

A brief summary of some of the diseases which may occur, or for which preventive measures should be taken, is given in Table 9.7. For a full description of these conditions, reference should be made to a more specialized textbook such as Gordon and Jordan (1982) or Sainsbury (1984).

Table 9.7
The main infectious diseases of the domestic fowl

Viruses		
Newcastle disease (fowl pest)[a]	CGA	Vaccination effective
Fowl plague (avian influenza)[a]	?	Vaccination effective
Infectious bronchitis	GA	Vaccination effective
Infectious bursal disease (Gumboro disease)	CG	Vaccination variable
Egg drop syndrome, 1976	A	No Vaccine
Infectious avian encephalomyelitis (crazy chick disease; epidemic tremors)	CA	Vaccination effective
Marek's disease (neural lymphomatosis)	GA	Vaccination effective
Leucosis (visceral lymphomatosis)	A	No vaccine
Infectious laryngotracheitis	A	Vaccine available
Bacteria and mycoplasma		
Avian mycoplasmosis	A	No vaccine. Blood test and eradicate
Salmonellosis (*S. gallinarium, S. typhimurium*)		No vaccines; antibiotic treatment
Yolk sac infection (*Escherichia, Staphylococcus, Clostridium*)		No vaccines; antibiotic treatment
Parasites		
Coccidiosis (*Eimeria* spp.)		Coccidiostatic drugs in feed
Helminths (roundworms, threadworms)		Anthelminthic drugs in severe infestation
Ectoparasites (lice, mites)		Insecticidal sprays in severe infestation

[a] Notifiable disease.
C, chick; G, Grower; A, adult.

Some non-specific disorders

Feather pecking and cannibalism

There is often considerable loss of feathers from hens kept intensively, both in cages and on litter and its various modifications. The extensive loss of neck feathers in caged hens is probably caused by abrasion on the wire fronts of cages and on feed hoppers. There is certainly a genetic factor involved, since some strains are more susceptible to feather loss than others; no dietary deficiencies have been demonstrated. Given adequate control of the house temperature, there is no reason to consider feather loss as harmful, but it is certainly unsightly and therefore looked upon, by some, as indicating poor welfare. Feathers are also lost, particularly under deep litter systems, by the ritual combats between pairs of hens by which the social hierarchy, or peck order, is established. Generally, the birds lowest in a peck order come to no harm if there is room to escape the attentions of the dominant birds and there is sufficient trough space for feeding. However, there are times when the pecking becomes persistent and aggressive, leading to the loss of feathers and to tissue laceration. This bullying behaviour can cause economic loss and at times becomes so intensive that substantial numbers of birds suffer extensive tissue damage and ultimately death. When this occurs it is referred to as cannibalism since a victim will literally be pecked to death.

There is no known cause of these outbreaks but their appearance is likened to a disease, because it spreads through a flock and then eventually subsides. Various vitamin or amino acid deficiencies have been suggested and so has the stimulus of the sight, smell or taste of blood. It is often associated with bright illumination and therefore dim lighting, sometimes with the use of red bulbs, has been tried as a preventive measure. Removing part of the beak was also in vogue for many years to minimize the tissue damage caused by pecking but there is little, if any, justification for this practice today.

Cage layer fatigue

This disorder not seen frequently today occurs in high producing hens at or soon after peak of laying. They become unable to stand but continue laying for a time until paralysis and death occur. The only lesion is the thinning of the leg bones. No infective agent is involved. A low calcium intake has been implicated but this is not the only factor and the aetiology remains obscure. It is, however, primarily a disorder found in battery-caged birds and those concerned with the management of this system should be aware of the possibility of metabolic disorders of this sort.

Public health

Food poisoning, caused by bacterial contamination, particularly *Salmonella* organisms, rarely arises with eggs or egg products. Normally these bacteria do not gain entry to the intact egg, and liquid egg, used by bakeries, is subject to compulsory pasteurization.

On-farm slaughter

It is necessary to kill poultry either as part of the poultry production cycle (disposing of unwanted male chicks or laying flocks which have completed their profitable laying period), which requires mass slaughter techniques, or as an occasional emergency procedure for small numbers of unhealthy birds at any time in their life.

At hatcheries, male chicks are best killed by placing them in containers into which a high concentration of carbon dioxide is introduced. This has the advantage of anaesthetizing before reaching a lethal concentration in the blood (MAFF, 1987b). Drowning or the use of toxic vapours such as carbon tetrachloride are not considered humane. Carbon dioxide can also be used for older birds. However, cervical dislocation is the best method for emergency slaughter both for these and for chicks, if only small numbers are involved, but it does require a measure of skill.

Welfare considerations

Poultry welfare is not a science and ultimately the criteria of welfare amount to subjective judgements as to what is or is not acceptable. The *Codes of Recommendations* contain much sound advice which any poultry farmer would adopt solely on economic grounds and indeed there is much truth in the view that production is a sound guide to welfare. Thus regular inspection, the provision of adequate food, water and light, careful handling and consideration of emergency precautions are all aspects of good management. A conflict between production and welfare may arise when the economics of the flock do not require the maximum production of the individual bird. Thus, by increasing stocking density, the total egg production from the enterprise is greater, despite the lower egg number per bird as a result of the greater overcrowding and stress.

The central issue of welfare is that of space, particularly the stocking density for battery cages. The revised *Codes* (MAFF, 1987) recommend that in deep litter pens 1 m^2 should support a total live-weight of 17 kg, as rearing birds or as layers; in practice, 10 light hybrid hens per m^2. In cages the recommendation is for 250 cm^2/kg live-weight during rearing, but no recommendation is made for laying hens and the European Commission has subsequently introduced a *legal* minimum of 450 cm^2/bird (p. 212). The British recommendation would have been for 600 cm^2, but it is not possible to give this the official recognition of a Code if it conflicts with European law.

The Farm Animal Welfare Council (1986) made their own assessment of the systems of housing for laying poultry in which they list the points for or against each system in terms of welfare and economics. The views of such a Council must command respect and, therefore, the main features of this report are given in Table 9.3.

References and further reading

Agricultural Research Council (1975). *Nutrient Requirements of Farm Livestock: Poultry*. London: ARC.

Dawkins, M. S. (1980). *Animal Suffering*. London: Chapman and Hall.

Duncan, I. J. H. (1987). The welfare of farm animals: an ethological approach. *Science Progress, Oxford* **71**: 317–326.

Elson, H. A. (1986). The economics of poultry welfare. In *Report on Second Symposium on Poultry Welfare*, pp. 244–253. Celle, West Germany: World's Poultry Science Association.

Farm Animal Welfare Council (1986). *Egg Production Systems – an Assessment*. Tolworth: FAWC.

Gordon, R. F. and Jordan, F. T. W. (eds) (1982). *Poultry Diseases*, 2nd edition. London: Baillière Tindall.

Michie, W. (1985). Development trials on lighting regimes for layers. *North of Scotland College of Agriculture, Technical Note 87*.

Ministry of Agriculture, Fisheries and Food (1974). *Poultry Nutrition*. RB 174. London: HMSO.

Ministry of Agriculture, Fisheries and Food (1976). *The Climatic Environment of Poultry Houses*. B 212. London: HMSO.

Ministry of Agriculture, Fisheries and Food (1977). *Incubation and Hatchery Practice*. RB 148. London: HMSO.

Ministry of Agriculture, Fisheries and Food (1978). *Artificial Insemination of Poultry*. B 213. London: HMSO.

Ministry of Agriculture, Fisheries and Food (1987a). *Disposal of Unwanted Day-old Chicks, Turkey Poults and Hatchery Waste*. L 568. London: HMSO.

Ministry of Agriculture, Fisheries and Food (1987b). *Codes of Recommendations for the Welfare of Livestock: Domestic Fowl*. L 703. London: HMSO.

Nesheim, M. C., Austic, R. E. and Card, L. E. (1979). *Poultry Production*, 12th edition. Philadelphia: Lea and Febiger.

Rothschild, M. (1986). *Animals and Man*. Oxford: Oxford University Press.

Sainsbury, D. (1984). *Poultry Health and Management*, 2nd edition. London: Granada.

Scott, M. L., Nesheim, M. C. and Young, R. J. (1982). *Nutrition of the Chicken*, 3rd edition. Ithaca: Scott Associates.

10 Broiler chicken

D. W. B. Sainsbury

Broiler chicken is the common name for young meat birds that are produced for the table. The average broiler weighs between 1.5 and 2.3 kg live-weight, but over the whole range broilers may be killed at a minimum of about 1.3 kg live-weight up to a maximum of about 4.5 kg live-weight. The heavier types are often called "roasters" or capons, although artificial caponization either by surgery* or by hormone administration,† is now illegal in the UK. Broiler chicken are carefully selected hybrids usually derived from two basic breeds, White Rock and Cornish Game. To produce a bird that weighs 2.3 kg takes only some 50 days, with a food conversion efficiency of 2.0–2.1.

Cockerels and pullets are frequently grown together "as hatched". However, it is now very popular to sex the chicks at 1 day old and grow the pullets and cockerels separately ("sexed growing" or "separate sex growing"), which has several advantages. The growth rate and feed conversion efficiency of the female falls off at lower weights than in males of the same strain, so ideally pullets should be marketed earlier. If a house is divided into two sections, holding equal numbers of either males or females, the pullets can be removed early at the lighter marketable weight, and the cockerels given the extra space. This system ensures more uniform batches for processing, with less damage at the factory. Keeping the sexes separate also reduces feed costs, in that males and females can each be fed the diet best suited to their requirements.

* Welfare of Livestock (Prohibited Operations) Regulations, 1982.

† EEC Directive 81/602.

Sexing of day-old chicks can be carried out in three ways:

1. Vent-sexing, which requires great skill and is slow and rather stressful.
2. Feather-sexing, in those crosses in which female day-olds have longer wing feathers than the males.
3. Colour-mating, in those crosses in which the males have white or yellow-coloured down while the females are a buff or red colour.

Environmental requirements

Temperature

Brooding

This first stage of rearing broilers invariably requires the application of an artificial heat source, usually known as a "brooder". A method that is widely used is to arrange a source of warmth in a confined area of the house, providing a temperature of about 35°C at 1 day old, and reducing this by 3°C a week as shown in Table 10.1. The area warmed should be as large as possible; up to one-third of the total house is recommended. This allows a wide distribution of birds over a large environmentally suitable area, a factor which is known to improve growth and reduce the likelihood of disease.

To ensure full use of the house, the ambient temperature is as important as the brooder temperature, a range of 25–30°C being associated with the best all-round performance. Below and above this range, weight gains and food conversion efficiencies are reduced. The best performance appears to be obtained if the house ambient temperature is reduced from 30°C during the first week to 27°C in the second and 24°C in the third week, and then to 21°C from the fourth week onwards. The worst results are associated with correct brooder temperatures accompanied by low house temperatures below 20°C, when the chicks are reluctant to venture away from the brooder heat to feed and drink. On the other hand, too high a temperature restricts appetite and retards activity and growth.

Table 10.1
Recommended brooder temperatures

Age (days)	Temperature (°C)
1–7	32–35
8–14	29–32
15–21	26–29
22–28	23–26
29–35	21–23
36 onwards	21

If the chicks are to be well distributed within the brooding area the temperature must be uniform, and draughts at floor level avoided. Overhead, largely radiant sources of heat give the most satisfactory results, because their fine thermostat control and adjustable height offer flexibility of arrangement (Fig. 10.1). They also serve the dual purpose of brooding and space heating.

As an alternative, however, blown hot air has its advocates because of its simplicity, low running costs and good space heating qualities. In this case, an initial temperature of 31°C is recommended for the whole house, which represents something of a compromise between the ideal brooder and house temperature. There should be a reduction of about 0.5°C per day until a level of about 21°C is reached. All changes should be made steadily and regularly to avoid stress to the birds. Sometimes attempts are made to blow air great distances from end to end of the house, which creates considerable draughts and uneven air temperatures. To compensate for this it is necessary to lift the temperature several degrees, which can be both costly and unsatisfactory for the productivity and health of the chicks. The correct procedure is to avoid this set of circumstances altogether either by ducting the hot air along the length or width of the house, or so integrating it with the ventilation system that the incoming air is heated as it comes into the house, perhaps through a central intake duct under the ridge taking a mixture of hot and fresh air combined. The very dry air conditions produced by hot air systems are not entirely favourable to the birds' health and well-being.

With a whole-house heating system the chicks should be evenly distributed with no concentration in one particular area. With brooders the distribution of chicks under each heater should be watched closely (see Fig. 10.1).

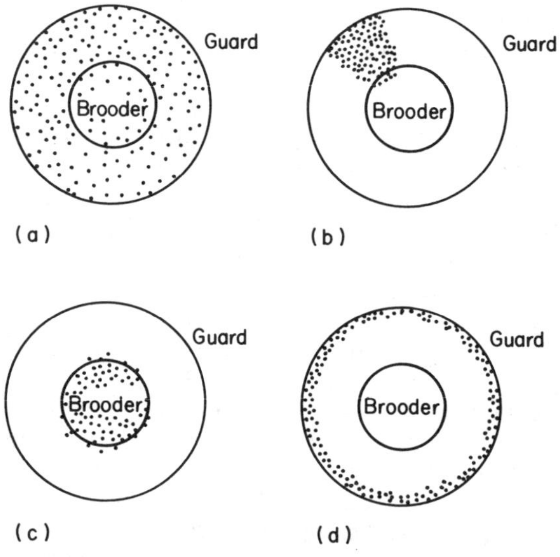

Fig. 10.1 Brooder conditions and chick behaviour. (a) Just right; (b) too draughty; (c) too cold; (d) too hot.

Post-brooding

The brooding period, as such, effectively ends at about 3 weeks of age, after which the birds are encouraged to use the whole area of the house. Brooding heat will still be necessary, other than in warm weather, to ensure favourable environmental conditions and the brooders are used as whole-house space heaters. Where radiant heaters are installed, they are raised towards the ceiling of the house to give a wider spread of heat. As a general rule, ambient house temperatures for broilers should be maintained at a minimum of 21°C.

Flooring and litter

A critical element in the rearing of broilers is to have a hygienic and "kind" surface for the birds to spend their lives on. Many types of flooring and litter have been tried but nothing has been found better than a well-built, damp-proofed concrete floor on which is placed about 100–150 mm of litter. The best litter is of soft wood shavings, but a close second is chopped straw. New clean litter is provided for each crop because the re-use of litter for more than one batch usually results in higher mortalities and an excessive production of ammonia. Management of the litter is very important, as there is a strong tendency for it to become wet and caked, especially around the drinkers; this is not only bad for the health of the birds but may also bruise the flesh and lead to carcasses being "down-graded". It is often necessary, therefore, to turn the litter physically, a time-consuming task if done manually, but which is often dealt with more satisfactorily by using a small electrically powered cultivator, which is quiet and fumeless.

Lighting programmes

Normally, broilers are given 23 h of lighting and 1 h of darkness in each 24 h, the latter being necessary to train the birds to darkness. If this is not done and the birds suddenly find themselves without light, they tend to panic, crowd into corners and suffocate. Alternatively, a programme of intermittent lighting, which is capable of giving improved growth and food conversion efficiency, can be used; A suitable example is as follows:

0–3 weeks	continuous lighting (with 1 h off in 24 h)
3–5 weeks	cycle 3 h on and 1 h off
5–7 weeks	cycle 2 h on and 2 h off
7 weeks onwards	cycle 1 h on and 3 h off

With such a programme, it is vital to provide the highest amount of feeder and drinker space, because when the lights come on after an "off" period there will be a strong demand by all birds simultaneously for food and water.

Initially, light intensity must be high to encourage the young chicks to seek food and water, but after about 5–6 days it can be reduced. It is important that the intensity is uniform across the area used by chicks, to encourage even use of the floor. Suitable intensities for broilers are as follows:

0–5 days	up to 30 lux
6–10 days	10–15 lux
11–30 days	reducing gradually to 5 lux
30 days onwards	reducing further, if required, to 2–3 lux

Lighting may be provided by fluorescent tubes, tungsten bulbs or long-life bulbs. It is desirable for all systems to be provided with a dimming mechanism, because the use of variations in the intensity of light can greatly help in controlling the behaviour of the birds. For example, if the birds are overactive, the light intensity will be reduced or, if they are ill and inclined to huddle together, increased light intensity (and warmth) will be beneficial.

Ventilation

Good ventilation is vital at all times to provide the birds with a constant and uniform supply of fresh air and to extract from the house the products of respiration and the moisture and gases arising from the litter and droppings, including air-borne droplets containing pathogenic microorganisms. While high ventilation rates are needed to keep the birds cool in hot summer conditions, the system must be capable of reducing these rates, in cold weather, to a level which will prevent the birds being chilled, while still expelling the by-products mentioned above.

Broilers can be kept in open, freely ventilated natural convection housing (Fig. 10.2) or, as is more usual, in controlled environment housing (Fig. 10.3).

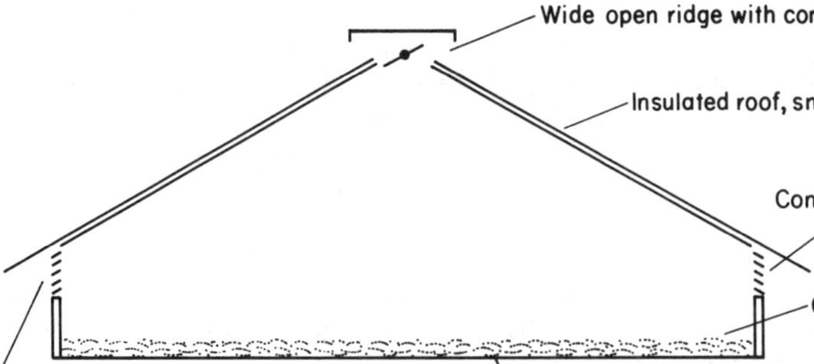

Fig. 10.2 Cross-section of freely ventilated natural convection housing suitable for broilers.

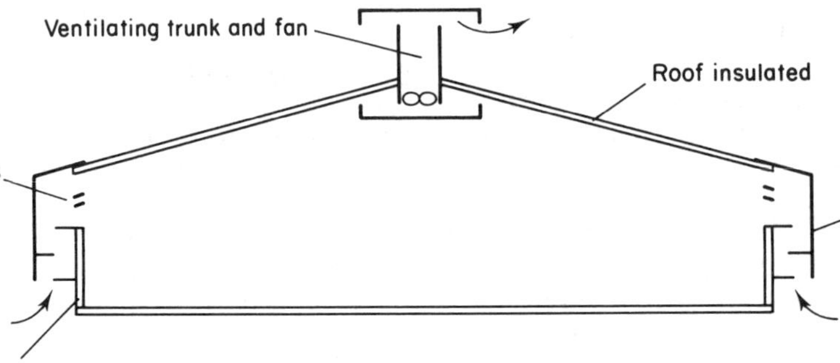

Fig. 10.3 Cross-section of typical conventionally ventilated, controlled environment house.

Naturally ventilated housing

Figure 10.2 shows the facilities for variation in the amount of ventilation and air movement in such houses, using controllable inlets and outlets. The roof should be insulated to a high standard, equivalent to not less than 150 mm of glass fibre or 100 mm of expanded lightweight plastic. An outer cladding in light-coloured heat reflective materials, is to be preferred.

The width of naturally ventilated houses should not exceed 13 m and the ridge height should not be more than 4 m, with a generous opening of up to 0.6 m at the ridge, suitably capped to prevent rain entry. A sharp-angled pitch to the roof aids ventilation. An overhang on the roof of up to 1 m beyond the eaves gives protection from the hot midday sun.

Inlet ventilators should extend along the whole length of the house on both sides, and can be controlled automatically by thermostats and motors with emergency "fail-safe" release for high temperatures.

In very hot weather, high air speeds are of great benefit to both birds and stockmen, increasing the heat loss by convection. It should be possible to open up much if not all of the sides of the house without restriction, especially in very warm areas. Siting of naturally ventilated buildings should allow maximum use to be made of winds flowing through the house when required.

Controlled environment housing

This system (Fig. 10.3) is the most widely used arrangement for broilers. The principal requirements of controlled environment housing are as follows:

1. A high standard of thermal insulation.
2. Mechanical ventilation to provide a maximum of 7 m^3/h/kg bodyweight.
3. Positive pressure systems to allow good control of air movement.
4. Ceiling or other circulating fans to help air circulation.
5. Reliable electricity supplies and stand-by generators.
6. Substantially built structures to reduce the diurnal variation in temperature, being slower to warm up by day or cool down at night.
7. Increased ventilation rate at night during warm periods to cool the entire structure and so delay the heating up of the house the following day.

Environmental cooling systems

During very hot weather, special measures may be necessary to avoid birds being adversely affected by over-heating. Measures currently in use include:

1. The introduction into the house of free-standing reciprocating circulation fans, usually mounted on the floor.
2. Trickling of water over the roof by means of a perforated plastic pipe.
3. Painting of external surfaces with reflective paint.
4. Provision of maximum "over-hang" at the eaves to keep the sun off the walls.
5. Use of construction materials which are not only good insulators but also have a good heat capacity.

Thermal insulation

The poultry industry is generally well aware of the several advantages of ensuring a high standard of thermal insulation. Apart from the merit of saving food and fuel, maximizing production, eliminating condensation and helping air-flow, good insulation generally stabilizes environmental conditions. However, in spite of the great amount of building which has been carried out using insulated construction, some serious errors are still made that have a strong bearing on the building's success and the health and well-being of the animals inside. There is a vast difference in this respect between the longevity of good and bad work. The main error is in the sealing of the insulation and of the cavity, if there is one, against damp penetration. The damp does not normally come from the outside, as the external cladding is usually adequate, but from vapour penetration from within the internal air space of the house, which will be warm and laden with moisture from the animals. If the sealing is ineffective, the moisture constantly penetrates the fabric and condenses somewhere within the cavity. This not only causes massive deterioration in almost any material, but also removes most of the insulating qualities of the materials. Thus, the vapour or moisture seal must be on the warm, inner side of the insulation. It may consist of special polythene sheets suitably overlapped at the joints. In other cases the materials themselves (such as polystyrene or polyurethane boards or sheets) are a barrier to the vapour, but the potential weakness here lies in the joints between the boards which must be sealed and protected efficiently.

Many of the more traditional methods of insulation leave a cavity between the inner and outer claddings. This becomes a happy hunting ground for vermin and insects, and it is best avoided. More recent methods of insulation in which materials, such as polyurethane, are bonded to the inner and outer claddings, avoid the cavity altogether and, as they are prefabricated in the factory, they make erection much easier. Also, provided the joints between the boards are moisture sealed, then there is every reason to recommend this type of structure.

While in temperate climates, such as in Western Europe, the principal problem is to conserve heat, there are also great advantages to be gained in warmer weather from good insulation. Good thermal insulation is becoming even more important with the current enthusiasm for natural ventilation.

It is essential that thermal insulation is used all round the building. Floors, walls and roof must be so dealt with that there are no weak points to serve as thermal "leaks" and so create damp areas of great harm to the animals.

Stocking density

Stocking density is critical to the health and welfare, and indeed production of broilers. If too few birds are placed in the house, environmental control, especially of warmth, may be difficult to maintain. If stocking density is too great, growth is impeded, access to food and water is impaired and the birds are much more likely to be adversely affected by disease. Also, the litter is almost certain to become wet and/or caked.

In "open" non-controlled environment houses the maximum stocking density should be 9–12 birds/m^2 (14–18 kg/m^2). In closed controlled environment houses a density of no more than 18–21 birds/m^2 (27–32 kg/m^2) is advised.

Emergencies

Intensively housed poultry in controlled environment houses are dependent on a reliable supply of electricity, heating equipment and other machinery in order to supply food and water and to keep the ventilation, heating and lighting functioning at all times. It is essential that fool-proof arrangements are installed to warn the stockman of any equipment failure and to provide alternative means of keeping the essential services going in any such emergency. For example, on a large site, it is usual to have a stand-by electrical generator which takes over automatically if the mains electricity fails. If this is not provided, then there must be provision for alternative natural ventilation and battery-operated lighting. Heating is usually by gas from a storage tank on site which is a reliable source of fuel unless the weather interrupts the supply vehicles. This problem is easily avoided if the tank is kept full when bad weather threatens.

Nutrition and feeding

Broilers are always fed *ad lib.* from 1 day old to finishing. A broiler "starter" ration, usually a "crumb", contains 22–23 per cent crude protein and may be fed to an age of 3–4 weeks, the later part being fed in the form of a small pellet. Thereafter, a "grower" pellet containing about 21 per cent protein is used, followed by "finisher" pellets of 19 per cent protein. If birds are taken beyond 50 days they will be finished on roaster or capon (finisher 2) diets with a further reduction of 1–2 per cent in protein. The essential nutrient levels of these rations are given in Table 10.2, and the mineral and vitamin requirements in Tables 10.3 and 10.4. The protein-rich foods used in formulation are largely white fishmeal, and various types of soya preparation; the cereal fraction lies between preparations of wheat, maize, barley and milo. The exact preparations to be incorporated in the finished ration will usually be calculated by a computer, which will be programmed to give a ration which is a safe compromise betweeen the need for economies of cost and the need for optimal biological efficiency.

Water management

Water makes up 60–70 per cent of the weight of the chicken and dehydration must be avoided at all times (see Table 10.5 for water consumption figures at different ages). At 1 day old, about 20 drinkers should be provided per 1000 birds, made up of six permanent, 400 mm-diameter, automatically filled, hanging drinkers positioned in the brooding area, supplemented by 14 chick founts. Water must be placed close to the chicks. The chick founts must be emptied, cleaned and re-filled daily with fresh uncontaminated water and on the third day the founts can be moved closer to the automatic drinkers and then removed altogether at 5 days. Drinker height should be maintained at the height of the bird's back throughout the growing period, being gradually raised to avoid excessive spillage. Careful adjustment should also be made to the depth of the water in the drinker, for the same reason.

Table 10.2
Recommended nutrient levels for broilers

	Level in rations (%)			
Nutrient	Starter	Grower	Finisher 1	Finisher 2
Protein	22	21	19	17.5
Lysine	1.26	1.20	1.06	0.90
Methionine	0.58	0.54	0.48	0.42
Methionine plus cystine	0.96	0.92	0.84	0.75
Tryptophane	0.22	0.21	0.20	0.19
Calcium	0.90	0.90	0.90	0.90
Available phosphorus	0.45	0.45	0.40	0.40
Salt	0.32	0.32	0.35	0.35
Sodium	0.17	0.17	0.17	0.17
Chloride	0.16	0.16	0.17	0.17
Potassium	0.40	0.40	0.40	0.40
Linoleic acid	1.50	1.30	1.20	1.10
Metabolizable energy (Kcal/kg)	3060	3100	3125	3125

Table 10.3
Recommended trace mineral mix for broilers

Element	Content (%)	Parts per million
Manganese	15	75
Zinc	12	60
Iron	4	20
Copper	2	10
Iodine	0.25	1.25
Selenium	0.03	0.15

Table 10.4
Recommended vitamin premix for broilers

Vitamin	Unit	Level
Vitamin A	million IUs[a]	13
Vitamin D3	million IUs	4
Vitamin E	thousand IUs	20
Vitamin K3	grams	2
Vitamin B1	grams	1
Vitamin B2	grams	8
Vitamin B6	grams	2
Vitamin B12	milligrams	10
Biotin	milligrams	150
Choline	grams	250
Folic acid	grams	1
Nicotinic acid	grams	30
Pantothenic acid	grams	10

[a] IU = International unit.

Feeder management

At the day-old stage, the "crumbs" are provided *ad lib.* on feeder pans, egg trays or chick-box lids, one being provided for each 100 chicks up to about 7 days of age. The birds are made accustomed to the trough pans or other automatic feeder systems as soon as possible, and these are used exclusively after the first few days. An allowance should be made of 25 mm of trough space per bird or 15–20 380 mm-diameter pans per 1000 birds.

Table 10.5
Water consumption in broilers

Age (days)	Consumption/1000 chicks/day (litres)
7	35
14	57
21	76
28	99
35	129
42	160
49	186
56	208

Hygiene

Probably the single most important factor in keeping poultry healthy is maintaining good hygiene. Healthy breeding stock and hygienic hatchery management ensure that newly hatched broiler chicks are free of disease and, if hygiene is good in the broiler house, an excellent chance of rearing healthy birds is ensured.

Hygiene implies dedicated attention to the standards of every process in the growing operation. The principal requirement is to prevent disease coming on to the site through visitors and vehicles. All such entries should be restricted to the minimum and, when they do come, there should be wheel dips, foot dips and protective clothing for the personnel.

"All-in, all-out" broiler management

This process entails the complete stocking of a site or unit with day-old chicks during a very "tight" period, and a subsequent complete depopulation at the end of the growing or production period. It has enormous advantages:

1. The clearing of all living animals off the site can be the biggest factor in eliminating most, if not all, organisms capable of causing disease. The more frequently this "depopulation" is practised, the more likely it is that any build-up of disease will be prevented.
2. Once the site has been completely cleared of birds it is possible to apply the most rigorous and effective programme of disinfection and fumigation (see below), so progressing towards an effective elimination of bacterial, viral, fungal and parasitic infection.
3. If all birds on a site are within a close age range this makes for a more uniform state of immunity to disease. Mixed ages lead to an immunological confusion, so that the effects of vaccines or treatment are less satsifactory, and it is certainly much more difficult to administer disease prevention programmes correctly.
4. There are sound and practical husbandry advantages. If a site is filled at one time, there will be no further unavoidable disturbance until the site is cleared. If units have birds coming and going at irregular intervals, the repeated disturbances can lead to health and behavioural upsets.
5. It may be considered an advantage if the operator takes a break in the exacting task of management. Maintenance of equipment can also be properly carried out between batches so that there are likely to be far fewer breakdowns.
6. In recent years, with the great increase in intensification, there have been serious problems caused by flies breeding in the manure and other organic matter around poultry sites that are in continuous use. Large sites often have poor arrangements for muck disposal. Smells can cause highly objectionable conditions for nearby residents. An "all-in, all-out" system can limit the feasible size of a unit, which is not a bad thing in itself but, above all, it can make it simpler to eliminate insects and rodents.

It may be emphasized that the merits of depopulation are in inverse proportion to the age of the birds: it is less important with adults because, by the time a bird has reached maturity, it may have achieved a satisfactory immunity to most diseases.

Hygiene programme for broiler sites

1. Remove all birds from the site.
2. Clean out the litter from all houses and remove it totally from the site.
3. Clean all the dust and dirt from the building, paying special attention to less obvious places, e.g. air inlets and fan ventilator boxes.
4. Place all equipment that can be removed (e.g. brooders, drinkers and feeders) on a clean area around the building for disinfection.
5. Wash down all interior surfaces of the house with a detergent disinfectant, preferably in a pressure washer.
6. Apply a specific, broad-spectrum disinfectant to soak all the lower parts of the house to destroy viruses, bacteria, mycoplasmas, fungi and parasites.
7. If there are litter beetles and other insects in the house, use an insecticide. Insects are known to act as carriers for all poultry infections.
8. Give a similar cleaning and disinfection treatment to all the equipment.
9. When the house interior is dry, set up the equipment and put in the litter. Then close the house and fumigate with formaldehyde gas after the house has been warmed to 20°C.

10. After 24 h, open up the house and ventilate to remove the remaining gas. The disinfected building is then ready for the chicks.

Health and disease

From the emergence of the intensive poultry industry some 40 years ago until the present time the mortality rate in broilers in the UK has steadily decreased from an average of well over 10 per cent to the current one of about 4 per cent. However, averages conceal the fact that at one time there were frequent occasions when mortalities rose to 40 per cent or more, whereas this would be very unlikely today. Indeed, nowadays, many good broiler growers will average mortalities of little over 2 per cent.

How has this excellent picture been obtained? First, careful environmental control with good heating and ventilation has played a fundamental part, together with well constructed and thermally insulated buildings. Secondly, the practice of an "all-in, all-out" policy for virtually all sites, combined with a thorough clean and disinfection in the empty period has gone far to eliminate the risk of a build-up of infection. Thirdly, the strategic use of drugs has played an essential part. All broilers are fed a coccidiostat at a low level for most of the growing period, and a growth promoter is also normally used. The growth promoters are antibiotics which are used for this purpose alone and have no therapeutic application or activity either in man or animals. Otherwise, drugs are used sparingly and the majority of flocks receive no treatment. The nutrition of the broiler has also played an enormous part in the maintenance of such good health and the various metabolic disorders affecting broilers are largely avoided now by ensuring the correct balance and quality of protein and energy and sufficient additives, especially vitamins and minerals. One of the results of this has been a dramatic reduction in the development of the skeletal abnormalities which once plagued the growing broiler.

However, if growth rates are forced excessively by the use of feed which has a higher energy than the bird can cope with, metabolic problems may result. Examples of these are poor (soft) bone development, abnormalities of the growth of the tendons, which may break, and "heart attacks", which kill the bird. If these do occur in a substantial number of birds, then adjustments to the ration by feeding a lower energy feed and relatively greater quantities of high-quality protein, minerals and vitamins, may be necessary.

Vaccines are used sparingly for broilers in the UK. Almost all birds receive an administration of live infectious bronchitis vaccine at 1 day old, usually given by a large particle spray in the hatchery. For most broilers this is the only vaccine used, but occasionally it is necessary to use Newcastle disease and Gumboro disease live vaccines, depending on the likely environmental challenge. Because the period of growth is so short, most of the immunological protection for broilers is obtained from ensuring that good maternal immunity is transferred passively to the chicks from the dams. Broiler growers are also usually careful to avoid infection being brought on to a site by visitors or deliveries of food and other necessities, referred to earlier. One risk which cannot be eliminated is that from air-borne infections. It is known that infective particles can travel many miles in the wind from diseased sites and, as broiler growing tends to be concentrated in limited areas, the risk is intensified. The first necessity in dealing with any disease condition in broilers is to *recognize* the trouble at once, *diagnose* rapidly and treat *speedily*, thereby the loss can usually be minimized.

General care and handling

Catching

Broilers are usually caught at night for slaughter the following day. Lights are dimmed to keep the birds calm and the catchers pick up three or four birds in each hand and place them in crates which are either outside the house on a lorry or, in the case of the "modular" system, are brought into the house in groups, on fork lift trucks. There are also some mechanical methods of gathering birds up from the floor, using plastic or rubber covered "arms" which slowly revolve and draw the birds towards an operator who can place them in a crate. These have at present a limited following, though further developments are taking place, and in due course they may become more generally used.

Transport

The transportation of the birds is another period of extreme importance and great care has to be taken to minimize stress at this time. Generally, broilers do not travel more than about 50 miles to the processing plant but, even so, severe losses can take place if the whole procedure from beginning to end is not handled with understanding and skill (see Appendix).

The stockman

The stockman is legally required to inspect his livestock and equipment at least once daily,* and in practice he will usually inspect both the stock and the equipment several times every day.

A major requirement in all animal accommodation is to provide a first class service room for the stockman. In here he should be able to carry out the necessary hygiene for himself and his colleagues, to store his equipment and medicines safely, have access to hot water and maintain records in an efficient way.

Welfare with reference to the *Codes of Recommendations*

The *Codes of Recommendations for the Welfare of Livestock: Domestic Fowls* (MAFF, 1987) have certain parts which are more specifically directed towards the welfare of broiler chicken. It is very important with broilers, where there are usually many thousands of birds in close contact, that ill-health is rapidly recognized and action is immediately taken to diagnose the ailment, treat the birds and hence limit the risk of contagion. Sometimes the most appropriate procedure after diagnosis is to remove the birds as quickly as possible to the processing plant for slaughter before there may be a greater degree of disease. Such action requires very careful professional and largely veterinary judgement, as with all slaughtering which is made in an emergency. The decision will be made on the basis of a careful balance between the economic and welfare aspects.

Fire precautions must be given a major priority – the risk is quite high. Most fires arise from electrical faults with short circuiting creating the greatest risk. The *Codes* detail the welfare risks from the failure of electricity and the breakdown of equipment, and the dangers of overcrowding are emphasized. Advice is given that the *maximum* stocking density for broilers should be 34 kg/m^2: this is good advice from the aspect of productivity as well as welfare.

Reference is made in the Preface to the *Codes* to the need to allow birds a proper exercise of their basic physiological needs. In general, broilers are able to do this and are ensured a comfortable existence if management is good and the environment and the litter are well maintained. There is no restriction on normal behaviour – broilers can exercise as much as they wish, dust-bathe and choose warmer areas in the house for their resting periods. The lighting should be adjusted to ensure there is no stress or undue excitement, and excessive sudden noise should be avoided to prevent panic which can lead to a pile-up and some birds being suffocated. The attendant must learn to move slowly among the stock to prevent any panic moves by the birds. Uniform and quiet conditions are essential and all changes in the environment should be made as gradually as possible.

The way ahead

In the past two decades, the methods of managing, housing and feeding broilers have not altered a great deal and nor do significant changes appear to be imminent. The number of broilers reared annually in the UK continues to increase steadily and now stands at about 450 million, which is some 8 birds per annum per

* The Welfare of Livestock (Intensive Units) Regulations, 1978.

head of the population. Attempts were made about 10 years ago to rear broilers in tiered cages but, because the capital cost was high and there was damage to the birds, especially of the breast muscles, leading to down-grading, the method found no favour and is no longer practised. A few attempts are currently being made to rear broilers either on free range or in straw yards in more "natural" conditions, but they have not been very successful because of the climatic stress these systems impose on the young growing chicks. There is, however, an increasing enthusiasm for the natural ventilation of broiler houses and for techniques that minimize the amount of vaccination or medication given to the birds. There is a real possibility that, within the next few years, improved methods of management and hygiene will enable broilers to be reared for the entire period of growth, some 45 days, without any routine medication. Some of the impetus for this trend has, without doubt, come from the official action, stimulated by consumer awareness, to enforce an absence of drug residues in the meat.

However, growers are also more aware of the fact that costs can be reduced by using a minimum of additives and placing more reliance on high standards of management and hygiene, together with less complicated systems of environmental control. There is also a trend towards smaller units with the birds sub-divided into smaller groups than hitherto. This makes practical use of the clear evidence that shows that the best results biologically come from farms of a modest size under the management of one good stockman. A further bonus in the case of small livestock farms is that it is easier to find a good use for the litter as manure on local land, thus removing the nuisance factor to the local human population.

Further reading

Agricultural Research Council (1975). *Nutrient Requirements of Farm Livestock: Poultry*. London: ARC.

Card, L. and Nesheim, M. (1972). *Poultry Production*. Philadelphia: Lea and Febiger.

Carter, F. (1976). *Intensive Poultry Management for Egg Production*. London: HMSO.

Charles, D. R. and Spencer, P. G. (1976). *The Climatic Environment of Poultry Houses*. MAFF Bulletin 212. London: HMSO.

Dobson, C., Charles, D. R., Emmans, G. C. and Rhys, I. W. (1970). *Housing and Environment*. MAFF Bulletin 56. London: HMSO.

Feltwell, R. and Fox, S. (1978). *Practical Poultry Feeding*. London: Faber and Faber.

Gordon, R. F. and Jordan, F. T. W. (eds) (1982). *Poultry Diseases, 2nd edition*. London: Baillière Tindall.

Hofstad, M. S., *et al.* (1972). *Diseases of Poultry*, 6th edition. Ames, Iowa: Iowa State University Press.

Ministry of Agriculture, Fisheries and Food (1987). *Codes of Recommendations for the Welfare of Livestock: Domestic Fowl*. L 703. London: HMSO.

Sainsbury, D. (1983). *Animal Health*. London: Collins.

Sainsbury, D. (1984). *Poultry Health and Management*, 2nd edition. London: Granada.

Sainsbury, D. and Sainsbury, P. (1988). *Livestock Health and Housing*, 3rd edition. London: Baillière Tindall.

Appendix: Code of Practice for the Catching and Transport of Live Chicken (as formulated by a major broiler company, and reproduced with their permission)

1. Catching

(a) Live chicken must be handled with extreme care at all times to ensure freedom from stress, injury and unnecessary suffering.

(b) Birds must never be roughly handled by driving, kicking or throwing and quiet movements are essential to prevent smothering, bruising or more serious damage.

(c) The top drawer of the module must be loaded first and filling should work downwards to avoid injury on closing the drawer.

(d) Birds must be lifted into the drawers carefully and never be thrown into them.

(e) The modules must be loaded on to the transporter with great care avoiding violent manoeuvres, e.g. heavy landings. All loaded module exit doors must be securely locked, standing of loaded modules must be kept to a minimum.

(f) It is essential to catch birds in one hand at a time and never in both hands at once.

(g) All staff involved with the catching and transportation should be instructed in the relevant Animal Welfare Legislation.

2. Transportation

(a) The vehicle and associated equipment must be inspected before every journey to make certain that all parts are mechanically sound and in good working order.

(b) Scheduling of loaded transporters into the plant should be such that transport and waiting times are kept to a minimum. All loads must be processed in strict delivery rotation.

(c) It must be established that the flock is in satisfactory health and is fit to travel.

(d) The weather conditions will determine the loading densities and a recommended schedule for modular crates is as follows:

Live weight (lbs)	Normal temperatures	Abnormally hot conditions
4.1 and below	30	27
4.2–4.5	27	24
4.6–5.0	24	21
5.1 and above	21	18

(e) In the event of temperatures near or below freezing or when it is cold and there is heavy rainfall and there is a journey in excess of about 20 miles, then the modules must be protected with appropriate sheeting.

(f) On arrival at the factory the vehicle must be unsheeted immediately, weighed and parked in the special holding bay.

(g) The Foreman in charge and his staff should be aware of welfare regulations and must be informed of the procedure to take whenever a breakdown occurs in the transport or there is some unavoidable hold-up.

3. Unloading

(a) All crates must be opened carefully and the birds removed with great care to avoid pain or injury.

(b) Any birds that break loose must be collected immediately.

(c) Any birds unfit or unsuitable for slaughter may be removed for slaughter elsewhere.

(d) All birds found to be dead on arrival must be placed in the receptacle provided.

(e) If there is a plant breakdown so loading cannot continue, the official Veterinary Surgeon or Senior Poultry Meat Inspector must be informed immediately and appropriate action taken after due consultation with the Factory Management and Agricultural Division (of the company).

11 Turkeys

C. Nixey

Origins

The turkey was originally domesticated by the Indians in the USA. There are at least five subspecies of wild turkeys. It seems likely that they have all made a genetic contribution to the domesticated turkey of today. The early settlers in America found the wild turkey easy to shoot. By 1900 only small populations remained of the once great flocks which roamed from the Great Lakes south to Central America. Due to protective measures by the Government, the number of wild turkeys is increasing again in the USA.

The turkey is a truly transatlantic bird in that settlers took the turkey back to Europe in the sixteenth century. These European domesticated turkeys were then reintroduced to the USA, probably in the nineteenth century and crossed with wild turkeys. Further development took place and there have been notable importations of stock, both to Europe and back to the USA, in the twentieth century. Compared to other poultry and farm animals, the turkey has been domesticated for only a short period. Its recent wild bird origins mean that it is more sensitive to environmental factors and managemental stress than most farm animals. The staff working with the birds therefore need to have a high level of stock managerial skills.

As has been outlined in Chapter 8, the turkey fulfils several markets in addition to the traditional turkey market at Christmas. To meet these different markets, strains of turkey differing markedly in growth rate and fecundity have been developed.

The first domesticated turkeys had either black or bronze feathers. However, because of the colour blemishes occurring in the plucked turkey with black or bronze-feathered birds, there has been a strong swing towards white-feathered birds, with the result that now only small numbers of coloured-feathered turkeys are produced for the specialist market of traditional farm fresh (TFF) turkeys.

Reproductive performance

It is normal for the age at onset of lay to be controlled by the daylength given to the turkey hens. First eggs are normally planned for 32 weeks of age. A normal economic laying cycle is 20–24 weeks. The rate of egg production is negatively correlated with bodyweight, i.e. the larger the strain of turkey, the fewer eggs it will tend to produce. To meet the market requirements, the strains available fall into three main types: the small type for the TFF market; the fast-growing strains for the large turkey market, and the dual-purpose strains which try to cater for all sections. Examples of typical reproductive data are shown in Table 11.1.

The reproductive results achieved are very dependent on the skill of the stockman looking after the breeders. Because of the size discrepancy between the sexes, and the broad conformation developed by genetic progress, natural mating is difficult or impossible (Fig. 11.1). It is almost universal for turkey breeder hens to be artificially inseminated, in order to obtain good fertility and to prevent damage to the females. Extremely high fertility results are achieved, but the results are very dependent on the skill of the inseminating team.

Table 11.1
Reproductive data for turkeys

	Type of bird		
	Small	Dual-purpose	Heavy
32-week live-weights (kg)			
female	6.8	8.6	11.4
male	18.0	25.0	27.0
Settable eggs per hen housed in 20 weeks	100	90	80
Mean live-germ % after 8 days of incubation	87.0	86.5	86.0
% Hatch of 8-day live-germs	89.0	88.0	84.0
% Hatch of eggs set	77.4	76.1	72.2
Poults per hen housed in 20 weeks	77.4	68.5	57.8

Fig. 11.1 Turkey stag and hen. Note the difference in size between the sexes. Photograph: British United Turkeys Ltd.

The first insemination should be done when around 60 per cent of the flock adopt the mating posture when the stockman enters the pen. This will also coincide with the first eggs being laid by the flock. It is normal to carry out three inseminations in the space of 10 days and then to go on to a weekly pattern of insemination, although equally good results can be obtained by a good insemination team carrying out inseminations every 10 days in the first 10 weeks of lay.

Figure 11.2 illustrates the weekly reproductive performance pattern obtained by the most popular dual-purpose breed currently used in the UK.

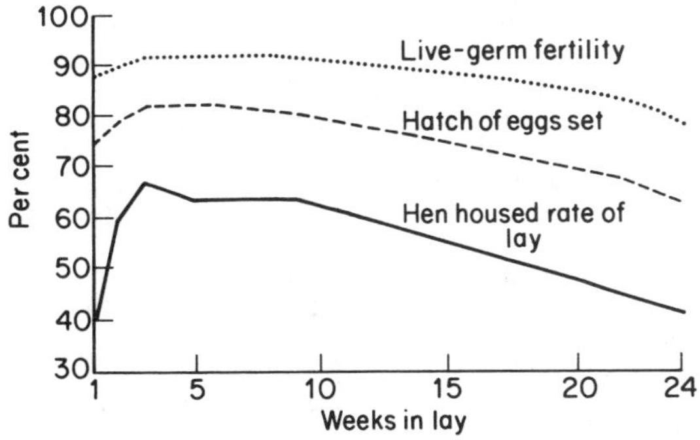

Fig. 11.2 Reproductive characteristics of large-type turkeys.

The turkey egg requires 28 days of incubation, which is longer than for the domestic fowl. The turkey egg is more sensitive to conditions during storage and incubation than the chicken egg. It is generally recommended that eggs are stored at a temperature of 13–16°C and a relative humidity of 80 per cent. Hatchability declines with increasing severity with storage beyond 7 days of age.

Eggs are incubated at a temperature of 37.5°C until 25 days, when they are transferred to the hatcher at a lower temperature of about 36.8°C. Variations in temperature around these figures will be seen, depending on the type of incubator.

Growing performance

Examples of typical growth rates for the three types of turkey are shown in Table 11.2. The performance achieved is greatly influenced by nutrition, and also by factors that influence food intake, such as housing, stress, pellet quality and temperature.

As has been mentioned earlier, the turkey carcass is increasingly being cut into joints, and examples of the likely yields from a large-type male are shown in Table 11.3. It will be seen that breast meat yield, which is the most valuable meat, increases steadily with age. However, if the ration is lacking in amino acids, or stress factors such as stocking density or lack of feed and water space adversely affect food intake, the breast meat is the first to be reduced. It is therefore greatly in the farmer's interest to reduce such stress factors to a minimum.

Nutritional requirements

Normally, commercial fattening feed formulations (Table 11.4) aim to produce maximum growth rates, whereas the aim when feeding future breeding stock (Table 11.5) is to produce a fit bird at start of lay for the most economic cost.

If turkeys older than 8 weeks of age are offered a choice of high-protein concentrated feed with vitamins and minerals included, and whole wheat or oats, they can balance their intake of each type of food so as to meet their requirements closely.

Turkey poults are normally fed feed in the form of crumbs for the first 3 or 4 weeks of life. They then move on to pellets 3.5 mm in diameter and approximately 4 mm in length. As they get older, the size of the pellet may be increased up to 4.5 mm in diameter.

Because of their rapid growth rate, turkeys are very sensitive to factors which depress appetite. One of these factors is the pellet quality, and dusty pellets will reduce growth rates. Farmers who are home-mixing their turkey food will often be feeding a mash. This may also reduce food intake. To maintain satisfactory inputs of nutrients, the nutrient density of the mash should be increased, with particular attention being given to the protein and metabolizable energy levels. Where the food is fed in a mash or pellet form, no benefit has been shown from feeding insoluble grit. However, this ingredient is essential if whole grain constitutes part of the diet.

A most important nutrient is water. The turkey's food intake is closely correlated with its water consumption. Turkeys consume approximately twice as much water by weight as they

Table 11.2
Examples of growth patterns

	Type of turkey					
	Small		Dual-purpose		Heavy	
	Live-weight (kg)	Food conv. rate	Live-weight (kg)	Food conv. rate	Live-weight (kg)	Food conv. rate
8 weeks						
females	2.02	1.95	2.98	1.92	3.30	1.91
males	2.43	1.85	3.62	1.82	4.10	1.81
12 weeks						
females	3.64	2.46	5.37	2.36	6.10	2.29
males	4.75	2.23	7.09	2.14	7.94	2.11
16 weeks						
females	5.05	3.04	7.46	2.88	8.76	2.75
males	7.10	2.64	10.60	2.49	11.99	2.44
20 weeks						
females[a]	5.97	3.76	8.81	3.51	10.84	3.28
males	9.37	3.11	14.00	2.92	15.88	2.85
24 weeks						
males	11.20	3.65	17.22	3.41	19.68	3.31

[a] Females not grown for slaughter beyond 20 weeks.

Table 11.3
Relative yields from a heavy-type male (% live-weight)

Age (weeks)	Live-weight (kg)	Evisceration loss	Giblets	Breast meat and skin	Thigh joint	Drum joint	Wing joint
14	9.98	18.7	7.1	27.3	13.4	10.9	9.8
16	11.99	17.9	6.5	29.0	13.9	10.7	9.3
18	13.94	17.4	6.0	30.6	14.1	10.5	8.9
20	15.88	17.2	5.7	32.1	14.2	10.1	8.3
22	17.79	17.1	5.4	33.6	14.2	9.8	7.7
24	19.68	16.9	5.2	34.8	14.2	9.6	7.2

Table 11.4
A typical commercial fattening feed programme

	Age (weeks)					
	0–4	4–8	8–12	12–16	16–20	20–24
Metabolizable energy (MJ/kg)	11.80	11.90	12.00	12.10	12.20	12.40
Crude protein (%)	29.00	26.50	22.50	19.00	17.00	15.00
Lysine (%)	1.85	1.60	1.33	0.95	0.85	0.75
Sulphur amino acids (%)	1.17	1.03	0.86	0.63	0.57	0.52
Calcium (%)	1.40	1.20	1.15	1.00	0.90	0.90
Available phosphorus (%)	0.80	0.70	0.65	0.55	0.50	0.45

Table 11.5
A typical future breeding stock feed programme

	Age (weeks)				
Female	0–4	4–8	8–12	12–29	29–
Male	0–4	4–8	9–12	12–	
Metabolizable energy (MJ/kg)	11.80	11.90	12.00	12.10	11.50
Crude protein (%)	25.00	21.00	18.00	12.50	16.00
Lysine (%)	1.50	1.16	0.93	0.58	0.75
Sulphur amino acids (%)	0.93	0.75	0.63	0.48	0.60
Calcium (%)	1.40	1.20	1.15	0.90	2.50
Available phosphorus (%)	0.80	0.70	0.65	0.50	0.45

do food, so it is very important that they have unrestricted access to water at all times. It is also important that the water is fresh, and uncontaminated with litter or faeces. As a rule of thumb, there should be one waterer per 100 birds and at least 3 cm of feeder space for each turkey. Feeders and waterers should be evenly distributed about the house.

The turkey has a simple digestive system, similar to that of the chicken (p. 201). There is, however, some evidence that it is able to utilize fibrous foods more efficiently. Because of its faster growth rate and lower fat content compared to the chicken, it does require a greater concentration of protein in its diet as shown in Table 11.4. The diet should be supplemented with vitamins and minerals. These are normally supplied in a premix by companies specializing in this field, for inclusion in the formulation. More information on the nutrition of turkeys can be obtained from MAFF, (1974, 1985).

Environmental requirements

The number of turkeys reared on pasture has dropped dramatically and is now of neglible proportions. The turkey can happily withstand cold, but dislikes wet conditions. The UK climate and soils make this country unsuitable for rearing turkeys outside. A major disease problem experienced on pasture is gapeworms (*Syngamus trachea*) which can remain present in the soil for several years after destocking.

Contrary to popular belief, no turkeys, either fattening or breeding stock, are grown commercially in cages. The physical size of the birds causes problems both to their feet and to the equipment. Egg numbers and shell quality are reduced when breeding hens have been accommodated in cages. Both on welfare and on economic grounds, it is very unlikely that cages will ever be used to any extent with turkeys.

The day-old turkey (poult) requires warm, draught-free conditions. In large flocks, this is usually provided by controlled environment housing, but on the general farm, turkeys can be brooded in converted farm buildings. At day-old, it is usual to confine poults to a warm area by encircling the brooder with a cardboard surround approximately 50 cm high, in a circle approximately 4 m in diameter. The spot heat brooder is normally gas-fired. The temperature in the centre of the area under the brooder is normally around 40°C, graduating out to a total house temperature of initially 29°C and reducing with age. The range of temperature allows the poult to choose its own comfort zone. The behaviour of the birds is a good indication of the suitability of the temperature. They should be moving around freely and not huddling, which indicates either that the temperature is too cold or there are draughts. It should be remembered that huddling will also occur when a disease problem is starting or is present. Panting and lying down with the wings spread would indicate that the temperature is too hot.

By 7 weeks of age, the turkeys should be fully feathered and weaned off artificial heating. Where the birds are being grown for early killing around 12 weeks of age, it would be normal for them to remain in controlled environment housing. On the general farm, where the birds are being grown for the Christmas market, or if the birds are being grown to 18–24 weeks of age, they may be moved into polebarn housing after 7 weeks of age in the summer and 8 weeks in the winter.

The *Codes of Recommendations for the Welfare of Livestock: Turkeys* (MAFF, 1987) suggest a maximum live-weight density of 39 kg/m^2 in controlled environment housing and 24.4 kg/m^2 in polebarn housing.

A polebarn house (Fig. 11.3) is a simple corrugated roofed structure supported by poles usually 3 m apart. The sides of the polebarn are enclosed with wire netting, or partly enclosed with corrugated sheeting with wire netting above. Natural ventilation occurs within the polebarn. In winter, the wind flow can be reduced by plastic curtaining. The birds usually receive only natural light. The floor may be of earth or concrete. The litter is usually straw in polebarns and wood shavings in controlled environment housing. In controlled environment houses, the light intensity and daylength, the ventilation and, indirectly, the temperature in the house, can be controlled. There is a wide range of lighting programmes practised. At one end of the spectrum, 2 h of light alternates with 2 h of darkness, while at the other extreme there is 23 h of dim light (5 lux) and 1 h of darkness. Dim light is most widely used, because it prevents feather-pecking and fighting, and this means that the birds do not have to be debeaked.

The optimum growing temperature for turkeys past the brooding period is between 16 and 20°C. Whenever possible, temperature should be maintained at the target level by controlling the ventilation rate. However, the first priority of ventilation must be to provide oxygen and remove carbon dioxide and excess moisture and other pollutants such as ammonia and smells. The limiting factor is usually the level of ammonia, which should not be above 15 p.p.m. The ammonia is easier to control when the litter is kept dry. If the litter in the house has a cracked hard surface, the house has been badly managed.

Future breeding stock have precise daylength requirements beyond 18 weeks of age. The turkey hen's requirement is similar to that of the wild birds. To commence lay, the breeder hen must first of all have experienced a winter type daylength, i.e. short day and long night for at least 6 weeks, prior to commencement of the long day (spring type) stimulatory light treatment. Given long daylengths (14 h), the turkey hen may reach sexual maturity and commence laying as early as 22 weeks of age. The resulting egg production is poor and egg size is small. The optimum age for starting lay is usually judged to be around 32 weeks of age.

The recommended lighting pattern is:

1. Prior to 18 weeks, the hens can receive natural daylengths, or 14 h light if in a controlled environment.
2. Between 18 and 29 weeks, the hens should receive 7 h of light which should have a light intensity of at least 50 lux.
3. From 29 weeks, the hens should receive 15 h of light per day.

The males do not have such sensitive light requirements and can receive either the normal seasonal daylength, or 14 h of artificial light until 26 weeks of age. They should then receive a minimum of 14 h of light for the rest of the breeding season.

The best egg production is obtained in polebarn housing rather than controlled environment housing. This may be due to the greater light intensity of natural light, or because the greater temperature variation between day and night acts as a stimulant. Breeder males should be kept in warm draught-proof housing.

General care and handling

Stockmanship

Although good stockmanship comes more naturally to some individuals than others, it is possible, with good training, to teach the right attitude and good habits which are at the heart of good stockmanship.

The most important time for this is in the early weeks of a bird's life. The day-old turkey (poult) would normally associate itself with its mother. In artificial incubators, this is not possible, so it looks for a substitute "mother", and it imprints itself to whatever it is most associated with in the first few days of life. It is therefore

Fig. 11.3 A Polebarn: (a) Exterior view; (b) Interior view. Photograph: British United Turkeys Ltd.

important for the future welfare and behaviour of the flock that the stockman spends a lot of time with the poults, inspecting them at regular intervals, replenishing food, etc., and just generally being around the house. He should always move unhurriedly and quietly. The resulting behaviour and placidness of the flock reflects the stockmanship practised early in their lives.

Handling and restraint

Due of their size, it is important that older turkeys are handled correctly. It is a cardinal sin to catch a turkey by one leg and then hang on while it flaps its wings. When the birds are small enough to do this, both legs should be grasped simultaneously with one hand and the bird lowered on to its breast and then lifted up, holding one wing with the other hand to prevent flapping. Large turkeys should be grasped with one hand on the shoulder of the wing farther from the catcher, while the other hand searches for the legs. The catching operations should be carried out with smooth movements so as not to unduly disturb the other birds, as this will cause them to panic and clamber on top of each other, causing scratching of the backs and sides. It is normal to build a catching area within the main house with the use of wire frames. Small numbers – 20–40 birds, depending on their size – should be driven into this catching area; larger numbers cause more stress, with birds flapping their wings and piling on top of each other.

Routine procedures

It is important that attention is paid to detail when carrying out routine procedures. Keen powers of observation are required when inspecting the birds. Birds should be visited at least daily,* preferably in the morning *and* afternoon. Before entering the house, the door should be knocked on to alert the birds, otherwise they may be frightened and agitated. Any abnormal behaviour of the flock or individuals should be noted and, if abnormal mortality occurs, a veterinary surgeon called. It is vital that the birds always have access to fresh food and clean water. This means that the feeders should never be allowed to become empty and, at the same time, care should be taken not to over-order from the compounder, to guard against food becoming stale in store. The drinkers should be swilled around and emptied each day, and scrubbed each week. The atmosphere in the house should be assessed on entering. The ventilation rates will need adjusting according to the climate. Where possible, the ventilation rate should be reduced in the evening to allow for the colder night temperatures.

Many operators are able to rear their turkeys throughout their life without debeaking them, by using low light levels to which the birds' eyes adjust, but which prevents them being attracted to blood and lessens the level of fighting which precipitates cannibalism. However, where the need for debeaking is anticipated, it is best carried out at 5–7 days of age using a sharp pair of secateurs. Although bleeding occurs, the poults will be seen eating again almost immediately. The alternative is to use a hot cauterizing blade which prevents bleeding but probably causes the bird more discomfort. The upper beak should be cut back to just in front of the nostril. It may be necessary at an older age to do further trimming, particularly at point of lay for breeding stock. This is important because there is a tendency for birds to peck around the eyes of others when they are "squatting", the position taken up when ready for mating. It is regrettable that the birds have to be debeaked, but the alternative, to risk severe damage from fighting and cannibalism, could involve considerable suffering by some birds within the flock.

It is usual to remove the snoods of day-old poults if they are potential future breeding stock. This is done by pinching the thumb and first fingernails together at the base of the snood and simply pulling it off. If allowed to develop, it is easily damaged when the birds fight, and can become a focus for infection.

On the breeding farm, the two main routine procedures are egg collecting and artificial insemination. Eggs should be collected eight times during the day, to reduce the number cracked while in the nestbox and to enable the eggs to be sanitized quickly after laying.

* This is a legal requirement under the Welfare of Livestock (Intensive Units) Regulations, 1978.

Artificial insemination is a skilled procedure. The technique of gathering semen from the male is a skill that can only be mastered by people with an aptitude for it or long training. The insemination of the hens requires attention to detail, and accuracy on the part of the inseminator and a smooth co-ordination between the team of three, which comprises the person catching, the person everting the oviduct, and the inseminator. A skilled team can inseminate in excess of 500 hens/h and achieve fertility levels of well over 90 per cent. It is important that the hens are disturbed as little as possible during insemination, because this could affect subsequent egg production. Obviously, catching and handling should be as gentle as possible. Birds should also be denied access to food and water for the shortest possible time. This means penning only small numbers at a time.

At some time, whether grown for fattening or breeding, the birds have to be transported for killing. The same principles for catching the birds apply. Careless handling at this stage can damage the birds in a few seconds and ruin several months' work.

Health and disease

In a normal healthy flock, all the birds will be active, giving an alert appearance with bright eyes. They should also be mobile and walk without trembling legs or wanting to sit down hastily. Sick turkeys will often stand with their head buried in their shoulders, with feathers fluffed up and their eyes closed. The normal place to find the poorest turkeys or ailing birds is along the outside of the house. Bullying of sick birds occurs, and it is therefore useful to have a spare pen which can be used as a "hospital" for sick or pecked birds. However, birds which are obviously suffering and likely to die in a few days should be slaughtered immediately by neck dislocation. It is important to dispose of these birds quickly and efficiently to avoid the risk of spreading disease. The same applies to birds which are already dead. This is best done by an incinerator on the farm or, alternatively, cadavers can be put into plastic bags for collection.

The national turkey flock is probably the healthiest it has ever been. Coccidiosis and blackhead are two once-common diseases which can be completely prevented by the addition of suitable preventive drugs to the feed for the first 8 weeks of life. In farms with a history of either disease, the drug may be included to 12 weeks of age. When turkeys are being killed at young ages, it is important that medication which could result in a residue in the meat should be carried out strictly according to the manufacturer's instructions regarding the withdrawal period prior to slaughter.

Mycoplasma gallisepticum, which causes respiratory problems, including swollen sinuses, *Mycoplasma synoviae* which causes similar symptoms and puffy joints, and *Mycoplasma meleagridis* which reduces hatchability and causes mortality and leg problems in the young turkey, are all problems of the past. These diseases, which were transmitted via the egg, have been eliminated from the parent breeding stock. There is still a residue of infection in the chicken population, so it is not advisable to rear chicken and turkeys in close proximity.

Fowl cholera can strike flocks and cause heavy mortality. Poor hygiene, particularly wet litter or muddy areas, can predispose the flock to the problem. If a sudden increase in mortality levels occurs, diseased birds should be sent quickly for post-mortem. Some measure of protection can be achieved by vaccination but, because there are several strains of bacteria, the vaccine may not give complete protection.

The turkey is susceptible to respiratory infections which are often caused by viruses. A particular cause for future concern is turkey rhino-tracheitis (TRT) which is caused by a virus which has not been identified with certainty. Newcastle disease is an example of a virus disease which is prevented by vaccination.

To minimize the risk of disease, the following guidelines should be followed:

1. Access to the farm by traffic and visitors should be minimized.
2. Protective clothing and boots should be worn when working with turkeys, and a separate set should be kept for use in each house.
3. Water drinkers should be kept clean.
4. Litter should be maintained in a dry and

clean condition.

5. Grass between houses should be kept short, and heaps of wood or other attractive places for vermin should be avoided.
6. Accurate records should be kept of mortality, and egg production if applicable, because they will provide an early indication of a problem, as will falls in feed consumption.
7. Dead birds should be disposed of effectively and daily.
8. Close and regular contact should be maintained with the birds so that changes in behaviour are quickly noticed.

The way ahead

It is likely that turkey production will continue to increase in the UK. It is possible that the number of general farms growing turkeys all the year round, to supply the local catering trade, will increase. For these to be successful, good stockmen are necessary. To become one, requires some aptitude and empathy for animals, together with the correct training. In the UK, short training courses on aspects of turkey production are run regularly by the Agricultural Training Board. The National Proficiency Test Council also offers proficiency testing for people working with turkeys.

References and further reading

Crawford, R. D. (1984). Turkey. In Mason, I. L. (ed.), *Evolution of Domesticated Animals*. New York: Longman.

Hale, E. B., Schleidt, W. M. and Schein, M. W. (1969). The behaviour of turkeys. In Hafez, E. S. E. (ed.), *The Behaviour of Domestic Animals*. London: Baillière, Tindall and Cassell.

Ministry of Agriculture, Fisheries and Food (1974). *Poultry Nutrition*. RB 174. London: HMSO.

Ministry of Agriculture, Fisheries and Food (1978). *Artificial Insemination in Poultry*. L 512. London: HMSO.

Ministry of Agriculture, Fisheries and Food (1983). *Turkey Production: Health*. RB 243. London: HMSO.

Ministry of Agriculture, Fisheries and Food (1985). *Turkey Production: Breeding and Husbandry*. RB 242. London: HMSO.

Ministry of Agriculture, Fisheries and Food (1987). *Codes of Recommendations for the Welfare of Livestock: Turkeys*. L 704. London: HMSO.

12 Ducks

P. J. Hearn and K. R. Gooderham

In nature, the duck is a water bird. Man has domesticated the wild duck and, by selective breeding and removing it from its natural habitat, has produced strains which he uses for both meat and egg production.

The ancestors of most of the ducks used in the UK industry can be traced back to just three breeds: Aylesbury and Pekin for meat production and Khaki Campbell for eggs. Table 12.1 gives some details of these breeds. As with chicken and turkeys, hybrid strains have been produced, containing the best characteristics of two or more separate breeds. Improvements in productivity have been obtained by genetic selection and hybrid vigour. By crossing the Aylesbury and Pekin breeds, duck producers have obtained a meat duck with excellent conformation and also an acceptable level of egg production.

The consumption of duck eggs is low in the UK, through a combination of their taste, which is slightly stronger than that of hens' eggs, and the fact that they have a reputation for being contaminated, particularly by salmonellae. This chapter refers mainly to meat production, but attention is drawn to certain aspects of egg production.

Although the duck industry in the UK is, in parts, equally as intensive as the chicken and turkey industries, there is little published research on many aspects of production relative to this aspect. However, most of the meat ducks in the UK will be killed, eviscerated, packed and marketed through processing plants. There is very little processing on a farm scale.

Breeding

The objective of a breeding programme for duck meat is to produce a duckling which will grow rapidly and reach a weight of about 3 kg by 7 weeks of age with a food conversion ratio of about 2.75 : 1. At the same time, egg production from the breeding females must be at a level sufficient to make the operation viable.

Unless replacement breeding stock are bought from a primary breeder, breeding females will be selected from commercially reared meat flocks at about 7 weeks of age. This selection will be on the basis of conformation, freedom from defects and general health. Drakes should be similarly selected, but from a different family line, to avoid in-breeding. The selection will be biased towards the market requirements of the production company concerned. Care should be taken not to select for a characteristic to the extent that it could affect the general health of the duck. For example, too much emphasis on body weight alone might lead to increased leg weakness.

Breeding stock thus selected will have been on a diet designed to ensure rapid growth. These ducklings will have to be slimmed down by feeding them a maintenance diet or they will be too fat when they reach sexual maturity. This could cause various disorders such as prolapse or internal laying. During this period, future breeding stock are often kept on range with no more than a rudimentary shelter. Although ducks are hardy, they should be protected from the worst excesses of the weather.

It is common for the ducks and drakes of

Table 12.1
Origin and physical characteristics of commercial breeds of ducks

	Origin	Plumage	Mean adult wt (kg)		Laying potential per annum
			Drakes	Ducks	
Aylesbury	Bucks	White	4.5	4.0	180
Pekin	China	Cream/white	4.0	3.6	200+
Khaki Campbell	Glos	Khaki	2.3	2.1	300+

future breeding stock to be reared together. Imprinting in ducks is strong and the absence of females during the latter stage of the drakes' rearing period may result in misdirected sexual activity and therefore infertility. If, for any reason, the sexes are reared separately, they should be grouped together for several weeks before eggs are expected. This will usually be at about 24 weeks of age, depending on the pattern of daylength received by the stock (see p. 246).

The optimum group size has been found to be about 400, although for egg producing stock, it is claimed that maximum egg output is obtained from groups as small as 25. The mating ratio is one drake to five or six ducks for meat strains, or one drake to eight ducks for egg strains. Eggs will be fertile about 3 days after the first mating and will remain at an acceptable level of fertility from that mating for up to 10 days. Although drakes will tread any duck, most matings will occur within a specific group of females.

Artificial insemination has been shown to be possible with ducks, both in terms of semen collection and insemination, but it is not practised on commercial farms. The frequency of insemination required to ensure a high level of fertility is two to four times that required for turkeys and, in contrast to turkeys, high levels of fertility are obtained from natural mating.

About 90 per cent of all eggs will be laid within 3 h of the start of the ducks' day. In semi-intensive systems, ducks should not be let out of their house until most of the eggs have been laid. Nest-boxes must be provided at the rate of one space for every five or six ducks. Some producers provide very rudimentary nesting areas by resting large pieces of wood or similar material against a wall to make a sheltered nesting area. Nest boxes are better at litter level, since ducks will not readily use boxes raised off the ground. Ducks selected for meat production will not fly but will climb or scramble up to a second level of nest boxes if these are provided. Single boxes measure about 40 cm^2, but longer boxes for several ducks can be provided. There should be a lip about 15 cm in height to retain the nesting material and eggs, because ducks will roll eggs out of nests with no lip.

Broodiness is not a great problem with ducks. Indeed, since virtually all eggs are laid early in the morning, any broodies remaining in the nests are easy to identify. They should be removed from the rest of the flock and kept in a wire-floored or slatted pen with food and water until all signs of broodiness have disappeared.

It is essential that nests are cleaned out and re-littered regularly, so that eggs may be collected in as clean a condition as possible. If eggs are to be washed, this should be done as soon as possible after collection. An automatic egg-washing machine will produce the best results, but with careful control of the water temperature and the strength of the sanitizer, successful hand-washing is possible.

The length of the cycle for the production of hatching eggs is determined by the flock replacement policy. One method is to allow the ducks to lay for as long as is economically viable. This should be for up to 30 weeks of lay. The flock can then go through a natural or enforced moult and then lay for a second cycle. An output of about 140 eggs per cycle can be expected from this method. After two or three cycles, the flock will be replaced.

The second system and one which allows for a more even output of hatching eggs, is for the flock to be induced to moult every 20 weeks or so, followed by a rest period of 10 weeks. Some breeding flocks will lay profitably for up to 3 years or so on this method.

Although the *Codes of Recommendations for the Welfare of Livestock: Domestic Fowl* proscribe the restriction of feed and water to induce a moult in laying hens, the equivalent *Codes* for ducks (MAFF, 1987) make no mention of induced moulting. However, the method of inducing a moult in ducks is less severe than that used for laying fowl. Daylength is reduced by eliminating artificial light, no food is given for 24 h and then about 140 g of good quality whole wheat per duck per day is provided. Water should be available throughout the moult period. Egg production will cease in about 3–4 days. The standard of management should be high during this period. Any signs of increased mortality or lameness should be noted and acted upon, possibly by increasing the level of feeding. When the ducks have lost about 20–25 per cent of their body weight, their feed allocation should be increased until they are fed nearly to appetite by about 3 weeks before the onset of the next production cycle. The feeding programme during the moulting period must be modified in the light of the condition of the ducks. In particular, care should be taken when moulting during cold weather, because the ducks will need more energy for maintenance of body temperature. If the body weight is too low, a maintenance ration, rather than whole grain, should be fed.

When eggs are required, the full breeding ration is fed and daylength is increased to at least 15 h/day. If eggs are required in midsummer, daylength may be increased to 20 h or more. A higher level of fertility will be achieved if drakes are given an increasing daylength starting about 3 weeks before the females. As the breeding stock will have been kept on range during the moulting period, this can be achieved simply by bringing the drakes into the light-controlled house earlier than the females. Imprinting will have been achieved during the earlier rearing period.

Eggs for hatching should be stored at about 15°C and 80 per cent relative humidity. The incubation period for duck eggs is 28 days. In well-managed breeding flocks, over 80 per cent of all eggs set can be expected to hatch, although this figure will decline as fertility wanes towards the end of a laying cycle. Further details on the handling, cleaning and incubation of duck eggs can be found in other publications (e.g. ADAS, 1980).

The larger integrated companies will have their own breeding flocks, hatcheries and growing farms. Some will sell day-old stock to independent growers, but there are some other smaller hatcheries from whom day-old ducklings can be bought.

Nutrition

Ducks have a simple alimentary system, broadly similar to that of the fowl and turkey. The one main difference is that the duck's oesophagus, although capable of considerable distension, does not possess a clearly identifiable crop.

In the wild, ducks are omnivorous, scavenging what they can from their environment. In the domesticated state, grass is the only feedstuff which is available but, unlike the goose, the duck is not a true grazing animal. Therefore a complete formulated food has to be provided.

Like fowls, ducks have the ability to adjust their feed intake in relation to the energy level of the feed. However, the adjustment is not perfect and there is an increase in energy intake as the energy level of the feed rises. Because of this energy level/intake relationship, recommendations on levels of other nutrients must be qualified by stating the energy level of the diet. There is little published research work on the nutritional requirements of the strains of duck used in the UK. Most recommendations are based on formulations which have been found to be successful in commercial practice. The figures in Table 12.2 combine research and commercial practice (ADAS, 1980; Dean, 1986).

Duck feeds can either be bought from a specialist animal feed compounder or mixed on the farm. In the latter case, care should be taken to use only good quality ingredients.

Growing ducklings

The starter feed should be fed either as crumbs or as 3-mm pellets. Particular care should be taken to ensure good pellet quality, because dust or fines in the feed may cause a build-up of meal around the bill. Pellet size can be increased to 5 mm for the grower ration, which is usually

Table 12.2
Analyses of typical duck diets

	Starter	Grower	Finisher	Breeder	Holding ration
ME (MJ/kg)	12.1	12.1	12.4	11.3	11.8
Crude protein (%)	20.0	15.5	13.3	17.0	11.3
Methionine (%)	0.44	0.33		0.33	
Lysine %	1.10	0.75		0.65	
Calcium (%)	0.60	0.60	0.5	2.75	0.2
Available phosphorus (%)	0.37	0.33		0.28	
Salt (%)	0.12	0.12		0.11	

fed from 3 weeks of age. Sometimes, a finisher ration is fed for the last 2 or 3 weeks of the growing cycle.

Replacement breeding stock

During the period between selection and maturity, future breeding stock may be either fed a maintenance ration, or given access to range and fed good quality whole wheat. Ducks will be expected to lose weight during this period and will only be fed at about 75 per cent of appetite.

Breeding stock

Current practice in feeding breeding stock is based on what has been found to work, rather than on research. Feed intake of the breeders' ration will range between 170 and 230 g/day, depending on ambient temperatures and the body weight of the duck.

Ducks tend to feed early in the morning and late at night. It is common practice, therefore, to feed a pelleted feed in the house at night only. Ducks will quite happily eat in darkness. When they are let out on range during the day, they just have access to the grass, as they may overeat and get too fat if they are given pellets or grain during the day as well. This system has the advantage that it will prevent feed being taken by wild birds.

Micronutrients

For the maintenance of good health and production efficiency, it is essential to feed a proprietary supplement containing the complete range of vitamins and trace elements appropriate to the class of stock.

Water

It is often forgotten that water is a nutrient. A constant supply of fresh, clean water is essential for all classes of ducks. On a weight basis, ducks will consume about five times as much water as food. They will waste a lot more if given the opportunity. Ducks are not supplied with water for swimming on commercial farms and they do not appear to suffer from this lack of opportunity. Further, the numerous parasitic diseases requiring aquatic intermediate hosts are avoided.

Environmental and housing requirements

Light

In common with other avian species, the age at sexual maturity of the duck can be manipulated by controlling the duration of light received daily. Breeding activity can be induced by a progressive increase in the daily light period. Ducklings reared in the spring will reach maturity about 2 or 3 weeks earlier than those reared in the autumn because of the change in natural daylength. Few breeding stock are kept in completely controlled environment housing, so any modification of the light pattern is limited to topping-up natural daylength when it starts to decline from mid-June onwards.

There is no reason to suppose that ducks react differently from fowl to light intensity, therefore it is recommended that the mean intensity of light in breeding houses should be at least 10 lux.

Temperature

Detailed information on the response of ducks to environmental temperature is lacking, but commercial experience suggests that high temperatures during the growing period considerably reduce feed intake and therefore weight gain. This can result in differences of up to 0.3 kg in weight between summer and winter crops.

Ventilation

The ventilation system should be capable of providing plenty of fresh air in growing and breeding houses, to prevent overheating. Ducks will pant in response to high temperatures. The onset of panting will depend on relative humidity, body weight, etc., but will occur at about 27°C.

Where fans are the sole means of ventilation, commercial experience suggests that one 610-mm fan is needed for every 400 ducks. When natural ventilation is relied on, large gaps along each wall and at the ridge are necessary. Whenever fan-powered ventilation is used, adequate provision should be made for ventilation in the event of a mains power failure. This should be in the form of, at least, a stand-by generator which is regularly tested, or even a fail-safe ventilation system. This would use drop-out panels to provide sufficient air to prevent distress to the ducks until mains power can be restored. An efficient alarm system is advisable.

Ducks respond to increased air movement by improved growth performance, so a recent development has been to install stirrer or paddle fans to create extra turbulence. The reason for this may be a dilution of atmospheric pollutants (disease-bearing dust particles, ammonia, etc.) or an increased stimulus to general activity. In brooder houses, these fans have the additional function of reducing costs by recirculating heated air.

Insulation

To reduce heat loss to acceptable levels, buildings should be insulated. A U-value of 0.5 $W/m^2/°C$ is advised for brooder houses, but rearing and breeding houses can be less well insulated.

Feeding systems

In larger duck units, automatic feeding systems can be used. These reduce the labour requirement and can improve efficiency by reducing wastage. Large bulk hoppers which are filled every few days are also popular. Each 100 ducks should have about 0.5 m each of feed and water trough space to 8 weeks of age, increasing to 0.6 m after 8 weeks.

Drinking systems

There is some controversy over the acceptability of some designs of drinker. The established view is that ducks need to be able to immerse their nostrils and even eyes in water, so drinkers must contain enough water to enable them to do this. There is no doubt that if ducks are given access to open troughs, they will dip in their heads and shake off excess water on to the surrounding area. However, some commercial experience suggests that this is not an essential function. Ducklings have been shown to grow well and healthily with access only to nipple drinkers where they can only receive a drop or two of water at a time. If nipple drinkers are used, they should be provided at the rate of one per 5 ducks at least. The advantage of this system to the producer is the greatly reduced wastage of water and therefore the potential for improved living conditions for the stock. In the absence of evidence and statutory requirements to the contrary, nipple and cup systems as used in the poultry industry are likely to increase in use. Indeed, the system is almost universally used in some of the important American duck-growing areas with no apparent welfare disadvantages. If open troughs or bell-type drinkers are used, regular cleaning should be carried out to prevent a build-up of sour mash.

To keep the litter as dry as possible, rearing and breeder houses should contain a separate, well drained drinker area. The drinkers can be positioned over a wire or slatted area to allow the considerable amount of wasted water to drain away.

Because of the lower stocking densities in duck houses, the possibility of water freezing in the header tank and supply pipes is greater than in some other poultry houses. Therefore, extra care should be taken in cold weather to check that the water supply is not impeded. For ducks kept on range, a bowser should be available to take water to the troughs in cold weather.

Litter

Litter materials for all classes of ducks include wood shavings and straw. The latter is preferred, because not only is straw cheaper and plentiful in duck-growing areas, but the open structure of a layer of straw is more able to absorb the ducks' faeces. Wood shavings tend to be paddled into a dirty crust by the ducks' webbed feet. Only good quality, clean straw should be used, since ducks are susceptible to the *Aspergillus* mould which occurs on badly stored straw. Barley straw is preferred for young stock, being softer than wheat, but wheat straw can be used for rearing and breeding houses. Wood shavings should be used in nest boxes and changed regularly.

Because ducks waste so much water and their faeces have a high moisture content, extra litter is often added every day. There is no great advantage in chopping the straw before it is spread. The litter will not work as it does in a broiler house. The built-up litter is removed once per crop in most cases, although sometimes it is kept for further crops. There are less disease risks in this practice with ducks than with other intensively kept poultry, but it may lead to a poorer environment through higher ammonia levels.

Management systems

The vast majority of ducks are kept on litter or a combination of litter and range, although tier-brooders can be used to rear ducks for the first few weeks (see below). A variety of buildings can be used, from converted general farm buildings to specially-built houses.

Several factors influence the number of ducks which can be kept in a given area. Practical guidelines are given in the *Codes of Recommendations for the Welfare of Livestock: Ducks* (MAFF, 1987) and these are summarized in Table 12.3. All recommendations on stocking densities for breeding stock are based on numbers of females only.

Brooding

Heat can be supplied by a variety of methods, but gas is widely used. Hover brooders with a chick capacity of 1500 will take about 1000 ducklings. The brooding temperature for ducks is about 30°C. The rate of temperature reduction depends where and when they move on to rearing quarters. The objective should be to match up the temperature at the end of the brooding period with the expected temperature at the start of the rearing period. This may be high if the ducks are to be kept in the same house, or low if they are to be moved to range. For the first few days, ducklings should be confined near the source of heat by a temporary barrier. Plenty of drinking and feeding points should be provided to ensure that the ducklings get a good start.

Ducks are hardier than chicken or turkeys and will survive without artificial heat from 10–12 days of age. By this time, the temporary barrier will have been removed and the ducks will have access to the full brooding area.

In addition to heat, the stock will require a supply of fresh air, but controlled both in quantity and direction. Efficient control of the ventilation rate is necessary for optimum environmental conditions.

An alternative brooding system, not widely used today, is wire-floored tier brooders. If properly managed, there are no welfare problems with this system. The ducklings are kept in groups of 25–100 on 12.5-mm-square mesh wire floors. Heat can be provided by electricity or gas. Compared with litter-based systems, labour, capital and maintenance costs per bird

Table 12.3
Recommended stocking densities for ducks under different management systems

	Age			
System	0–10 days	10–21 days	21–56 days	Adults
		Ducks/m^2		
Slatted, perforated or mesh floors	50	25	8	5
Solid littered floors	36	14	7	3
		Ducks/ha		
Grass runs	Poorly grassed runs 2500 Well-grassed runs 4000			4000 + house

are higher. Tier brooders are therefore uneconomic for large numbers of birds.

Rearing

By about 3 weeks of age, ducks can be moved to their rearing quarters, where they will remain until going for processing or being selected for breeding. The practice of keeping stock in the same house throughout the growing period is not as widespread as it is in the broiler and turkey industries. Most ducks will be moved to cheaper buildings or to range. Comparative costings of these two systems are hard to obtain. The choice depends on what buildings are available. However, most producers would prefer to keep ducks intensively during the winter months.

Range rearing is still very popular, particularly in parts of Norfolk where the well-drained soil makes for ideal conditions. Ducklings need only rudimentary shelter; straw bales and corrugated iron sheets are sufficient. Fences need to be only about 1 m high to keep the ducks in. The installation of fox-proof fencing is not usually considered worthwhile, particularly as the range should be rested from stock periodically to enable the sward to recover. However, if some protection against foxes, badgers or dogs is required, then the addition of an electrified strand of wire about 25 cm off the ground and 25 cm out from a wire fence will prove a useful deterrent.

Breeding

A wide variety of houses can be suitable for breeding ducks. Although common practice, it is not essential to incorporate range into the breeding system. Good levels of production can be obtained by keeping breeding ducks intensively throughout the cycle, but in these circumstances, stocking densities in the house should be decreased from 3 to about 2 ducks/m^2.

Although there may be some advantages in having a controlled environment house for breeding ducks through being able to control the light pattern, this is only viable if the rearing house is similarly controlled. Most breeding houses will be similar to turkey pole-barns with rudimentary ventilation control and limited insulation.

General care and handling

As with all forms of intensive livestock, the art of stockmanship can be summed up as an understanding of the animal and its requirements, coupled with attention to the detail of whatever production system is used. Staff should be able to monitor the living conditions of the ducks under their care to ensure that the required environment is provided. Often, the first sign of ill-health is a lowering of production efficiency, either growth rate or egg production. Monitoring these factors by measurement or inspection should be part of the daily routine.

Fig. 12.1 A house for breeding ducks. There is a raised drinking area (far end), nest boxes (centre) and feed troughs (left). Courtesy of Buxted Ducklings Ltd.

Handling and transport

Whereas poultry are frequently caught and carried by their legs, ducks should not be handled in this way. The preferred method is to place one hand on each side of the duck's body, over the wings. Only one duck should be picked up at a time, except when dealing with very young ducklings. With older ducks, it may sometimes be necessary to catch them by the neck. If this is the case, they should be supported by placing one hand under the body.

It is often preferable to drive or herd ducks if they have to be moved over relatively short distances. For longer distances, ducks can be driven into flat-bed trailers rather than caught, crated and uncrated.

If ducks are badly handled, or herded too quickly or frequently, they can become temporarily lame and production may be affected.

Behaviour

Ducks, particularly young ducks, exhibit a behaviour pattern similar to flocks of birds in flight. For no obvious reason, ducks will start to move around their living space in a swirling motion, with more and more ducks joining in. Just as suddenly this will stop, only to start again in a short while in another direction. As the ducks get older, this behaviour is seen less and less and will disappear by sexual maturity. Older growing ducks and adult breeding stock of the meat strains are usually fairly docile. They tend to walk rather than run.

Perhaps in a reference to the wild state when they climb out of water on to a bank, ducks will readily climb (though not jump very far) on to any straw bale or anything similar left in their pens. This should be prevented as they may injure themselves by falling off high vantage

Fig. 12.2 Breeding ducks on range. Note the paddocks used for rotating ducks over the available area. Courtesy of Buxted Ducklings Ltd.

points. Ducks are more inquisitive than chicken and will readily explore their surroundings.

Ducks are not normally aggressive birds. However, off-colour breeding ducks may be bullied by drakes. These ducks should be culled or removed to a sick-pen as soon as possible.

Ducks do not need any form of perch, since their webbed feet are not designed for clasping. They simply squat down to rest.

Routine procedures

Flight prevention

Some of the lighter strains of duck, particularly the egg-laying varieties, will readily fly. Flight can be prevented by trimming the flight feathers of one wing only. Surgical pinioning must not be practised (Welfare of Livestock (Prohibited Operations) Regulations, 1982).

Sexing

Ducklings can be vent-sexed at 1 day-old. Drakes are relatively easy to sex, but ducks often require a second look. Unless the sexer is very careful, the extra pressure which may be necessary can cause a ruptured yolk-sac.

In the male, there is a phallus or penis which is easily visible at the time of hatching, being 2–3 mm in length. In the adult it is considerably larger and is used as an intromittent organ in mating. Occasionally, this organ can become infected and will not retract. This condition can result in the eventual death of the male.

Up to about 3 weeks of age, ducklings of both sexes cheep in a similar way to chickens. From that age onwards, the sexes differ. Females develop the familiar quack, but males have a more muted, tremulous burble. This is because the male syrinx is distended into a large cartilaginous bulla. So distinct is the difference that ducks can be voice-sexed relatively easily from about 5 weeks onwards.

The adult male plumage includes a few curled feathers on the tail and this can be used as a visual method of sexing.

Bill trimming

Outbreaks of feather-pulling in young ducklings may be caused by a variety of factors, including overcrowding or lack of feed or water. Attempts should be made to improve the management, but as a last resort, it may be necessary to trim the ducks' bills. If this is done at about 1 week of age, it is necessary to remove only about 1 mm of the downward directed part of the bill rim so that down or feathers cannot be gripped. More of the rim should be removed if the procedure is carried out on older birds. If it is done correctly with an electrically heated hot blade, no bleeding occurs and it only needs to be done once. The procedure should be carried out by skilled personnel or under their direction.

Marking

It is often advisable to be able to identify particular groups of ducks, so some form of identification is necessary. Two methods are used, web-slitting and wing-banding. There are two webs on each foot, so several combinations of marks are possible. An incision can be made with a sharp knife or scissors. Visible blood vessels should be avoided. The marking will persist for life. Wing-bands are usually used for pedigree breeding, where it is necessary to be able to identify individual birds. Bands or tags can be attached to the flap of skin which stretches across the inner part of the wing. Particular care is necessary when tagging ducklings at 1 day old to prevent the wing-tag from entering tissue other than skin. All tags should be checked a few days after they have been applied, because the small wing may be able to pass through the loop of the tag.

Emergency slaughter

Even in the most well-managed units, the occasional sick or injured duck will be found. The preferred method of humane killing is by neck dislocation. The legs should be grasped firmly in the left hand and the index and third fingers of the right hand placed on either side of the back of the skull. With the left hand held at chest height, the right hand should be pushed quickly down by the right thigh. If the operation is performed correctly, it should be possible to feel a gap between the vertebrae in the duck's neck.

Guidance on the general care and handling of ducks can be found in the *Codes of Recommendations* (MAFF, 1987).

Health and disease

The stockman should be aware of the signs of good health in the duck. He should look for almost continuous feeding and drinking activity and vigorous movements if the ducks are disturbed. On closer inspection, the duck should exhibit clear eyes, clean feathers and skin and a general alertness.

Many diseases, both specific and non-specific, result from inadequate husbandry. Attention to the husbandry details listed in this section can reduce or prevent much clinical disease and mortality. The most important diseases are mentioned briefly. More detail can be found in Gooderham (1982).

Duck viral hepatitis (DVH) is an acute disease of duckling usually under 3 weeks of age. Losses can be minimized by vaccinating the parents or the duckling at 1 day old with a live vaccine, but this should not be resorted to unless the disease is endemic. Duck virus enteritis can cause heavy losses in adult as well as younger ducks. Typically, it is diagnosed by the combination of lesions in the oesophagus and cloaca. It can be prevented by the use of an attenuated vaccine.

Chlamydiosis (ornithosis) is a scheduled disease under the Psittacosis or Ornithosis Order, 1953. This is because symptomless carriers can transmit the infection to man (psittacosis). This is especially so when dry-plucking ducks. Infection can also precipitate secondary bacterial diseases in ducks.

Pasteurella anatipestifer causes a bacterial septicaemic disease, especially from 2 to 6 weeks of age. Its effects can be minimized or prevented by avoiding environmental stresses. Sulphonamides in drinking water or streptomycin/dihydrostreptomycin by injection are useful for treatment. *Escherichia coli* causes a similar septicaemic disease which is common up to 8 weeks of age. The prevention and treatment are similar to that for *Pasteurella anatipestifer*. *Pasteurella multocida* and *Erysipelothrix* also cause septicaemic disease, the former especially in older birds.

Aspergillosis is a common respiratory disease in all ages of birds and is caused by breathing in fungal spores from mouldy feed or litter. Mycotoxicosis is caused by toxins produced by a number of different moulds in feed or feed ingredients. Prevention of these conditions is by providing wholesome feed and litter and good ventilation.

Amyloidosis is a chronic disease mainly affecting the liver. It is most common in adult birds. The incidence is reduced if damage to the feet is avoided and birds are prevented from becoming excessively fat.

Salpingitis, which is a lesion of the oviduct occurring in younger birds, may result in death at sexual maturity. Healthy rearing reduces these losses.

The way ahead

The introduction of *Codes of Recommendations for the Welfare of Livestock: Ducks* (MAFF, 1987) has set a base-line for general care and management. This is to be welcomed, since the duck industry is likely to follow the chicken and turkey industries on the road to further intensification. Improved genetic selection and a better knowledge of nutritional requirements will lead to improved growth performance of meat ducks.

Although wire-floored growing houses have been tried and some are in use, they are not widespread. If pressure is put on growing costs, then there may be a further development of this system. One of the biggest problems associated with it is likely to be the disposal of effluent.

The development of methods of controlling the environment automatically in intensive growing and breeding houses should lead to improved conditions for the stock.

Increased sales of duck meat may be more likely to come through the further development of portioned and cooked products than from the sale of oven-ready carcasses. This may lead to the development of specific strains of duck with the required characteristics for particular products.

References and further reading

Agricultural Development and Advisory Service (1980). *Ducks and Geese*. RB 70. London: HMSO.

Dean, W. F. (1986). Nutrient requirements of meat-type ducks. In Farrell, D. J. and Stapleton, P. (eds), *Duck Production Science and World Practice*. Armidale: University of New England.

Gooderham, K. R. (1982). Diseases of the domestic duck. In Gordon, R. F. and Jordan, F. T. W. (eds), *Poultry Diseases*. London: Baillière Tindall.

Farrell, D. J. and Stapleton, P. (eds). *Duck Production Science and World Practice*. Armidale: University of New England.

Ministry of Agriculture, Fisheries and Food (1987), *Codes of Recommendations for the Welfare of Livestock: Ducks*. London: HMSO.

Index